Android 经典应用实例

张德丰　编著

清华大学出版社
北京

内 容 简 介

本书是一本 Android 开发应用书籍,内容由浅入深,循序渐进;以经典、丰富的实例为引导,介绍 Android 的具体应用,让读者快速地了解 Android,并掌握 Android 的应用。全书共分 7 章,主要包括 Android 基本知识、Android 界面开发、Android 深入开发、Android 动态效果、Android 通信服务、Android 手机功能、Android 媒体应用内容。

本书具有内容全、覆盖广、实例多、上手易、图文全的特点,对每个实例进行剖析,让读者对 Android 应用开发有一个全面的认识。

本书适合 Android 初学者,也可作为程序开发人员的 Android 开发参考书。

本书封面贴有清华大学出版社防伪标签,无标签者不得销售。
版权所有,侵权必究。侵权举报电话: 010-62782989　13701121933

图书在版编目(CIP)数据

Android 经典应用实例/张德丰编著. --北京:清华大学出版社,2015
ISBN 978-7-302-40082-0

Ⅰ. ①A… Ⅱ. ①张… Ⅲ. ①移动终端-应用程序-程序设计 Ⅳ. ①TN929.53

中国版本图书馆 CIP 数据核字(2015)第 089625 号

责任编辑:刘　星　王冰飞
封面设计:刘　键
责任校对:徐俊伟
责任印制:李红英

出版发行:清华大学出版社
　　　　　网　　址: http://www.tup.com.cn, http://www.wqbook.com
　　　　　地　　址: 北京清华大学学研大厦 A 座　　邮　编: 100084
　　　　　社 总 机: 010-62770175　　邮　购: 010-62786544
　　　　　投稿与读者服务: 010-62776969, c-service@tup.tsinghua.edu.cn
　　　　　质 量 反 馈: 010-62772015, zhiliang@tup.tsinghua.edu.cn
　　　　　课 件 下 载: http://www.tup.com.cn, 010-62795954
印　刷　者:北京富博印刷有限公司
装　订　者:北京市密云县京文制本装订厂
经　　　销:全国新华书店
开　　　本: 185mm×260mm　　印　张: 26.75　　字　数: 668 千字
版　　　次: 2015 年 6 月第 1 版　　　　　　印　次: 2015 年 6 月第 1 次印刷
印　　　数: 1~2000
定　　　价: 59.50 元

产品编号: 064392-01

前言

　　Android 是一个开放式手机和平板电脑的操作系统，是基于 Linux 平台的开源手机操作系统的名称，目前的发展势头十分迅猛。Android 平台由操作系统、中间件、用户界面和应用软件组成。

　　Android 市场的软件数量在以惊人的速度增长。目前以 Google Android 为平台的产品研发势头正旺，可以预见，Google Android 系统的产品在不久的将来将成为市场的主流。

　　在 Android 推出之前，移动开发领域的发展一直处于不温不火的局面，Android 的推出为移动互联网开发领域吹进一股清新的风。本书结合作者的编程经验，以 Android 的经典实例为引导，详尽地介绍了 Android 的应用，协助读者掌握应用重点，享受开发乐趣。

　　对于 Java 语言而言，Android 系统给了 Java 一个新的机会。在过去的岁月中，Java 语言作为服务器端编程语言，已经取得了极大的成功，Java EE 平台发展得非常成熟，而且一直是计算机、移动、银行、证券、电子商务应用的首选平台、不争的王者。但在客户端应用开发方面，Java 语言一直表现不佳。Android 是一个非常优秀的手机、平板电脑操作系统，它将会逐渐蚕食传统的桌面操作系统，而 Android 平台应用的开发语言就是 Java，这意味着 Java 语言将可以在客户端应用开发上大展拳脚。

　　对于 Java 开发者来说，Android 应用开发既是一个挑战，也是一个机遇。挑战是掌握 Android 应用开发需要重新投入学习成本；机遇是 Android 系统是一个新的发展趋势，这必定带来更多的就业机会与创业机会。

　　而本书的编写也具有其自身的特点。

1. 内容全

　　本书通过众多经典实例对 Android 常用的知识点进行详细的介绍，同样对每个实例进行剖析，让读者对 Android 应用开发有一个全面的认识。

2. 覆盖广

　　本书通过经典实例详细而又全面地介绍了 Android 的应用开发，内容深入浅出。

3. 实例多

　　为了让读者快速熟悉并掌握 Android，本书对每一个知识点都通过一个经典实例来说明，让读者在领略到 Android 的功能强大之外，同时也掌握了 Android 的实际开发应用，具有极高的参考价值。

4. 上手易

　　本书每个实例都具有详细的操作步骤，编程思路清晰，语言通俗易懂。通过本书可以轻松上手 Android 应用开发，同时也可满足实际企业中的要求。

5．图文全

本书每一个知识点的经典实例都给出相应的运行效果图，这对读者掌握这一知识点起到了很大的作用，并让读者直观地享受视觉效果。

全书共分为 7 章，其主要内容如下。

第 1 章介绍 Android 的魅力，主要包括 Android 的特点、Android 平台架构、Android 开发环境搭建、Android 的基本组件等内容。

第 2 章介绍 Android 界面开发实例，主要包括常用的布局实例、文本类实例、按钮类实例以及条类控件实例等内容。

第 3 章介绍 Android 深入开发实例，主要包括 Android 视图实例、温馨的提示实例、友好界面实例以及温馨消息对话框实例等内容。

第 4 章介绍 Android 动态效果实例，主要包括基本二维图形实例、绘制路径实例、图像的动画效果实例、图像的特效处理实例以及图像渲染实例等内容。

第 5 章介绍 Android 通信服务实例，主要包括电话拨号实例、电子邮箱实例、天气预报实例、Wi-Fi 实例以及在线查询实例等内容。

第 6 章介绍 Android 手机功能实例，主要包括振动器实例、闹钟实例、手电筒实例、备忘录实例以及万年历实例等内容。

第 7 章介绍 Android 媒体应用实例，主要包括视频/音频播放实例、视频/音频录制实例、Android 定位实例以及城市定位实例等内容。

本书适合 Android 初学者，适合初、中级程序员，也可作为程序开发人员的 Android 开发参考书。

本书主要由张德丰编写，此外参加编写的还有刘志为、栾颖、周品、曾虹雁、邓俊辉、邓秀乾、邓耀隆、高泳崇、李嘉乐、李旭波、梁朗星、梁志成、刘超、刘泳、卢佳华、张棣华、张金林、钟东山、李伟平、宋晓光和何正风。

由于作者的水平有限，加之时间紧凑，书中难免会存在不足之处，敬请广大读者批评指正。

作　者

2015 年 1 月

目录

第1章 领略 Android 的魅力 ········· 1
- 1.1 Android 平台的概述 ········· 1
- 1.2 Android 的特点 ········· 2
- 1.3 Android 平台体系 ········· 3
- 1.4 Android 开发环境 ········· 4
  - 1.4.1 Android 系统需求 ········· 4
  - 1.4.2 JDK 安装 ········· 5
  - 1.4.3 Eclipse 环境 ········· 6
  - 1.4.4 Android 的 SDK ········· 8
  - 1.4.5 Android 的 AVD ········· 9
- 1.5 Android 应用结构分析 ········· 13
- 1.6 Android 应用程序权限说明 ········· 16
- 1.7 Android 基本组件介绍 ········· 17
  - 1.7.1 Activity 与 View ········· 17
  - 1.7.2 Service ········· 18
  - 1.7.3 BroadcastReceiver ········· 18
  - 1.7.4 ContentProvider ········· 19
  - 1.7.5 Intent 与 IntentFilter ········· 19
- 1.8 应用程序生命周期 ········· 20
- 1.9 Activity 生命周期 ········· 22
  - 1.9.1 Activity 的周期 ········· 22
  - 1.9.2 Activity 的状态 ········· 22
- 1.10 Android 进程与线程 ········· 24
  - 1.10.1 进程 ········· 24
  - 1.10.2 线程 ········· 24
  - 1.10.3 线程安全的方法 ········· 24

第2章 Android 界面开发实例 ········· 26
- 2.1 常用的布局实例 ········· 26

2.1.1 计算器实例 ……………………………………………………………… 26
2.1.2 奖牌排行榜实例 …………………………………………………………… 32
2.1.3 标语与排行实例 …………………………………………………………… 34
2.1.4 登录界面实例 …………………………………………………………… 37
2.1.5 霓虹灯实例 …………………………………………………………… 39
2.1.6 显示信息实例 …………………………………………………………… 42
2.2 文本类实例 …………………………………………………………………… 45
2.2.1 文字说明实例 …………………………………………………………… 45
2.2.2 接收信息实例 …………………………………………………………… 47
2.2.3 自动搜索实例 …………………………………………………………… 50
2.3 按钮类实例 …………………………………………………………………… 53
2.3.1 按钮测试实例 …………………………………………………………… 53
2.3.2 图片说明实例 …………………………………………………………… 57
2.3.3 程序开闭实例 …………………………………………………………… 62
2.3.4 城市选择实例 …………………………………………………………… 64
2.3.5 确定选择实例 …………………………………………………………… 66
2.3.6 个人性格选择实例 …………………………………………………………… 71
2.4 计时实例 …………………………………………………………………… 74
2.5 条类控件实例 …………………………………………………………………… 77
2.5.1 进度提示实例 …………………………………………………………… 77
2.5.2 音量大小调节实例 …………………………………………………………… 81
2.5.3 等级打分实例 …………………………………………………………… 83
2.6 手机图片查看实例 …………………………………………………………………… 85
2.7 色彩选择实例 …………………………………………………………………… 89
2.8 手机模拟时钟实例 …………………………………………………………………… 91
2.9 记录购书时间实例 …………………………………………………………………… 94

第3章 Android 深入开发实例 …………………………………………………………………… 98

3.1 Android 视图实例 …………………………………………………………………… 98
3.1.1 左右浏览影片实例 …………………………………………………………… 98
3.1.2 单击显示控件实例 …………………………………………………………… 100
3.1.3 选择条目实例 …………………………………………………………… 103
3.1.4 显示文本列表实例 …………………………………………………………… 107
3.1.5 实现多选条目实例 …………………………………………………………… 109
3.1.6 实现单选条目实例 …………………………………………………………… 112
3.1.7 自定义 ListView 控件实例 …………………………………………………………… 114
3.1.8 制作相片集实例 …………………………………………………………… 117
3.1.9 手机图片查看器实例 …………………………………………………………… 120
3.1.10 九宫布局实例 …………………………………………………………… 123
3.1.11 带图片文字的 ListView 实例 …………………………………………………………… 127

3.1.12　手机浏览网页实例 …………………………………………………… 130
　　3.1.13　手势滑动实例 ………………………………………………………… 131
　　3.1.14　多个标签栏实例 ……………………………………………………… 134
3.2　温馨的提示实例 …………………………………………………………………… 140
　　3.2.1　"通知单"实例 ………………………………………………………… 141
　　3.2.2　手机消息提醒实例 ……………………………………………………… 145
3.3　友好界面实例 ……………………………………………………………………… 148
　　3.3.1　选项菜单实例 …………………………………………………………… 148
　　3.3.2　在菜单中添加单、多选功能实例 ……………………………………… 152
　　3.3.3　添加常用操作实例 ……………………………………………………… 155
3.4　温馨消息对话框实例 ……………………………………………………………… 158
　　3.4.1　单击弹出一个对话框实例 ……………………………………………… 159
　　3.4.2　Android 9 种对话框实例 ……………………………………………… 160
　　3.4.3　日期选择对话框实例 …………………………………………………… 169
　　3.4.4　时间日期选择对话框实例 ……………………………………………… 172

第 4 章　Android 动态效果实例 …………………………………………………… 175

4.1　基本二维图形实例 ………………………………………………………………… 175
4.2　绘制路径实例 ……………………………………………………………………… 179
4.3　绘制路径文本实例 ………………………………………………………………… 181
4.4　电影式播放实例 …………………………………………………………………… 183
4.5　平面贴图实例 ……………………………………………………………………… 187
4.6　图像淡入淡出实例 ………………………………………………………………… 189
4.7　图像变大变小实例 ………………………………………………………………… 192
4.8　图像移动实例 ……………………………………………………………………… 195
4.9　动画综合实例 ……………………………………………………………………… 197
4.10　图像特效实例 ……………………………………………………………………… 201
4.11　图像扭曲实例 ……………………………………………………………………… 203
4.12　图像渲染实例 ……………………………………………………………………… 206
4.13　示波器实例 ………………………………………………………………………… 211

第 5 章　Android 通信服务实例 …………………………………………………… 215

5.1　还原桌面实例 ……………………………………………………………………… 215
5.2　数据交换实例 ……………………………………………………………………… 217
5.3　查询星座实例 ……………………………………………………………………… 223
5.4　Intent 发短信实例 ………………………………………………………………… 227
5.5　发送短信实例 ……………………………………………………………………… 230
5.6　隐式 Intent 实例 …………………………………………………………………… 233
5.7　电话拨号实例 ……………………………………………………………………… 239
5.8　自定义拨打电话实例 ……………………………………………………………… 241

5.9 邮箱实例 …… 247
5.10 保护视力实例 …… 251
5.11 天气预报实例 …… 254
5.12 通信组件实例 …… 257
5.13 Wi-Fi 实例 …… 261
5.14 查看手机信息实例 …… 268
5.15 读取 SIM 卡参数实例 …… 273
5.16 查询电池剩余量实例 …… 277
5.17 通讯录实例 …… 279
5.18 在线查询实例 …… 291
5.19 自动朗读实例 …… 295

第 6 章 Android 手机功能实例 …… 299

6.1 振动器实例 …… 299
6.2 闹钟实例 …… 302
6.3 计算器实例 …… 305
6.4 手电筒实例 …… 313
6.5 备忘录实例 …… 316
6.6 手机状态提醒实例 …… 320
6.7 来电自动短信回复实例 …… 322
6.8 万年历实例 …… 326
6.9 存储卡查询实例 …… 332
6.10 RSS 阅读器实例 …… 335

第 7 章 Android 媒体应用实例 …… 348

7.1 MediaPlayer 播放音频实例 …… 348
7.2 SoundPool 播放音频实例 …… 360
7.3 MediaRecorder 录制音频实例 …… 363
7.4 VideoView 播放视频实例 …… 369
7.5 SurfaceView 播放视频实例 …… 371
7.6 摄像头实例 …… 375
7.7 录制视频实例 …… 380
7.8 SensorManager 开发传感器实例 …… 387
7.9 Android 定位实例 …… 389
7.10 城市定位实例 …… 393

附录 A 网上参考资源 …… 415

参考文献 …… 417

第 1 章

领略 Android 的魅力

Android 一词的本义是指"机器人",同时也是 Google 于 2007 年 11 月 5 日宣布的基于 Linux 平台的开源手机操作系统的名称,该平台由操作系统、中间件、用户界面和应用软件组成,号称首个为终端打造的真正开放和完整的移动软件。

1.1 Android 平台概述

Android 主要的使用场合就是移动通信领域,也就是 Android 开发出来的软件在手机或计算机上运行,随时随地可以很方便地处理数据和管理数据。要学习 Android,首先需要掌握以下一些基本知识。

(1) Android 是由 Google 公司进行设计与开发的移动通信平台。

(2) Android 平台基于 Linux 操作系统,所以内存的分配、线程的调度以及作业的执行都由底层 Linux 操作系统进行处理。

(3) Android 平台是开放源代码的。

(4) Android 单词的中文翻译是"机器人"的意思。

(5) Android 平台被很多知名的通信运营商支持,所以 Android 逐渐成为了一个通信平台的标准。

(6) Android 相当于一个操作系统,在这个操作系统上可以运行任何 Android 支持的软件。Android 相当于 Windows XP,所以安装 Android 系统的手机称为"智能手机"。

(7) Android 支持多个任务环境。

(8) Android 平台应用比较好的手机制造商有 HTC 和三星等。

(9) Android 是基于 WebKit 引擎的浏览器。

(10) Android 具有 2D 和 3D 软件开发能力,其中 3D 图形库基于 OpenGL ES 1.0 标准。

(11) Android 数据存储使用 SQLite,它是一个文件型数据库。

(12) Android 支持常用的图形及视频格式(MPEG4、H.264、MP3、AAC、AMR、JPG、PNG 和 GIF)。

1.2 Android 的特点

Android 是由操作系统、用户界面和应用程序组成，允许开发人员自由获取和修改源代码，也就是说，这是一套具有开源性质的手机终端解决方案。其具有如下特点：
- 开放性；
- 平等性；
- 无界性；
- 方便性；
- 丰富性。

下面对这 5 个特点进行详细介绍。

1. 开放性

提到 Android 的优势，首先想到一定是其真正的开放，其开放性包含底层的操作系统以及上层的应用程序等，Google 与开放手机联盟合作开发 Android 的目的就是建立标准化、开放式的移动单击软件平台，在移动产业内形成一个开放式的生成系统。

Android 的开放性也同样会使大量的程序开发人员投入 Android 程序的开发中，这将为 Android 平台带来大量新的应用。

2. 平等性

在 Android 系统上，所有应用程序完全平等，系统默认自带的程序与自己开发的程序没有任何区别，程序开发人员可以开发个人喜爱的应用程序并替换掉系统程序，来构建个性化的 Android 手机系统，这些功能在其手机平台上是没有的。

在开发之初，Android 平台就被设计成由一系列应用程序组成的平台，所有的应用程序都运行在一个虚拟机上面。该虚拟机提供了系统应用程序之间和硬件资源通信的 API。而除了该虚拟机，其他应用全部平等。

3. 无界性

Android 平台的无界性表现在应用程序之间的无界性，开发人员可以很轻松地将自己开发的程序与其他应用程序进行交互，例如你的应用程序需要播放声音模块，而正好你的手机中已经有一个成熟的音乐播放器，此时你就不需要再重复开发音乐播放功能，只需要简单地加上几句话即可将成熟的音乐播放功能添加到自己的程序中。

4. 方便性

在 Android 平台中开发应用程序是非常方便的，如果你对 Android 平台比较熟悉，想开发一个功能全面的应用程序并不是什么难事。Android 平台为开发人员提供了大量的实用库及方便的工具，同时也将 Google Map 等强大的功能集成了进来，开发人员只需要简单地调用几何代码可将强大的地图功能添加到自己的程序中。

5. 丰富性

由于平台的开放性，众多的硬件制造商推出各种各样、千奇百怪的产品，但这些产品功能上的差异并不影响数据的同步与软件的兼容。例如，原来在诺基亚手机上的应用程序。

1.3 Android 平台体系

Android 系统是以 Linux 系统为基础的，Google 将其按照功能特性划分为 4 层，自下而上分别是 Linux 内核、中间件、应用程序框架和应用程序，如图 1-1 所示。

图 1-1 Android 系统框架图

1. 应用程序

Android 系统内置了一些常用的应用程序，包括 Home 视图、联系人、电话和浏览器等。这些应用程序和用户自己编写的应用程序是完全并列的，同样都是采用 Java 语言编写的。而且，用户可以根据需要增加自己的应用程序，或者替换系统自带的应用程序。

2. 应用程序框架

应用程序框架提供了程序开发人员的接口，这是与 Android 程序员直接相关的部分。开发者可以用它开发应用，其中包括以下内容。

- 丰富而又可扩展的视图(Views)：可以用来构建应用程序，它包括列表(lists)、网格(grids)、文本框(text boxes)、按钮(buttons)，甚至可嵌入的 Web 浏览器。
- 内容提供器(Content Providers)：使得应用程序可以访问另一个应用程序的数据(如联系人数据库)，或者共享它们自己的数据。
- 资源管理器(Resource Manager)：提供非代码资源的访问，如本地字符串、图形、布局文件(layoutfiles)。
- 通知管理器(Notification Manager)：使得应用程序可以在状态栏中显示自定义的提示信息。
- 活动管理器(Activity Manager)：用来管理应用程序生命周期并提供常用的导航回退功能。

3. 中间件

中间件包括两部分：核心库（libraries）和 Android 运行时环境（Android runtime）。

1) 核心库

核心库中主要包括一些 C/C++核心库，方便开发者进行应用的开发。

- 系统 C 库(libc)：专门为基于 embedded linux 的设备定制的。
- 媒体库：支持多种常用的音频、视频格式回放和录制，同时支持静态图像文件。编码格式包括 MPEG4、H.264、MP3、AAC、AMR、JPG 和 PNG。
- SurfaceManager：对显示子系统的管理，并且为多个应用程序提供了 2D 和 3D 图层的无缝融合。
- WebKit/LibWebCore：Web 浏览引擎，支持 Android 浏览器和一个可嵌入的 Web 视图。
- SGL：底层的 2D 图形引擎。
- 3D libraries：基于 OpenGL ES 1.0 APIs 实现的 3D 引擎。
- FreeType：位图（bitmap）和矢量（vector）字体显示。
- SQLite：轻型关系型数据库引擎。

2) Android 运行时环境

Android 运行时环境主要包括：

- Android 核心库：提供了 Java 库的大多数功能。
- Dalvik 虚拟机：依赖于 linux 内核的一些功能，例如线程机制和底层内存管理机制。同时虚拟机是基于寄存器的，Dalvik 采用简练、高效的 byte code 格式运行，它能够在低资耗和没有应用相互干扰的情况下并行执行多个应用，每一个 Android 应用程序都在它自己的进程中运行，都拥有一个独立的 Dalvik 虚拟机实例。Dalvik 虚拟机中可执行文件为.dex 文件，该格式文件针对小内存使用做了优化。所有的类都经由 Java 编译器编译，然后通过 SDK 中的 dx 工具转化成.dex 格式由虚拟机执行。

4. Linux 内核

Android 平台运行在 Linux 2.6 之上，其 Linux 内核部分相当于手机硬件层和软件层之间的一个抽象层。Android 的内核提供了显示驱动、摄像头驱动、闪存驱动、键盘驱动、Wi-Fi 驱动、音频驱动和电源管理等多项功能。此外，Android 为了让 Android 程序可以用于商业目的，将 Linux 系统中受 GNU 协议约束的部分进行了取代。

1.4 Android 开发环境

在搭建环境前，需要了解安装开发工具所需要的硬件和软件配置条件。

1.4.1 Android 系统需求

本节介绍使用 Android SDK 进行开发所需的硬件和软件需求。对于硬件方面，要求 CPU 和内存尽量大。Android SDK 全部下载大概需要占用 4.5GB 硬盘空间。由于开发过程中需要反复重启模拟器，而每次重启都会消耗几分钟的时间（视机器配置而定），因此使用高配置的机器能节约不少时间。

支持 Android SDK 的操作系统及其要求如表 1-1 所示。

表 1-1　Android SDK 对操作系统的要求

操 作 系 统	要　　　求
Windows	Windows XP(32 位)
	Vista(32 或 64 位)
	Windows 7(32 位或 64 位)
Mac OS	10.5.8 或更新(仅支持 x86)
Linux(在 Ubuntu 的 10.04 版测试)	需要 GNU C Library(glibc)2.7 或更新 在 Ubuntu 系统上，需要 8.04 版或更新 64 位版本必须支持 32 位应用程序

对于开发环境，除了常用的 Eclipse IDE，还可以使用 Intelli J IDEA 进行开发。对于 Eclipse 在下载 Android SDK 时就自带相兼容的版本。

1.4.2　JDK 安装

在 Windows 平台上，搭建 Android 开发环境，首先下载并安装与开发环境相关的软件资源，这些资源主要包括 JDK、Eclipse、Android SDK 和 Development Tools 插件(ADT 插件)。

在 Android 平台上，所有应用程序都是使用 Java 语言编写的，所以要安装 Java 开发包 JDK(Java SE Development Kit)，JDK 是 Java 开发时所必需的软件开发包。

安装 JDK 的过程比较简单，运行该程序后，根据安装提示选择安装路径，将 JDK 安装到指定的文件夹即可，默认安装目标为 C:\Program Files\Java\jdk1.6.0_10(jdk-6u10-rc2-bin-b32-windows-i586-p-12_sep_2008)。

JDK 安装完毕后，要进一步设置 Java 的环境变量，即设置 bin 和 lib 文件夹的路径。其操作步骤如下(计算机操作系统为 Windows 7)：

（1）右击"计算机"图标，在弹出的快捷菜单中选择"属性"选项，在弹出的"系统"对话框中，单击"高级系统设置"按钮，弹出"系统属性"对话框，如图 1-2 所示。

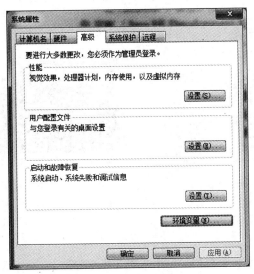

图 1-2　"系统属性"对话框

（2）在"系统属性"对话框的"高级"选项卡中，单击"环境变量"按钮，弹出"环境变量"对话框，如图1-3所示。

（3）选中"系统变量"区域的Path变量，单击"编辑"按钮，弹出"编辑系统变量"对话框，如图1-4所示。

图1-3 "环境变量"对话框　　　　　　　　　图1-4 环境变量Path设置

（4）在该对话框的"变量值"文本框中添加C:\Program Files\Java\jdk1.6.0_10\bin，然后单击"确定"按钮即可完成设置。这样即设置了bin文件夹的路径。

（5）在"环境变量"对话框的"系统变量"区域，单击"新建"按钮，弹出"新建系统变量"对话框，如图1-5所示。

（6）在图1-5中的"变量名(N)"右侧文本框中输入classpath，在"变量值(V)"右侧文本框中输入C:\Program Files\Java\jdk1.6.0_10\lib，即可设置lib文件夹的路径。

完成以上操作后，一个典型的Java开发环境便设置好了。在正式开始下一步前先验证Java开发环境的设置是否成功。

在Windows 7系统中单击"开始"按钮，在弹出的窗口中选择"运行"，在运行框中输入cmd并按Enter键，即可打开CMD窗口，在窗口中输入java - version，则可显示所安装的Java版本信息，如图1-6所示。

图1-5 新建环境变量classpath　　　　　　　图1-6 JDK安装成功页面

1.4.3 Eclipse环境

从Android 4.4版本开始，下载的Android软件包中包括Eclipse软件，在官方网站下载相应Android软件包并解压即可看到Eclipse软件启动器，双击该软件，打开效果如图1-7所示。

图 1-7　Eclipse 启动界面

启动 Eclipse 开发环境桌面，将会看到选择工作空间的提示，如图 1-8 所示。

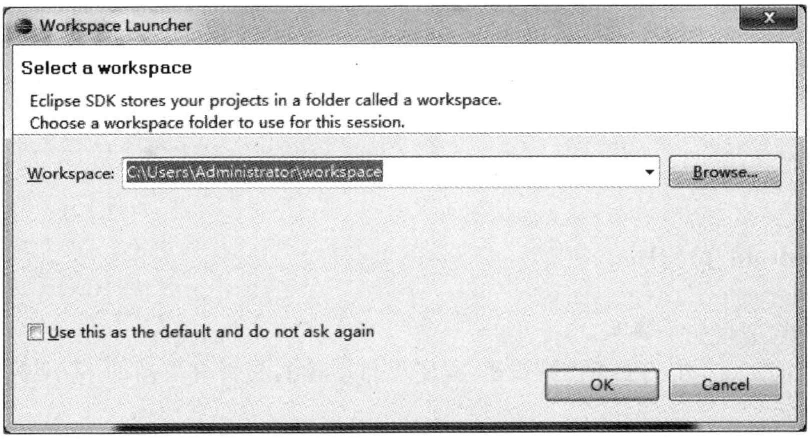

图 1-8　选择工作空间

接着单击图 1-8 中的 OK 按钮，即完成 Eclipse 的安装，系统进入 Eclipse 初始欢迎界面，如图 1-9 所示。接着单击图 1-9 左上角的"欢迎"按钮，即可进入 Eclipse 的开发环境界面，如图 1-10 所示。

图 1-9　Eclipse 欢迎界面

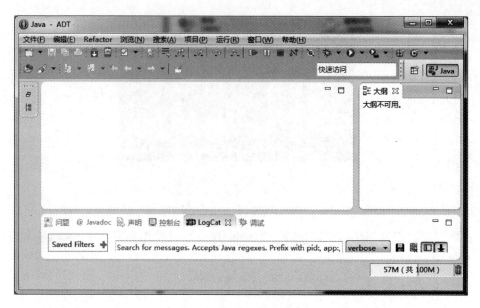

图 1-10　Eclipse 的开发界面

1.4.4　Android 的 SDK

安装 SDK 的操作步骤如下。

(1) 单击图 1-10 中的 ![btn] 快捷按钮,程序将自动检测是否有更新的 SDK 数据包可下载,如图 1-11 所示。

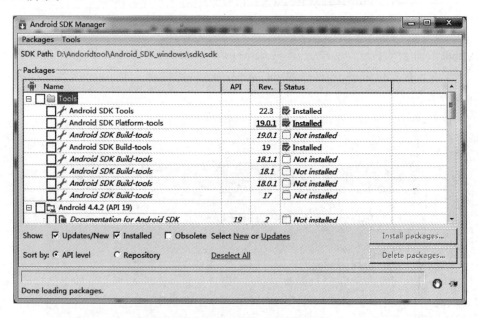

图 1-11　运行 SDK Manager.exe 执行文件

(2) 对于所要更新的内容,如果只要尝试一下 Android 4.4.2,那么只需要选择 Android 4.4.2 (API 19),然后单击 Install X packages 按钮安装就可以了。如果要在此 SDK 上开发应用程序和游戏应用,那么需要接受并遵守所有许可内容(Accept All),并单击 Install 按钮。

（3）接着将 SDK tools 目录的完整路径设置到系统变量 Path 中，这样便于在后面调用 Android 命令时，无须输入绝对路径。设置系统变量 Path 的方法与 JDK 的环境变量值操作一致，在 Path 环境变量的"变量值(V)"文本框中添加；D:\Androidtool\Android_SDK_windows\sdk\sdk\tools；即可，如图 1-12 所示。

图 1-12 设置 Android SDK 环境变量

最后检查 Android SDK 是否安装成功，能够正常运行。在 Windows 7 系统中单击"开始"按钮，在弹出的窗口中选择"运行"，在运行框中输入 cmd 并按 Enter 键，即可打开 CMD 窗口。在窗口中输入 android-h，则可显示所安装的 Android SDK 信息，如图 1-13 所示。

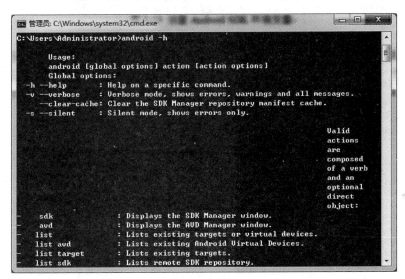

图 1-13 Android SDK 安装成功信息

1.4.5 Android 的 AVD

AVD 全称为 Android Virtual Device，是 Android 运行的虚拟设备，它是 Android 的模拟器识别。建立的 Android 要运行，必须创建 AVD，每个 AVD 上可以配置很多的运行项目。创建 AVD 时，可以配置的选项有模拟影像大小、触摸屏、轨迹球、摄像头、屏幕分辨率、键盘、GSM、GPS、Audio 录放、SD 卡支持和缓存大小等。

设置 AVD 的操作步骤如下。

（1）单击图 1-10 中的 ![icon] 快捷按钮，即可启动 Android AVD，弹出图 1-14 所示的 Android Virtual Device Manager 窗口。

（2）单击图 1-14 右侧的 New 按钮，弹出一个新的 Create new Android Virtual Device（AVD）对话框，如图 1-15 所示。在该对话框中可以设置模拟器的配置，主要包括如下几项。

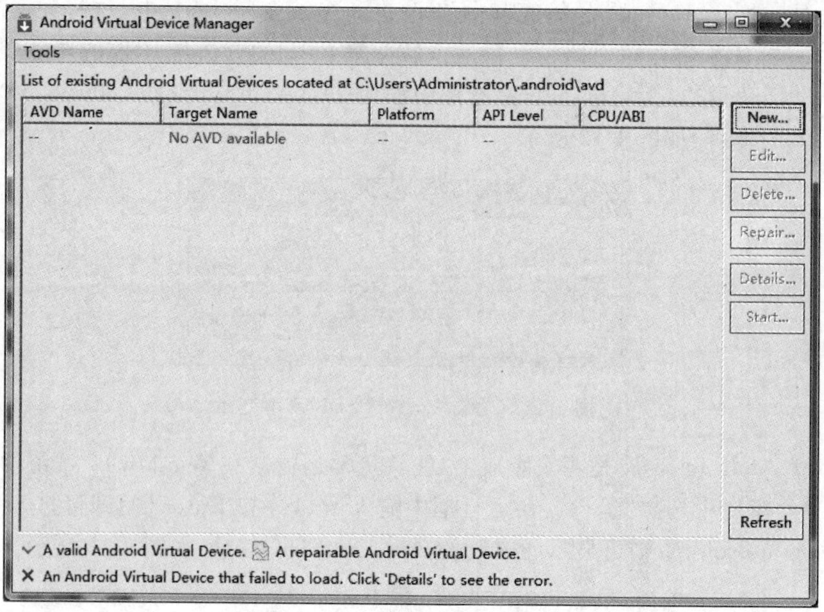

图 1-14　Android Virtual Device Manager 窗口

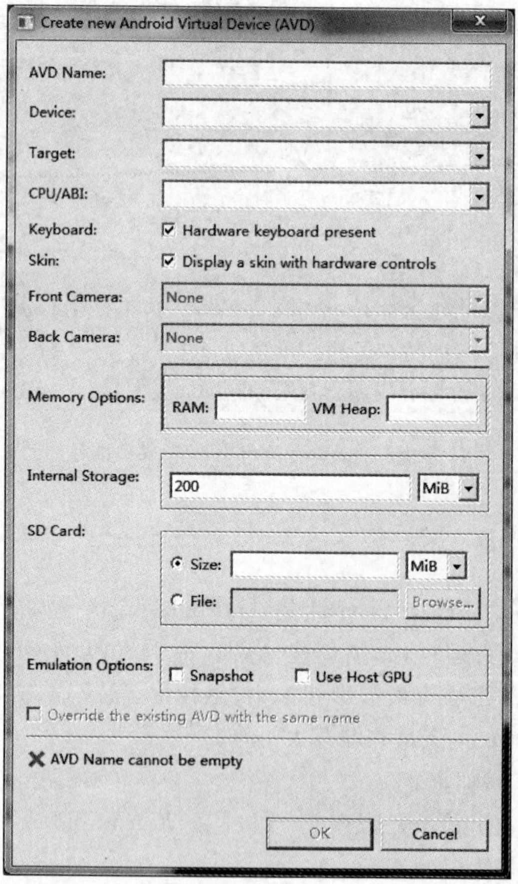

图 1-15　新建 AVD 时的 emulate 设置

- AVD Name：创建 AVD 的名称。可以在文本框中输入所要创建的 AVD 的名称，注意名称中不能有空格符。
- Target：选择 Android 版本和 API 的等级。单击右边的下拉按钮，选择相应的 Android 版本和 API 的等级。
- SD Card：设置 SD 卡。在 Size 文本中指定 SD 卡大小。另外，也可以在 File 文本框设置已有的 SD 卡镜像文件的路径。
- Skin：设置模拟器的外观和屏幕分辨率。单击 Built-in 右边的下拉按钮，可以选择默认的 HVGA（320×480）、QVGA（240×320）、WVGA（480×800 或 480×854）和 WQVGA（240×400 或 240×320）几种，在此选择默认的 HVGA（320×480）。另外，单击 Resolution 项，还可以自定义分辨率。不同版本的 Android 所设置的 Skin 参数有所不同。
- Hardware：设置模拟器支持的硬件设备的属性，包括影像大小、触摸屏、轨迹球、摄像头、屏幕分辨率、键盘、GSM、GPS、Audio 录放、SD 卡支持和缓存区大小等。单击该区域右边的 OK 按钮，在弹出的对话框中可以设置各项的属性。

（3）设置好模拟器的参数后，单击图 1-16 下边的 New 按钮即可创建一个 AVD。创建好的 AVD 将会显示在图 1-16 所示的 Android Virtual Device Manager 窗口的文件列表中。

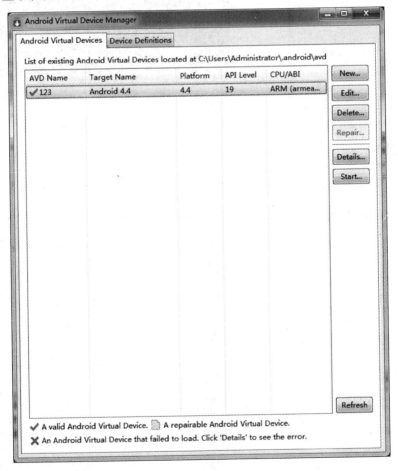

图 1-16　创建新的 AVD 界面

（4）选中所创建的 AVD 选项，单击右侧的 Start 按钮，弹出图 1-17 所示的 Launch Options 窗口。

（5）单击 Launch Options 窗口下的 Launch 按钮，即成功启动 AVD，效果如图 1-18 所示。

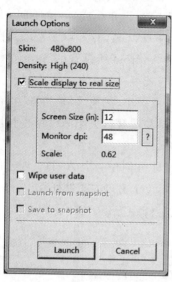

图 1-17　Launch Options 窗口

图 1-18　AVD 界面

使用同样的操作可以根据需要创建多个 AVD 模拟器。这样做的好处是，可以模拟程序在不同的 Android 版本上运行的兼容性。

图 1-18 右侧的各个控制按钮名称及其功能如表 1-2 所示。

表 1-2　AVD 的控制按钮功能

模拟器 AVD 的模拟按键	相应的图标	功　　能
音量渐小按钮		控制音量大小
电源按钮		设置电话模式，AVD 开关
音量增加按钮		控制音量大小
上/下/左/右按钮		确定按钮
中心按钮		上/下/左/右移动焦点
Home 按钮		返回主界面
Menu 按钮		打开应用程序菜单
查询按钮		在手机内部或上网查询
返回按钮		返回上一级界面

1.5 Android 应用结构分析

Android 的应用项目主要由以下部分组成。
- src 文件：项目源文件都保存在这个目录中。
- R.java 文件：这个文件是 Eclipse 自动生成的，应用开发者不需要去修改里面的内容。
- Android Library：这个为应用运行的 Android 库。
- assets 目录：主要放置多媒体等一些文件。
- res 目录：主要放置应用程序会用到的资源文件。
- drawable 目录：主要放置应用程序会用到的图片资源。
- layout 目录：主要放置用到的布局文件。这些布局文件都是 XML 文件。
- value 目录：主要放置字符串（strings.xml）、颜色（colors.xml）和数组（arrays.xml）。
- Androidmanifest.xml：相当于应用的配置文件。在这个文件中，必须声明应用的名称，应用所用到的 Activity、Service 以及 Receiver 等。

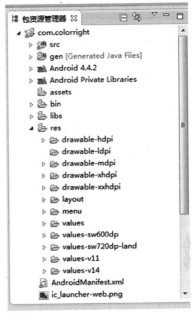

图 1-19 Android 应用工程文件组成

在 Eclipse 中，一个基本的 Android 项目的目录结构如图 1-19 所示。

1. src 目录

与一般的 Java 项目一样，src 目录下保存的是项目的所有包及源文件（.java），res 目录下包含了项目中的所有资源。例如，程序图标（drawable）、布局文件（layout）和常量（value）等。不同的是，在 Java 项目中没有 gen 目录，也没有每个 Android 项目都必须有的 AndroidManifest.xml 文件。

.java 格式文件是在建立项目时自动生成的，这个文件为只读模式，而 R.java 文件是定义该项目所有资源的索引文件。先来看看 Helloworld 项目的 R.java 文件，代码如下：

```
package fs.helloworld;
public final class R {
    public static final class attr {
    }
    public static final class dimen {
        public static final int activity_horizontal_margin = 0x7f040000;
        public static final int activity_vertical_margin = 0x7f040001;
    }
    public static final class drawable {
        public static final int ic_launcher = 0x7f020000;
    }
    public static final class id {
        public static final int action_settings = 0x7f080000;
    }
    public static final class layout {
```

```java
        public static final int main = 0x7f030000;
    }
    public static final class menu {
        public static final int main = 0x7f070000;
    }
    public static final class string {
        public static final int action_settings = 0x7f050001;
        public static final int app_name = 0x7f050000;
        public static final int hello_world = 0x7f050002;
    }
    public static final class style {
        public static final int AppTheme = 0x7f060001;
    }
}
```

从上述代码中,可以看到文件定义了很多常量,并且会发现这些常量的名字都与 res 文件夹中的文件名相同,这再次证明.java 文件中所存储的是该项目所有资源的索引。有了这个文件,在程序中使用资源将变得更加方便,可以很快地找到要使用的资源,由于这个文件不能手动编辑,所以当用户在项目中加入新的资源时,只需要刷新一下该项目,.java 文件便自动生成所有资源的索引。

2. res 目录

在 res 目录下包含该项目所使用到的资源文件,这里面的每一个文件或者资源都将在 R.java 文件中进行索引定义。主要包括如下几类。

- 图片文件:分别提供高分辨率(drawable-hdpi)、低分辨率(drawable-ldpi)、中分辨率(drawable-mdpi)、超高分辨率(drawable-xhdpi)和超高清分辨率(drawable-xxhdpi)的图片文件。
- 布局文件:在 layout 目录下,默认只有一个 main.xml,用户也可以添加更多的布局文件。
- 字符串:在 values 目录下的 strings.xml 文件中。

打开 main.xml 布局文件。代码为:

```xml
<RelativeLayout xmlns:android = "http://schemas.android.com/apk/res/android"
    xmlns:tools = "http://schemas.android.com/tools"
    android:layout_width = "match_parent"
    android:layout_height = "match_parent"
    android:paddingBottom = "@dimen/activity_vertical_margin"
    android:paddingLeft = "@dimen/activity_horizontal_margin"
    android:paddingRight = "@dimen/activity_horizontal_margin"
    android:paddingTop = "@dimen/activity_vertical_margin"
    tools:context = ".MainActivity" >
    <TextView
        android:layout_width = "wrap_content"
        android:layout_height = "wrap_content"
        android:text = "@string/hello_world" />
</RelativeLayout>
```

在该布局文件中,首先定义了相对布局,内部只有一个文本框控件。这个控件显示了内容引用 string 文件中 hello 变量。

其中,

- \<RelativeLayout\>\</RelativeLayout\>:相对版面配置,在这个标签中,所有元件都

是按相对排队排成的。
- android:layout_width：定义当前视图在屏幕上所占的宽度，fill_parent 即填充整个屏幕。
- android:layout_height：随着文字栏位的不同而改变这个视图的宽度或高度。
- android:paddingBottom：指屏幕界面底部的填充方式。
- android:paddingLeft：指屏幕界面左侧的填充方式。
- android:paddingRight：指屏幕界面右侧的填充方式。
- android:paddingTop：指屏幕界面顶部的填充方式。
- tools:context：该布局文件所调用的 Activity 内容。

在上述布局代码中，使用一个 TextView 来配置文件标签 Widget(构件)，其中设置的属性 android:layout_width 为整个屏幕的宽度，android:layout_height 可以根据文字改变高度，而 android:text 则设置了这个 TextView 要显示的文字内容，这里引用@string 中的 hello 字符串，即 String.xml 文件中的 hello 所代表的字符串资源。Hello 字符串的内容"HelloWorld、HelloAndroid"这就是用户在 HelloAndroid 项目运行时看到的字符串。

Strings.xml 文件的代码为：

```xml
<?xml version = "1.0" encoding = "utf-8"?>
<resources>
    <string name = "app_name">Hello World</string>
    <string name = "action_settings">Settings</string>
    <string name = "hello_world">Hello world!</string>
</resources>
```

3. AndroidManfest.xml 文件

在文件 AndroidManfest.xml 中包含该项目中所使用的 Activity、Service 和 Receiver，以下代码为"HelloWorld"项目中的 AndroidManfest.xml 文件。

```xml
<?xml version = "1.0" encoding = "utf-8"?>
<manifest xmlns:android = "http://schemas.android.com/apk/res/android"      //根节点
    package = "fs.helloworld"                          //包名
    android:versionCode = "1"
    android:versionName = "1.0" >
    <uses-sdk
        android:minSdkVersion = "8"
        android:targetSdkVersion = "18" />             //SDK 版本
    <application                                       //图标和应用程序名称
        android:allowBackup = "true"
        android:icon = "@drawable/ic_launcher"
        android:label = "@string/app_name"
        android:theme = "@style/AppTheme" >
        <activity
            android:name = "fs.helloworld.MainActivity"              //默认启动的 Activity
            android:label = "@string/app_name" >       //Activity 名称
            <intent-filter>
                <action android:name = "android.intent.action.MAIN" />
                <category android:name = "android.intent.category.LAUNCHER" />
            </intent-filter>
        </activity>
    </application>
</manifest>
```

1.6 Android 应用程序权限说明

一个 Android 应用可能需要权限才能调用 Android 系统的功能；一个 Android 应用也可能被其他应用调用，因此它也需要声明调用自身所需要的权限。

1. 声明所拥有的权限

通过为＜manifest…/＞元素添加＜uses-permission…/＞子元素即为自身声明权限。

例如在＜manifest…/＞元素中添加如下代码：

```
<!-- 声明需要打电话权限 -->
<uses-permission android:name="android.permission.CALL_PHONE"/>
```

2. 声明所需的权限

通过为应用的各组件元素，如＜activity…/＞元素添加＜uses-permission…/＞子元素即可声明调用该程序所需要的权限。

例如在＜activity…/＞元素中添加如下代码：

```
<!-- 声明需要发送短信的权限 -->
<uses-permission android:name="android.permission.SEND_SMS"/>
```

通过以上介绍可看出，＜uses-permission…/＞元素的用法不难，但到底有多少权限呢？实际上，Android 提供了大量的权限，这些权限都位于 Manifest.permission 类中。一般常用的权限如表 1-3 所示。

表 1-3 Android 系统的常用权限

权　　限	说　　明
ACCESS_NETWORK_STATE	允许应用程序获取网络状态信息的权限
ACCESS_WIFL_STATE	允许应用程序获取 Wi-Fi 网络状态信息的权限
BATTERY_STATS	允许应用程序获取电池状态信息的权限
BLUETOOTH	允许应用程序连接匹配的蓝牙设备的权限
BLUETOOTH_ADMIN	允许应用程序发现匹配的蓝牙设备的权限
BROADCAST_SMS	允许应用程序广播收到短信提醒的权限
CALL_PHONE	允许应用程序拨打电话的权限
CAMERA	允许应用程序使用照相机的权限
CHANGE_NETWORK_STATE	允许应用程序改变网络连接状态的权限
CHANGE_WIFL_STATE	允许应用程序 Wi-Fi 网络连接状态的权限
DELETE_CACHE_FILES	允许应用程序的删除缓存文件权限
DELETE_PACKAGES	允许应用程序删除安装包的权限
FLASHLIGHT	允许应用程序访问闪光灯的权限
INTERNET	允许应用程序打开网络 Socket 的权限
MODIFY_AUDIO_SETTINGS	允许应用程序修改全局声音设置的权限
PROCESS_OUTGOING_CALLS	允许应用程序监听、控制、取消呼出电话的权限
READ_CONTACTS	允许应用程序读取用户的联系人数据的权限
READ_HISTORY_BOOKMARKS	允许应用程序读取历史书签的权限
READ_OWNER_DATA	允许应用程序读取用户数据的权限
READ_PHONE_STATE	允许应用程序读取电话状态的权限

续表

权　限	说　明
READ_PHONE_SMS	允许应用程序读取短信的权限
REBOOT	允许应用程序重启系统的权限
RECEIVE_MMS	允许应用程序接收、监控、处理彩信的权限
RECEIVE_SMS	允许应用程序接收、监控、处理短信的权限
RECORD_AUDIO	允许应用程序录音的权限
SEND_SMS	允许应用程序发送短信的权限
SET_ORIENTATION	允许应用程序旋转屏幕的权限
SET_TIME	允许应用程序设置时间的权限
SET_TIME_ZONE	允许应用程序设置时区的权限
SET_WALLPAPER	允许应用程序设置桌面壁纸的权限
VIBRATE	允许应用程序访问振动器的权限
WRITE_CONTACTS	允许应用程序写入用户联系人的权限
WRITE_HISTORY_BOOKMARKS	允许应用程序写历史书签的权限
WRITE_OWNER_DATA	允许应用程序写用户数据的权限
WRITE_SMS	允许应用程序写短信的权限

1.7 Android 基本组件介绍

Android 应用通常由一个或多个基本组件组成，而最常用的组件为 Activity。事实上 Android 应用还可能包括 Service、BroadcastReceiver 和 ContentProvider 等组件。

1.7.1 Activity 与 View

Activity 是 Android 应用中负责与用户交互的组件，其与 Swing 编程中的 JFrame 控件的区别在于：JFrame 本身可以设置布局管理器，不断地向 JFrame 中添加组件，但 Activity 只能通过 setContentView(View) 显示指定组件。

View 组件是所有 UI 控件、容器控件的基类，View 组件就是 Android 应用中用户实实在在看到的部分。但 View 组件需要放到容器组件中，或者使用 Activity 将它显示出来。如果需要通过某个 Activity 把指定 View 显示出来，调用 Activity 的 setContentView() 方法即可。

setContentView() 方法可接收一个 View 对象作为参数，如以下代码所示：

```
//创建一个线性布局管理器
LinearLayout layout = new LinearLayout(this);
//设置该 Activity 显示 layout
super.setContentView(layout);
```

上面代码创建了一个 LinearLayout 对象（它是 ViewGroup 的子类，LinearLayout 又是 View 的子类），接着调用 Activity 的 setContentView(layout) 把这个布局管理器显示出来。

setContentView() 方法也可以接收一个布局管理资源的 ID 作为参数。代码为：

```
//设置 Activity 显示 main.xml 文件定义的 View
setContentView(R.layout.main);
```

从这个角度来看,大致可把 Activity 理解为 Swing 中的 JFrame 组件。当然,Activity 可完成的功能比 JFrame 更多。

Activity 为 Android 应用提供了可视化用户界面,如果该 Android 应用需要多个用户界面,那么这个 Android 应用将包含多个 Activity,多个 Activity 组成 Activity 栈,当前活动的 Activity 位于栈顶。

Activity 包含了一个 setTheme(int resid)方法来设置其窗口的风格,如果希望窗口不显示标题,则以对话框形式显示窗口,都可通过该方法实现。

1.7.2 Service

Service 是 Android 系统中的一种组件,它跟 Activity 的级别差不多,但是它不能自己运行,只能在后台运行,并且可以和其他组件进行交互。Service 是没有界面的长生命周期的代码。Service 是一种程序,它可以运行很长时间,但是它却没有用户界面。这么说有点枯燥,来看个例子。打开一个音乐播放器的程序,这个时候如果想上网,那么,打开 Android 浏览器,虽然已经进入浏览器这个程序,但是,歌曲播放并没有停止,而是在后台继续一首接着一首的播放。其实这个播放就是由播放音乐的 Service 进行控制。当然这个播放音乐的 Service 也可以停止,例如,当播放列表里边的歌曲都结束,或者用户按下了停止音乐播放的快捷键等。Service 可以在多场合的应用中使用,例如播放多媒体的时候用户启动了其他 Activity 这个时候程序要在后台继续播放,例如检测 SD 卡上文件的变化,再或者在后台记录地理信息位置的改变等,而服务是藏在后头的。

开启 Service 有两种方式。

(1) Context.startService():Service 会经历 onCreate→onStart(如果 Service 还没有运行,则 Android 先调用 onCreate(),然后调用 onStart();如果 Service 已经运行,则只调用 onStart(),所以一个 Service 的 onStart 方法可能会重复调用多次);如果调用 StopService 即直接进入 onDestroy,如果是调用者自己没有调用 StopService 的话,Service 会一直在后台运行。该 Service 的调用者再启动起来后可以通过 stopService 关闭 Service。注意,多次调用 Context.startservice()不会嵌套(即使会有相应的 onStart()方法被调用),所以无论同一个服务被启动多少次,一旦调用 Context.stopService()或者 StopSelf(),它都会被停止。补充说明:传递给 StartService()的 Intent 对象会传递给 onStart()方法。调用顺序为:onCreate → onStart(可多次调用) →onDestroy。

(2) Context.bindService():Service 会经历 onCreate()→onBind(),onBind 将返回给客户端一个 IBind 接口实例,IBind 允许客户端回调服务的方法,例如得到 Service 运行的状态或其他操作。这个时候调用者(Context,例如 Activity)会和 Service 绑定在一起,当 Context 退出了,Srevice 就会调用 onUnbind→onDestroyed 相应退出,所谓绑定在一起即为"共存亡"了。

1.7.3 BroadcastReceiver

在 Android 中,Broadcast 是一种广泛运用的在应用程序之间传输信息的机制。而 BroadcastReceiver 是对发送出来的 Broadcast 进行过滤接收并响应的一类组件。可以使用 BroadcastReceiver 让应用对一个外部的事件做出响应。这是非常有意思的,例如,当电话呼入这个外部事件到来的时候,可以利用 BroadcastReceiver 进行处理。例如,当下载一个程序成功完成的时候，仍然可以利用 BroadcastReceiver 进行处理。BroadcastReceiver 不能生成 UI，也就是说对于用户来说是不透明的，用户是看不到的。BroadcastReceiver 通过 NotificationManager 通知用户这些事情发生了。BroadcastReceiver 既可以在 AndroidManifest.xml 中注册，也可以在运行时的代码中使用 Context.registerReceiver() 进行注册。只要是注册了，当事件来临的时候，即使程序没有启动，系统也在需要的时候启动程序。各种应用还可以通过使用 Context.sendBroadcast() 将它们自己的 Intent Broadcasts 广播给其他应用程序。

1.7.4 ContentProvider

对于 Android 应用而言，它们必须相互独立，各自运行在自己的 Dalvik 虚拟机实例中，如果这些 Android 应用之间需要实现实时的数据交换。例如开发一个发送短信的程序，当发送短信时需要从联系人管理应用中读取指定联系人的数据——这就需要多个应用程序之间进行实时的数据交换。

Android 系统为这种跨应用的数据交换提供了一个标准：ContentProvider。当用户实现自己的 ContentProvider 时，需要实现如下抽象方法。

- insert(Uri,ContentValues)：向 ContentProvider 插入数据。
- delete(Uri,ContentValues)：删除 ContentProvider 中指定数据。
- udpate(Uri,ContentValues,String,String[])：更新 ContentProvider 中指定数据。
- query(Uri,String[],String,String[],String)：从 ContentProvider 查询数据。

通常与 ContentProvider 结合使用的是 ContentResolver，一个应用程序使用 ContentProvider 暴露自己的数据，而另一个应用程序则通过 ContentResolver 来访问数据。

1.7.5 Intent 与 IntentFilter

严格地说，Intent 并不是 Android 应用的组件，但它对于 Android 应用的作用非常大，它是 Android 应用内不同组件之间通信的载体。当 Android 运行时需要连接不同的组件时，通常就需要借助于 Intent 来实现。Intent 可以启动应用中另一个 Activity，也可以启动一个 Service 组件，还可以发送一个广播消息来触发系统中的 BroadcastReceiver。也就是说，Activity、Service 和 BroadcastReceiver 3 种组件之间的通信都以 Intent 作为载体，只是不同组件使用 Intent 的机制略有区别而已。

- 当需要启动一个 Activity 时，可调用 Context 的 startActivity(Intent intent) 方法，该方法中的 Intent 参数封装了需要启动的目标 Activity 的信息。
- 当需要启动一个 Service 时，可调用 Context 的 startService(Intent intent) 方法或 bindService(Intent service,ServiceConnection conn,int flags) 方法，这两个方法中的 Intent 参数封装了需要启动的目标 Service 的信息。
- 当需要触发一个 BroadcastReceiver 时，可调用 Context 的 sendBroadcast(Intent intent)、sendStickyBroadcast(Intent intent) 或 sendOrderedBroadcast(Intent intent, String receiverPermission) 方法发送广播消息，这 3 个方法中的 Intent 参数封装了需要触发的目标 BroadcastReceiver 的信息。

通过介绍可看出，Intent 封装了当前组件需要启动或触发的目标组件的信息，Intent 也可称为"意图"。实际上 Intent 对象中封装了大量关于目标组件的信息。

当一个组件通过 Intent 表示启动或触发另一个组件的"意图"后,这个意图可分为两类。
- 显式 Intent:显式 Intent 明确指定需要启动或者触发的组件的类名。
- 隐式 Intent:隐式 Intent 只是指定需要启动或者触发的组件应满足怎样的条件。

对于显式 Intent 而言,Android 系统无须对该 Intent 做任何解析,系统直接找到指定的目标组件,启动或触发它即可。

对于隐式 Intent 而言,Android 系统需要对该 Intent 进行解析,解析出它的条件,然后再去系统中查找与之匹配的目标组件。如果找到符合条件的组件,就启动或触发它们。

那么 Android 系统怎样判断调用组件是否符合隐式 Intent 呢?这就需要靠 IntentFilter 来实现,被调用组件可通过 IntentFilter 声明自己所满足的条件——也就是声明自己到底能处理哪些隐式 Intent。

1.8 应用程序生命周期

程序也如同自然界的生物一样,有自己的生命周期。应用程序的生命周期即程序的存活时间。Android 是构建在 Linux 之上的开源移动开发平台,在 Android 中,多数情况下每个程序都是在各自独立的 Linux 进程中运行的。当一个程序或其某些部分被请求时,它的进程就"出生"了,当这个程序没有必要再运行下去且系统需要回收这个进程的内存用于其他程序时,这个进程即为"死亡"了。可看出,Android 程序的生命周期是由系统控制的而非程序自身直接控制。这和人们编写桌面应用程序时的思维有一些不同,一个桌面应用程序的进程也是在其他进程或用户请求时被创建的,但是往往是在程序自身收到关闭请求后执行一个特定的动作(例如从 main 函数中返回)而导致进程结束。要想做好某种类型的程序或者某种平台下的程序的开发,最关键的就是要弄清楚这种类型的程序或整个平台下的程序的一般工作模式并熟记于心。在 Android 中,程序的生命周期控制就是属于这个范畴。

开发者必须理解不同的应用程序组件,尤其是 Activity、Service 和 Intent Receiver。了解这些组件是如何影响应用程序的生命周期的,这非常重要。如果不正确地使用这些组件,可能会导致系统终止正在执行重要任务的应用程序进程。

一个常见的进程生命周期漏洞的例子是 Intent Receiver(意图接收器),当 Intent Receiver 在 onReceive 方法中接收到一个 Intent(意图)时,其会启动一个线程,然后返回。一旦返回,系统将认为 Intent Receiver 不再处于活动状态,因而 Intent Receiver 所在的进程也就不再有用了(除非该进程中还有其他的组件处于活动状态)。因此,系统可能会在任意时刻终止该进程以回收占用的内存。这样进程中创建出的那个线程也将被终止。解决这个问题的方法是从 Intent Receiver 中启动一个服务,让系统知道进程中还有处于活动状态的工作。为了使系统能够正确地决定在内存不足时终止进程,Android 根据每个进程中运行的组件及组件的状态把进程放入一个 Importance Hierarchy(重要性分组)中。进程的类型按重要程度排序。

一个 Android 程序的进程是何时被系统结束的呢?通俗地说,一个即将被系统关闭的程序是系统在内存不足(low memory)时,根据"重要性层次"选出来的"牺牲品"。一个进程的重要性是根据其中运行的部件和部件的状态决定的。各种进程按照重要性从高到低排列如图 1-20 所示。

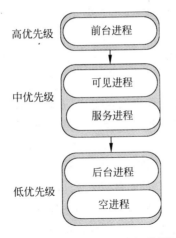

图 1-20　Android 进程优先级效果图

1. 前台进程

前台进程是与用户正在交互的进程，也是 Android 系统中最重要的进程。处于前台进程一般包含以下 4 种情况。

- 进行中的 Activity 正在与用户进行交互。
- 进程服务被 Activity 调用，而且这个 Activity 正在与用户进行交互。
- 进程服务正在执行生命周期中的回调函数，如 onCreate()、onStart() 或 onDestroy()。
- 进程的 BroadcastReceiver 正在执行 onReceive() 函数。

Android 系统为多任务操作系统，当系统中的多个前台进程同时运行时，如果出现资源不足的情况，此时 Android 内核将自动清除部分前台进程，保证最主要的用户界面能够及时响应操作。

2. 可见进程

在屏幕上显示，但是不是在前台的程序。例如一个前台进程以对话框的形式显示在该进程前面。这样的进程也很重要，它们只有在系统没有足够内存运行所有前台进程时，才会被结束。

3. 服务进程

这样的进程在后台持续运行，例如后台音乐播放、后台数据上传下载等。这样的进程对用户来说一般很有用，所以只有当系统没有足够内存来维持所有的前台和可见进程时，才会被结束。

4. 后台进程

这样的程序拥有一个用户不可见的 Activity。这样的程序在系统内存不足时，按照 LRU 的顺序被结束。

5. 空进程

这样的进程不包含任何活动的程序部件。系统可能随时关闭这类进程。

从某种意义上讲，垃圾收集机制把程序员从"内存管理噩梦"中解放出来，而 Android 的进程生命周期管理机制把用户从"任务管理噩梦"中解放出来。见过一些 Nokia S60 用户和 Windows Mobile 用户要么因为长期不关闭多余的应用程序而导致系统变慢，要么因为不及时查看应用程序列表而影响使用体验。Android 使用 Java 作为应用程序 API，并且结合其独特的生命周期管理机制同时为开发者和使用者提供最大程度的便利。

1.9 Activity 生命周期

在 Android 中,一般用系统管理来决定进程的生命周期。有时因为手机所具有的一些特殊性,所以人们需要更多地去关注各个 Android 程序部分的运行时生命周期模型。所谓手机的特殊性,主要有如下两点。

(1) 在进行手机应用时,大多数情况下只能在手机上看到一个程序的一个界面,用户除了通过程序界面上的功能按钮在不同的窗体间切换,还可以通过 Back(返回)键和 Home(主)键来返回一个窗口,而用户使用 Back 键或者 Home 键的时机是非常不确定的,任何时候用户都可以使用 Back 键或 Home 键来强行切换当前的界面。

(2) 通常手机上一些特殊的事件发生也会强制改变当前用户所处的操作状态,例如,无论任何情况,在手机来电时,系统都会优先显示电话接听界面。

1.9.1 Activity 的周期

Activity 生命周期是指应用程序的 Activity 从启动到销毁的全过程。Activity 的生命周期是在 Android 应用程序设计最重要的内容,直接关系到用户程序的界面和功能。

每一个活动的状态是由它在活动栈中所处的位置决定的,活动栈是当前所有正在运行的进程的后进先出的集合。当一个新的活动启动时,当前的前台屏幕就会移动到栈顶。如果用户使用 Back(返回)按钮返回到刚才的活动,或者前台活动被关闭了,那么栈中的下一个活动就会移动到栈顶,变为活动状态。图 1-21 说明了这个过程。

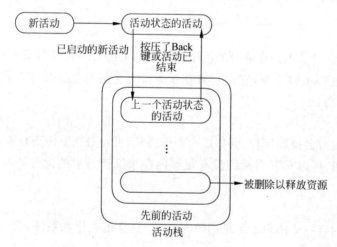

图 1-21 活动栈流程图

1.9.2 Activity 的状态

随着活动的创建和销毁,它们会按照图 1-21 所示的过程,从栈中移进移出。在这个过程中,它们也经历了活动、暂停、停止和非活动 4 种可能的状态,如图 1-22 所示。

- 活动状态:当一个活动位于栈顶的时候,它是可见的、被关注的前台活动,这时它可以

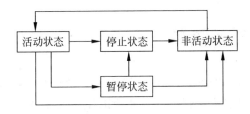

图 1-22　Activity 的状态

接收用户输入。Android 将会不惜一切代价来保持它处于活动状态,并根据需要来销毁栈下面部分的活动,以保证这个活动拥有它所需要的资源。当另一个活动变为活动状态时,这个活动就将被暂停。

- 暂停状态：在某些情况下,活动是可见的,但是没有被关注,此时它就处于暂停状态。当一个透明的或者非全屏的活动位于某个处于活动状态的活动之前时,这个透明的或者非全屏的活动就会达到这个状态。当活动被暂停的时候,它仍然会被当作近似于活动状态的状态,但是它不能接收用户的输入事件。在极端情况下,当一个活动变得完全不可见的时候,它就会变为停止状态。
- 停止状态：当一个活动不可见的时候,它就处于停止状态。此时,活动仍然会停留在内存中,保存所有的状态和成员信息,然而当系统的其他地方要求使用内存的时候,它们就会成为被清除的首要候选对象。在一个活动停止的时候,保存数据和当前的 UI 状态是很重要的。一旦一个活动被退出或者关闭,它就会变为非活动状态。
- 非活动状态：当一个活动被销毁之后,在它启动之前就处于非活动状态。处于非活动状态的活动已经从活动栈中移除了,因此,在它们可以被显示和使用之前,需要被重新启动。

在 Android 系统中,采用"栈"结构来管理 Activity,这是一种"后进先出"的原则,如图 1-23 所示。当一个 Activity 被启用时,将执行入栈操作。位于栈顶的 Activity 处于活动状态,其他 Activity 则处于暂停状态或者停止状态。当 Activity 关闭时,将执行出栈操作,从而改变成非活动状态。当 Android 系统资源紧张时,Android 内核也会终止部分长久没有响应的 Activity,使之为非活动状态,从而释放系统资源。

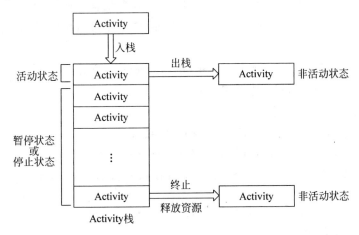

图 1-23　Activity 的栈结构

1.10 Android 进程与线程

当某个组件第一次运行的时候,Android 启动了一个进程。默认所有的组件和程序运行在这个进程和线程中。也可以安排组件在其他的进程或者线程中运行。

1.10.1 进程

组件运行的进程由 ManifestFile 控件负责,组件中的节点＜activity＞、＜service＞、＜receiver＞和＜provider＞都包含了一个 Process 属性。通过 Process 属性,可以设置组件运行的进程,既可以配置组件在一个独立进程中运行,也可以配置多个组件在同一个进程运行,甚至可以配置多个程序在同一个进程中运行(前提是这些程序共享一个 User ID 并给定同样的权限)。另外,在＜application＞节点中也包含了 Process 属性,可以用来设置程序中所有组件的默认进程。

所有的组件在此进程的主线程中实例化,系统对这些组件的调用从主线程中分离。并非每个对象都会从主线程中分离。一般来说,响应如 View.onKeyDown()用户操作的方法和通知的方法也在主线程中运行。这就表示,组件被系统调用的时候不应该长时间运行或者阻塞操作(如网络操作或者计算大量数据),因为这样会阻塞进程中的其他组件。可以把这类操作从主线程中分离。

当更加常用的进程无法获取足够内存,Android 可能会关闭不常用的进程。下次启动程序的时候会重新启动进程。

当决定哪个进程需要被关闭的时候,Android 会考虑哪个对用户更加有用。如 Android 会倾向于关闭一个长期不显示在界面的进程来支持一个经常显示在界面的进程。是否关闭一个进程取决于组件在进程中的状态。

1.10.2 线程

即使为组件分配了不同的进程,有时候也需要再分配线程。例如用户界面需要很快对用户进行响应,因此某些费时的操作,如网络连接、下载或者非常占用服务器时间的操作应该放到其他线程。

线程通过 Java 的标准对象 Thread 创建,在 Android 中提供了如下管理线程的方法。
- Looper:在线程中运行一个消息循环。
- Handler:传递一个消息。
- HandlerThread:创建一个带有消息循环的线程。

Android 会让一个应用程序在单独的线程中,指导它创建自己的线程。除了上述方法外,通过使用应用程序组件,如 Activity、Service 和 Broadcast Receiver,可以在主线程中实现实例化操作。

1.10.3 线程安全的方法

了解了进程和线程的基本知识后,开发人员很有必要了解线程安全方面的知识。在某些情况下，程序中的方法可能不止调用一个线程，这样在多个线程协同交互工作时，需要特别注意线程安全的问题。例如有一个方法正在调用在 IBinder 中的接口对象，此对象中的方法的程序启动了和 IBinder 对象相同的进程，这个方法就相当于在 IBinder 的进程中执行。但是，如果此时调用者发起另外一个进程，这个方法需要在另外一个线程中运行。因为这个线程和 IBinder 对象在同一个线程池中，所以它不会在进程的主线程中运行。假如应用中的一个 Service 从主线程调用 onBind() 方法，onBind() 返回的对象中的方法会从线程池中调用。这样就造成了一个服务被多个客户端请求的情景，不止一个线程池会在同一时间调用 IBinder 中的方法，所以此时 IBinder 必须保证线程安全。如果不安全，则会影响多个线程，从而影响和 IBinder 相关的所有应用。

第 2 章 Android 界面开发实例

Android 应用程序的界面开发,主要包括界面显示和事件处理两方面内容。对于界面显示,可以通过两个大类实现:一个是通过 ViewGroup 类进行整体布局;另一个是通过 View 类进行控件使用。对于事件处理则包括回调事件和监听事件等。

2.1 常用的布局实例

本节主要介绍 Android 几种常用的布局。

2.1.1 计算器实例

在本实例中,通过 LinearLayout 线性布局的应用,构建一个小型的计算器界面。该计算器实现整数间的加减乘除四则运算,并通过本计算器的实现,演示了 LinearLayout 线性布局的具体应用。

本实例中实现小型计算器软件的开发。该计算器通过 0~9 这 10 个数字按钮,"加"、"减"、"乘"、"除"和"等于"5 个运算按钮以及清空按钮实现对本实例的操控,并通过一个文本框显示计算的结果。应用本实例可以进行整数间的加、减、乘、除四则运算。其具体实现步骤如下。

(1) 在 Eclipse 中创建一个 Android 应用项目,命名为 LinearLayou_test。

(2) 打开 res\layout 目录下的 main.xml 布局文件,在文件中声明 4 个 LinearLayout 布局,并分别在这些线性布局中声明 Button 控件及 TextView 控件。代码为:

```
<?xml version = "1.0" encoding = "utf - 8"?>
<LinearLayout xmlns:android = "http://schemas.android.com/apk/res/android"
    android:orientation = "vertical"
    android:layout_width = "fill_parent"
    android:layout_height = "fill_parent"
    android:paddingTop = "5dip"
    android:background = " # 000000">
    <TextView
        android:id = "@ + id/tv"
        android:layout_width = "fill_parent"
        android:layout_height = "40dip"
```

```xml
            android:layout_marginRight = "5dip"
            android:layout_marginLeft = "5dip"
            android:background = "#FFFFFF"
            android:gravity = "center_vertical|right"
            android:textSize = "30dip"
            android:textColor = "#fffccc">
</TextView>
<LinearLayout
        android:orientation = "horizontal"
        android:layout_width = "fill_parent"
        android:layout_height = "wrap_content"
        android:paddingTop = "5dip">
    <Button
            android:text = "7"
            android:textSize = "25dip"
            android:id = "@ + id/Button07"
            android:layout_width = "80dip"
            android:layout_height = "wrap_content"/>
    <Button
            android:text = "8"
            android:textSize = "25dip"
            android:id = "@ + id/Button08"
            android:layout_width = "80dip"
            android:layout_height = "wrap_content"/>
    <Button
            android:text = "9"
            android:textSize = "25dip"
            android:id = "@ + id/Button09"
            android:layout_width = "80dip"
            android:layout_height = "wrap_content"/>
    <Button
            android:text = " + "
            android:textSize = "25dip"
            android:id = "@ + id/ButtonJia"
            android:layout_width = "80dip"
            android:layout_height = "wrap_content"/>
</LinearLayout>
<LinearLayout
            android:orientation = "horizontal"
            android:layout_width = "fill_parent"
            android:layout_height = "wrap_content"
            android:paddingTop = "5dip">
    <Button
            android:text = "4"
            android:textSize = "25dip"
            android:id = "@ + id/Button04"
            android:layout_width = "80dip"
            android:layout_height = "wrap_content"/>
    <Button
            android:text = "5"
            android:textSize = "25dip"
            android:id = "@ + id/Button05"
            android:layout_width = "80dip"
            android:layout_height = "wrap_content"/>
    <Button
```

```xml
            android:text = "6"
            android:textSize = "25dip"
            android:id = "@ + id/Button06"
            android:layout_width = "80dip"
            android:layout_height = "wrap_content"/>
    <Button
            android:text = " - "
            android:textSize = "25dip"
            android:id = "@ + id/ButtonJian"
            android:layout_width = "80dip"
            android:layout_height = "wrap_content"/>
</LinearLayout>
<LinearLayout
    android:orientation = "horizontal"
    android:layout_width = "fill_parent"
    android:layout_height = "wrap_content"
    android:paddingTop = "5dip">
    <Button
            android:text = "1"
            android:textSize = "25dip"
            android:id = "@ + id/Button01"
            android:layout_width = "80dip"
            android:layout_height = "wrap_content"/>
    <Button
            android:text = "2"
            android:textSize = "25dip"
            android:id = "@ + id/Button02"
            android:layout_width = "80dip"
            android:layout_height = "wrap_content"/>
    <Button
            android:text = "3"
            android:textSize = "25dip"
            android:id = "@ + id/Button03"
            android:layout_width = "80dip"
            android:layout_height = "wrap_content"/>
    <Button
            android:text = " * "
            android:textSize = "25dip"
            android:id = "@ + id/ButtonCheng"
            android:layout_width = "80dip"
            android:layout_height = "wrap_content"/>
</LinearLayout>
<LinearLayout
    android:orientation = "horizontal"
    android:layout_width = "fill_parent"
    android:layout_height = "wrap_content"
    android:paddingTop = "5dip">
    <Button
        android:text = "0"
        android:textSize = "25dip"
        android:id = "@ + id/Button00"
        android:layout_width = "80dip"
        android:layout_height = "wrap_content"/>
    <Button
        android:text = "c"
```

```xml
                android:textSize = "25dip"
                android:id = "@ + id/ButtonC"
                android:layout_width = "80dip"
                android:layout_height = "wrap_content"/>
            <Button
                android:text = " = "
                android:textSize = "25dip"
                android:id = "@ + id/ButtonDengyu"
                android:layout_width = "80dip"
                android:layout_height = "wrap_content"/>
            <Button
                android:text = "/"
                android:textSize = "25dip"
                android:id = "@ + id/ButtonChu"
                android:layout_width = "80dip"
                android:layout_height = "wrap_content"/>
        </LinearLayout>
</LinearLayout>
```

(3) 打开 src\fs.linearlayout 目录下的 MainActivity.java 文件,在文件中实现简单的四则运算和清空功能。代码为:

```java
package fs.linearlayout_test;
import android.app.Activity;
import android.os.Bundle;
import android.view.View;
import android.widget.Button;
import android.widget.TextView;
import android.view.View.OnClickListener;
public class MainActivity extends Activity
{
    TextView tv;
    int[] buttons;                                  //数字按钮数组
    int result;
    int result0;
    int result1;
    Button buttonC;                                 //按钮对象声明
    Button buttonJia;
    Button buttonJian;
    Button buttonCheng;
    Button buttonChu;
    Button buttonDengyu;
    String str1;                                    //旧输入的值
    String str2;                                    //新输入的值
    int flag = 0;              //计算标志位,0 第一次输入;1 加;2 减;3 乘;4 除;5 等于
    Button temp;
    @Override
    public void onCreate(Bundle savedInstanceState)
    {
        super.onCreate(savedInstanceState);
        setContentView(R.layout.main);              //跳转到 main 界面
        initButton();
        //清空按钮的单击事件监听器
        buttonC.setOnClickListener
        (
            new OnClickListener()
```

```java
        {
            public void onClick(View v)
            {
                str1 = "";
                str2 = "";                              //清空记录
                tv.setText(str1);
                flag = 0;
            }
        }
    );
    //监听
    for( int i = 0;i < buttons.length;i++)
    {
        temp = (Button)findViewById(buttons[i]);
        temp.setOnClickListener
        (                                               //为 Button 添加监听器
            new OnClickListener()
            {
                public void onClick(View v)
                {
                    str1 = tv.getText().toString().trim();
                    str1 = str1 + String.valueOf(((Button)v).getText());    //获得新输入的值
                    System.out.println("str1" + ":::" + str1);
                    tv.setText(str1);
                }
            }
        );
    }
    buttonListener(buttonJia,1);
    buttonListener(buttonJian,2);
    buttonListener(buttonCheng,3);
    buttonListener(buttonChu,4);
    buttonDengyu.setOnClickListener
    (
        new OnClickListener()
        {
            public void onClick(View v)
            {System.out.println(str1);
                result1 = Integer.parseInt(str1);
                if(flag == 1)
                {
                    result = result0 + result1;
                    System.out.println(result0 + ":" + result1);
                }
                else if(flag == 2)
                {
                    result = result0 - result1;
                }
                else if(flag == 3)
                {
                    result = result0 * result1;
                }
                else if(flag == 4)
                {
                    result = (int)(result0/result1);
```

```
            }
            String str = (result + "").trim();
            System.out.println(str);
            tv.setText(str);
        }
    }
);
}
//初始化按钮
public void initButton()
{                                                //初始化控件资源
    tv = (TextView)this.findViewById(R.id.tv);   //获取文本框控件对象
    str1 = String.valueOf(tv.getText());str2 = "";  //初始化运算输入数值
    buttonC = (Button)this.findViewById(R.id.ButtonC);  //获得计算按钮的按钮对象
    buttonJia = (Button)this.findViewById(R.id.ButtonJia);
    buttonJian = (Button)this.findViewById(R.id.ButtonJian);
    buttonCheng = (Button)this.findViewById(R.id.ButtonCheng);
    buttonChu = (Button)this.findViewById(R.id.ButtonChu);
    buttonDengyu = (Button)this.findViewById(R.id.ButtonDengyu);
    buttons = new int[]
    {                                            //记录数值按钮的 id
        R.id.Button00,R.id.Button01,R.id.Button02,R.id.Button03,R.id.Button04,
        R.id.Button05,R.id.Button06,R.id.Button07,R.id.Button08,R.id.Button09
    };
}
//按钮监听
public void buttonListener(Button button,final int id)
{
    button.setOnClickListener
    (
        new OnClickListener()
        {
            public void onClick(View v)
            {
                String str = tv.getText().toString().trim();
                result0 = Integer.parseInt(str);
                tv.setText("");
                flag = id;
            }
        }
    );
}
}
```

运行程序,效果如图 2-1 所示。

在以上实例中实现计算器的开发,主要运用了 LinearLayout 线性布局的相关知识。线性布局是最简单的布局之一,其几个常用的属性主要如下。

- android:layout_weight:指不同的控件在 Activity 中占有体积大小的比例。
- android:paddingLeft:指左内边距,即控件内文字离控件左边边界的距离。其他的类推。
- android:gravity:指控件内文字相对于控件本身的方向属性,长度为 dip,与像素独立的长度。
- android:background:为控件内文字颜色的背景色,颜色采用 rgb 时前面需用#号。

图 2-1 计算器

- android:textSize：为文本的大小，单位为 pt，即磅。
- android:id：为该控件的 id，即在此处可以设置控件的 id。
- android:layout_width：为控件本身的宽度属性，其他的类似。

除此之外，线性布局中可以使用 gravity 属性设置控件的对齐方式，如表 2-1 所示。

表 2-1 gravity 可取的属性及说明

属 性 值	描 述
top	不改变控件大小，对齐到容器顶部
bottom	不改变控件大小，对齐到容器底部
left	不改变控件大小，对齐到容器左侧
right	不改变控件大小，对齐到容器右侧
center_vertical	不改变控件大小，对齐到容器纵向中央位置
center-horizontal	不改变控件大小，对齐到容器横向中央位置
center	不改变控件大小，对齐到容器中央位置
fill_vertical	若有可能，纵向拉伸以填满容器
fill_horizontal	若有可能，横向拉伸以填满容器
fill	若有可能，纵向横向同时拉伸以填满容器

2.1.2 奖牌排行榜实例

本实例中，通过 TableLayout 表格布局的应用，构建了一个奖牌排行榜的程序，并通过本实例向读者演示 TableLayout 表格布局的具体应用。本实例的具体实现步骤如下。

（1）在 Eclipse 中创建一个 Android 应用项目，命名为 FrameLayout_test。

（2）打开 res\layout 目录下的 main.xml 布局文件，在文件中定义 TableLayout 布局，并在帧布局中声明对应的控件。代码为：

```
<TableLayout xmlns:android = "http://schemas.android.com/apk/res/android"
    xmlns:tools = "http://schemas.android.com/tools"
```

```xml
    android:layout_width = "match_parent"
    android:layout_height = "match_parent"
    android:stretchColumns = "1"
    android:background = "#aabbcc">
<TableRow>
    <TextView
        android:text = "国家"
        android:background = "#848484"
        android:padding = "2dip"/>
    <TextView
        android:text = "金牌"
        android:background = "#ff0000"
        android:padding = "2dip" />
    <TextView
        android:text = "银牌"
        android:background = "#00ff00"
        android:padding = "2dip"/>
    <TextView
        android:text = "铜牌"
        android:background = "#0000ff"
        android:padding = "2dip"/>
</TableRow>
<TableRow>
    <TextView
        android:text = "中国"
        android:background = "#848484"
        android:padding = "2dip"/>
    <TextView
        android:text = " * "
        android:background = "#ff0000"
        android:padding = "2dip"/>
    <TextView
        android:text = " ** "
        android:background = "#00ff00"
        android:padding = "2dip"/>
    <TextView
        android:text = " *** "
        android:background = "#0000ff"
        android:padding = "2dip"/>
</TableRow>
<TableRow>
    <TextView
        android:text = "美国"
        android:background = "#848484"
        android:padding = "2dip"/>
    <TextView
        android:text = " * "
        android:background = "#ff0000"
        android:padding = "2dip"/>
    <TextView
        android:text = " ** "
        android:background = "#00ff00"
        android:padding = "2dip"/>
    <TextView
        android:text = " *** "
```

```
            android:background = "#0000ff"
            android:padding = "2dip"/>
    </TableRow>
</TableLayout>
```

运行程序,效果如图 2-2 所示。

图 2-2 表格布局

表格布局有点类似表单的意思,可以在 Activity 中建立多行,每一行又可以设置为多列,所以看起来横竖条理比较清晰,因此叫作表格布局。表格布局各控件属性与线性布局类似,其几个属性如下。

- 用 TableRow 来增加一行,然后该行内各列依次并排。
- android:padding 指的是内边距的 4 个方向都采用同样的间距。
- android:stretchColumns 属性表示当该行属性设置为填充屏幕时,指定将哪一列拉伸。

2.1.3 标语与排行实例

混合布局的原理是:大的 Layout 嵌入小的 Layout,而且小的 Layout 又可以嵌入不同的 Layout 中。

本实例主要是将线性布局与表格布局混合起来显示的,即总的布局为垂直方向上的线性布局,上面那个布局内部又是垂直方向的布局,下面那个布局也是一个线性布局,不过里面嵌入了一个表格布局,所以总共有 4 个布局。其具体实现步骤如下。

(1) 在 Eclipse 中创建一个 Android 应用项目,命名为 Linear_Table_test。

(2) 打开 res\layout 目录下的 main.xml 布局文件,在文件中实现线性布局与表格布局的混合。代码为:

```
<LinearLayout xmlns:android = "http://schemas.android.com/apk/res/android"
    xmlns:tools = "http://schemas.android.com/tools"
    android:layout_width = "match_parent"
    android:layout_height = "match_parent"
    android:orientation = "vertical"
```

```xml
android:background = "#aabbcc">
<LinearLayout
    android:layout_width = "fill_parent"
    android:layout_height = "fill_parent"
    android:orientation = "vertical"
    android:layout_weight = "1" >
    <TextView
        android:id = "@+id/London"
        android:layout_width = "fill_parent"
        android:layout_height = "wrap_content"
        android:text = "伦敦奥运"
        android:textSize = "5pt"
        android:background = "#00ff00"
        android:gravity = "center_horizontal"
        android:padding = "10pt"
        android:layout_weight = "1"/>
    <TextView
        android:id = "@+id/China"
        android:layout_width = "fill_parent"
        android:layout_height = "wrap_content"
        android:text = "中国加油!!!"
        android:textSize = "8pt"
        android:background = "#ff00ff"
        android:layout_weight = "3"/>
</LinearLayout>
<LinearLayout
    android:layout_width = "fill_parent"
    android:layout_height = "fill_parent"
    android:layout_weight = "3">
    <TableLayout
        android:layout_width = "match_parent"
        android:layout_height = "match_parent"
        android:stretchColumns = "1">
        <TableRow>
            <TextView
            android:text = "国家"
            android:background = "#848484"
            android:padding = "2dip"/>
            <TextView
                android:text = "金牌"
                android:background = "#ff0000"
                android:padding = "2dip"/>
            <TextView
                android:text = "银牌"
                android:background = "#00ff00"
                android:padding = "2dip"/>
            <TextView
                android:text = "铜牌"
                android:background = "#0000ff"
                android:padding = "2dip"/>
        </TableRow>
        <TableRow>
            <TextView
            android:text = "中国"
            android:background = "#848484"
```

```xml
                android:padding = "2dip"/>
            <TextView
                android:text = " * "
                android:background = " # ff0000"
                android:padding = "2dip"/>
            <TextView
                android:text = " ** "
                android:background = " # 00ff00"
                android:padding = "2dip"/>
            <TextView
                android:text = " *** "
                android:background = " # 0000ff"
                android:padding = "2dip"/>
        </TableRow>
        <TableRow>
            <TextView
                android:text = "美国"
                android:background = " # 848484"
                android:padding = "2dip"/>
            <TextView
                android:text = " * "
                android:background = " # ff0000"
                android:padding = "2dip"/>
            <TextView
                android:text = " ** "
                android:background = " # 00ff00"
                android:padding = "2dip"/>
            <TextView
                android:text = " *** "
                android:background = " # 0000ff"
                android:padding = "2dip"/>
        </TableRow>
    </TableLayout>
  </LinearLayout>
</LinearLayout>
```

运行程序,效果如图 2-3 所示。

图 2-3 线性布局与表格布局

2.1.4 登录界面实例

本实例通过 RelativeLayout 相对布局的应用,构建了一个登录界面。通过本实例的实现,可了解 RelativeLayout 相对布局的具体应用。

相对布局 RelativeLayout 允许子元素指定它们相对于其父元素或兄弟元素的位置,这是实际布局中最常用的布局方式之一。它灵活性大,当然属性也多,操作难度也大,属性之间产生冲突的可能性也大,使用相对布局时要多做些测试。

本实例的具体实现步骤如下。

(1) 在 Eclipse 中创建一个 Android 应用项目,命名为 RelativeLayout_test。

(2) 打开 res\layout 目录下的 main.xml 布局文件,在文件中定义一个相对布局,在布局中声明对应的控件。代码为:

```xml
<RelativeLayout xmlns:android = "http://schemas.android.com/apk/res/android"
    xmlns:tools = "http://schemas.android.com/tools"
    android:layout_width = "match_parent"
    android:layout_height = "match_parent"
    android:paddingBottom = "@dimen/activity_vertical_margin"
    android:paddingLeft = "@dimen/activity_horizontal_margin"
    android:paddingRight = "@dimen/activity_horizontal_margin"
    android:paddingTop = "@dimen/activity_vertical_margin"
    tools:context = ".MainActivity"
    android:background = "#aabbcc">
    <TextView
        android:id = "@+id/user"
        android:layout_width = "wrap_content"
        android:layout_height = "wrap_content"
        android:text = "账号:" />
    <EditText
        android:id = "@+id/userEdit"
        android:layout_width = "fill_parent"
        android:layout_height = "wrap_content"
        android:layout_toRightOf = "@id/user"/>
    <TextView
        android:id = "@+id/password"
        android:layout_width = "wrap_content"
        android:layout_height = "wrap_content"
        android:layout_alignParentLeft = "true"
        android:layout_below = "@id/userEdit"
        android:text = "密码:" />
    <EditText
        android:id = "@+id/passwordEdit"
        android:layout_width = "fill_parent"
        android:layout_height = "wrap_content"
        android:layout_below = "@id/userEdit"
        android:layout_toRightOf = "@id/password"/>
    <Button
        android:id = "@+id/ok"
        android:layout_width = "wrap_content"
        android:layout_height = "wrap_content"
        android:layout_below = "@id/passwordEdit"
        android:layout_alignParentRight = "true"
```

```
            android:layout_marginLeft = "10px"
            android:text = "登录" />
        <Button
            android:layout_width = "wrap_content"
            android:layout_height = "wrap_content"
            android:layout_alignBaseline = "@ + id/ok"
            android:layout_alignBottom = "@ + id/ok"
            android:layout_alignLeft = "@ + id/password"
            android:text = "取消" />
</RelativeLayout>
```

运行程序,效果如图 2-4 所示。

图 2-4　相对布局

进行相对布局时可能用到的属性有很多,但都不复杂,首先看属性值为 true 或 false 的属性,主要常用属性如下。

① 相对于给定 ID 控件。

- android:layout_above：将该控件的底部置于给定 ID 的控件之上；
- android:layout_below：将该控件的底部置于给定 ID 的控件之下；
- android:layout_toLeftOf：将该控件的右边缘与给定 ID 的控件左边缘对齐；
- android:layout_toRightOf：将该控件的左边缘与给定 ID 的控件右边缘对齐；
- android:layout_alignBaseline：将该控件的 baseline 与给定 ID 的 baseline 对齐；
- android:layout_alignTop：将该控件的顶部边缘与给定 ID 的顶部边缘对齐；
- android:layout_alignBottom：将该控件的底部边缘与给定 ID 的底部边缘对齐；
- android:layout_alignLeft：将该控件的左边缘与给定 ID 的左边缘对齐；
- android:layout_alignRight：将该控件的右边缘与给定 ID 的右边缘对齐.

② 相对于父组件。

- android:layout_alignParentTop：如果为 true,将该控件的顶部与其父控件的顶部对齐；
- android:layout_alignParentBottom：如果为 true,将该控件的底部与其父控件的底部

对齐；
- android:layout_alignParentLeft：如果为 true，将该控件的左部与其父控件的左部对齐；
- android:layout_alignParentRight：如果为 true，将该控件的右部与其父控件的右部对齐。

③ 居中。
- android:layout_centerHorizontal：如果为 true，将该控件的置于水平居中；
- android:layout_centerVertical：如果为 true，将该控件的置于垂直居中；
- android:layout_centerInParent：如果为 true，将该控件的置于父控件的中央。

④ 指定移动像素。
- android:layout_marginTop：上偏移的值；
- android:layout_marginBottom：下偏移的值；
- android:layout_marginLeft：左偏移的值；
- android:layout_marginRight：右偏移的值。

2.1.5 霓虹灯实例

本实例中，通过 FrameLayout 帧布局的应用，构建霓虹灯效果。并通过本实例，演示 FramLayout 帧布局的具体应用。本实例主要运用了 FrameLayout 帧布局的相关知识。

帧布局是最简单的布局之一，采用帧布局的容器中无论放入多少个控件，默认情况下控件都对齐到容器的左上角。如果控件一样大，同一时刻只能见到最上面的一个控件。

本实例的具体实现步骤如下。

(1) 在 Eclipse 中创建一个 Android 应用项目，命名为 Frame_test。

(2) 打开 res\layout 目录下的 main.xml 布局文件，在文件中定义帧布局，并在布局中声明相应的控件。代码为：

```xml
<FrameLayout xmlns:android = "http://schemas.android.com/apk/res/android"
    android:layout_width = "fill_parent"
    android:layout_height = "fill_parent"
    android:background = "#aabbcc">
<TextView
        android:id = "@+id/view01"
        android:layout_width = "wrap_content"
        android:layout_height = "wrap_content"
        android:layout_gravity = "center"
        android:width = "320px"
        android:height = "320px"
        android:background = "#f00" />
<TextView
        android:id = "@+id/view02"
        android:layout_width = "wrap_content"
        android:layout_height = "wrap_content"
        android:layout_gravity = "center"
        android:width = "280px"
        android:height = "280px"
        android:background = "#0f0" />
<TextView
```

```
            android:id = "@ + id/view03"
            android:layout_width = "wrap_content"
            android:layout_height = "wrap_content"
            android:layout_gravity = "center"
            android:width = "320px"
            android:height = "320px"
            android:background = "#00f" />
<TextView
            android:id = "@ + id/view04"
            android:layout_width = "wrap_content"
            android:layout_height = "wrap_content"
            android:layout_gravity = "center"
            android:width = "240px"
            android:height = "240px"
            android:background = "#ff0" />
<TextView
            android:id = "@ + id/view05"
            android:layout_width = "wrap_content"
            android:layout_height = "wrap_content"
            android:layout_gravity = "center"
            android:width = "200px"
            android:height = "200px"
            android:background = "#f0f" />
<TextView
            android:id = "@ + id/view06"
            android:layout_width = "wrap_content"
            android:layout_height = "wrap_content"
            android:layout_gravity = "center"
            android:width = "160px"
            android:height = "160px"
            android:background = "#0ff" />
</FrameLayout>
```

(3) 在 res/value 目录下新建一个 color.xml 文件,用于实现霓虹灯颜色。代码为:

```
<?xml version = "1.0" encoding = "utf - 8"?>
<resources>
    <color name = "color1">#f00</color>
    <color name = "color2">#0f0</color>
    <color name = "color3">#00f</color>
    <color name = "color4">#ff0</color>
    <color name = "color5">#f0f</color>
    <color name = "color6">#0ff</color>
</resources>
```

(4) 打开 src\fs.frame_test 包下的 MainActivity.java 文件,在文件中实现霓虹灯。代码为:

```
package fs.frame_test;
import java.util.Timer;
import java.util.TimerTask;
import android.os.Bundle;
import android.os.Handler;
import android.os.Message;
import android.app.Activity;
import android.util.Log;
```

```java
import android.view.Menu;
import android.widget.TextView;
public class MainActivity extends Activity {
    private String TAG = "androidtest1";
    private int currentcolor = 0;
    final int[] colors = new int[]{
            R.color.color1,
            R.color.color2,
            R.color.color3,
            R.color.color4,
            R.color.color5,
            R.color.color6
    };
    final int[] names = new int[]{
        R.id.view01,
        R.id.view02,
        R.id.view03,
        R.id.view04,
        R.id.view05,
        R.id.view06
    };
    TextView[] views = new TextView[names.length];
    Handler handler = new Handler()
    {
        @Override
        public void handleMessage(Message msg)
        {
            if(msg.what == 0x123)
            {
                for(int i = 0;i < names.length;i++)
                {
                        views[i].setBackgroundResource(colors[(i + currentcolor) % names.
                        length]);                           //i+后,每次颜色往后循环一个
                }
                currentcolor++;
            }
            super.handleMessage(msg);
        }
    };
    @Override
    protected void onCreate(Bundle savedInstanceState)
    {
        super.onCreate(savedInstanceState);
        setContentView(R.layout.main);

        for(int i = 0;i < names.length;i++)
        {
            views[i] = (TextView)findViewById(names[i]);
        }
        new Timer().schedule(new TimerTask()
        {
            @Override
            public void run()
            {
                handler.sendEmptyMessage(0x123);
```

```
            }
        },0,500);
    }
}
```

运行程序,效果如图 2-5 所示。

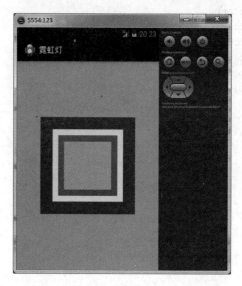

图 2-5 霓虹灯

FrameLayout 继承自 ViewGroup,除了继承自父类的属性和方法,FrameLayout 类中包含了自己特有的属性和方法,主要有:
- android:foreground:设置绘制在所有子控件之上的内容;
- android:foregroundGravity:设置绘制在所有子控件之上内容的 gravity。

提示:在 FrameLayout 中,子控件是通过栈来绘制的,所以后添加的子控件会被绘制在上层。

2.1.6 显示信息实例

本实例中,通过 AbsoluteLayout(绝对布局)的应用,构建一个登录界面,并通过实例的实现演示 AbsoluteLayout 的具体应用。

AbsoluteLayout 是指屏幕中所有控件的摆放由开发人员通过设置控件的坐标来指定,控件容器不再负责管理其子控件的位置。由于子控件的位置和布局都通过坐标来指定,AbsoluteLayout 类中并没有开发特有的属性和方法。

本实例的具体实现步骤如下。

(1) 在 Eclipse 中创建一个 Android 应用项目,命名为 AbsoluteLayout_test。

(2) 打开 res\layout 目录下的 main.xml 布局文件,在文件中定义一个 AbsoluteLayout 布局,并在布局中声明对应的控件。代码为:

```
<?xml version = "1.0" encoding = "utf - 8"?>
< AbsoluteLayout xmlns:android = "http://schemas.android.com/apk/res/android"
        android:id = "@ + id/AbsoluteLayout01"
        android:layout_width = "fill_parent"
```

```xml
        android:layout_height = "fill_parent"
        android:background = "#aabbcc" >
<!-- 声明一个绝对布局 -->
    <TextView
        android:id = "@+id/TextView01"
        android:layout_width = "wrap_content"
        android:layout_height = "wrap_content"
        android:layout_x = "20dip"
        android:layout_y = "20dip"
        android:text = "用户名"/>
<!-- 声明一个TextView控件 -->
    <TextView
        android:id = "@+id/TextView02"
        android:layout_width = "wrap_content"
        android:layout_height = "wrap_content"
        android:layout_x = "20dip"
        android:layout_y = "80dip"
        android:text = "密码"/>
<!-- 声明一个TextView控件 -->
    <EditText
        android:id = "@+id/EditText01"
        android:layout_width = "180dip"
        android:layout_height = "wrap_content"
        android:layout_x = "80dip"
        android:layout_y = "20dip" />
<!-- 声明一个EditText控件 -->
    <EditText
        android:id = "@+id/EditText02"
        android:layout_width = "180dip"
        android:layout_height = "wrap_content"
        android:layout_x = "80dip"
        android:layout_y = "80dip"
        android:password = "true"/>
<!-- 声明一个EditText控件 -->
    <Button
        android:id = "@+id/Button02"
        android:layout_width = "wrap_content"
        android:layout_height = "wrap_content"
        android:layout_x = "210dip"
        android:layout_y = "140dip"
        android:text = "取消"/>
<!-- 声明一个Button控件 -->
    <ScrollView
        android:id = "@+id/ScrollView01"
        android:layout_width = "250dip"
        android:layout_height = "150dip"
        android:layout_x = "10dip"
        android:layout_y = "200dip">
<!-- 声明一个ScrollView控件 -->
        <EditText
            android:id = "@+id/EditText03"
            android:layout_width = "fill_parent"
            android:layout_height = "wrap_content"
            android:gravity = "top"
            android:singleLine = "false" />
```

```xml
<!-- 声明一个 EditText 控件 -->
    </ScrollView>
    <Button
        android:id = "@+id/Button01"
        android:layout_width = "wrap_content"
        android:layout_height = "wrap_content"
        android:layout_x = "14dp"
        android:layout_y = "138dp"
        android:text = "确定" />
</AbsoluteLayout>
```

(3) 在 res\values 目录下新建一个 color.xml 文件，用于实现程序中将会用到的颜色资源。代码为：

```xml
<?xml version = "1.0" encoding = "utf-8"?>
<resources>
    <color name = "red">#fd8d8d</color>
    <!-- 声明名为 red 的资源 -->
    <color name = "green">#9cfda3</color>
    <!-- 声明名为 green 的资源 -->
    <color name = "blue">#8d9dfd</color>
    <!-- 声明名为 blue 的资源 -->
    <color name = "white">#FFFFFF</color>
    <!-- 声明名为 white 的资源 -->
    <color name = "black">#000000</color>
    <!-- 声明名为 black 的资源 -->
</resources>
```

(4) 打开 src\fs.absolutelayout_test 包下的 MainActivity.java 文件，在文件中实现当在用户名及密码编辑框中输入对应的信息时，单击界面中的"确定"按钮即把相应的信息显示在滚动条中的编辑框中，当单击界面中的"取消"按钮时，即清空用户名密码编辑框中的信息。代码为：

```java
package fs.absolutelayout_test;
import android.app.Activity;
import android.os.Bundle;
import android.view.View;
import android.widget.Button;
import android.widget.EditText;
public class MainActivity extends Activity {
    @Override
    public void onCreate(Bundle savedInstanceState) {              //重写 onCreate 方法
        super.onCreate(savedInstanceState);
        setContentView(R.layout.main);                             //设置当前屏幕
        final Button OkButton = (Button) findViewById(R.id.Button01);       //获取"确定"按钮对象
        final Button cancelButton = (Button) findViewById(R.id.Button02);   //获取"取消"按钮对象
        final EditText uid = (EditText) findViewById(R.id.EditText01);      //获取用户名文本框对象
        final EditText pwd = (EditText) findViewById(R.id.EditText02);      //获取密码文本框对象
        final EditText log = (EditText) findViewById(R.id.EditText03);      //获取登录日志文本对象
        OkButton.setOnClickListener(                                        //为按钮添加 OnClickListener
            new View.OnClickListener() {
                public void onClick(View v) {                               //重写 onClick 方法
                    String uidStr = uid.getText().toString();               //获取用户名文本框的内容
                    String pwdStr = pwd.getText().toString();               //获取密码文本框的内容
                    log.append("用户名：" + uidStr + " 密码：" + pwdStr + "\n");
```

```
            }
        });
        cancelButton.setOnClickListener(              //为按钮添加 OnClickListener
            new View.OnClickListener() {
                public void onClick(View v) {          //重写 onClick 方法
                    uid.setText("");                   //清空用户名文本框内容
                    pwd.setText("");                   //清空密码文本框内容
                }
            });
    }
}
```

运行程序,效果如图 2-6(a)所示,当输入对应的信息时,单击"确定"按钮,如图 2-6(b)所示。

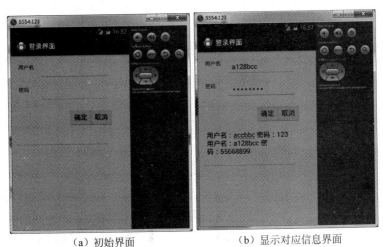

(a)初始界面　　　　　　　　(b)显示对应信息界面

图 2-6　登录界面

2.2　文本类实例

在 Android 中提供了各种类型的文本控件用于实现信息的编写及录入,下面给予介绍。

2.2.1　文字说明实例

本实例中利用 TextView 控件显示图片,并利用单行文本框与多行文本框显示文字。通过本实例可以演示 TextView 控件的具体应用。

其具体实现步骤如下。

(1) 在 Eclipse 中创建一个 Android 应用项目,命名为 TextView_test。

(2) 打开 res\layout 目录下的 main.xml 布局文件,在文件中声明几个 TextView 控件。代码为:

```
<?xml version = "1.0" encoding = "utf - 8"?>
<LinearLayout xmlns:android = "http://schemas.android.com/apk/res/android"
    android:orientation = "vertical"
    android:layout_width = "fill_parent"
```

```xml
        android:layout_height = "fill_parent"
        android:background = "#000000">
    <TextView
        android:layout_width = "wrap_content"
        android:layout_height = "wrap_content"
        android:text = "文本框实例"
        android:autoLink = "email"
        android:height = "50px" />
    <TextView
        android:layout_width = "wrap_content"
        android:id = "@+id/textView1"
        android:text = "带图片的TextView"
        android:drawableTop = "@drawable/ic_launcher"
        android:layout_height = "wrap_content" />
    <TextView
        android:id = "@+id/textView2"
        android:textColor = "#0f0"
        android:textSize = "20px"
        android:text = "多行文本：吃饱饭后，一戒吸烟，二戒洗澡，三戒生气，四戒松裤带，五戒刷牙，六戒上厕所，七戒喝酒，八戒你知道了吗？"
        android:width = "300px"
        android:layout_width = "wrap_content"
        android:layout_height = "wrap_content" />
    <TextView
        android:id = "@+id/textView3"
        android:textColor = "#f00"
        android:textSize = "20px"
        android:text = "单行文本：吃饱饭后，一戒吸烟，二戒洗澡，三戒生气，四戒松裤带，五戒刷牙，六戒上厕所，七戒喝酒，八戒你知道了吗？"
        android:width = "300px"
        android:singleLine = "true"
        android:layout_width = "wrap_content"
        android:layout_height = "wrap_content" />
</LinearLayout>
```

运行程序，效果如图2-7所示。

TextView控件的功能是向用户显示文本内容，同时可选择性地让用户编辑文本。从功能上来说，一个TextView就是一个完整的文本编辑器，只不过其本身被设置为不允许编辑，其子类EditText被设置为允许用户对内容进行编辑。在TextView中有其自己的属性，如下所示。

- android:autoLink：设置是否显示为可单击的链接，可选值（none/web/email/phone/map/all）。
- android:drawableBottom：在text的下方输出一个drawable（图片）。
- android:drawableLeft：在text的左边输出一个drawable（图片）。
- android:drawableRight：在text的右边输出一个drawable（图片）。
- android:drawableTop：在text的正上方输出一个drawable（图片）
- android:drawablePaddingL：设置text与drawable（图片）的间隔，与drawableLeft、drawableRight、drawableTop、drawableBottom一起使用，可设置为负数，单独使用没有效果。
- android:ellipsize：设置当文字过长时，该控件该如何显示。可设置如下属性值：

图 2-7　文本框显示多样式的文字

"start"省略号显示在开头；"end"省略号显示在结尾；"middle"省略号显示在中间；"marquee"以跑马灯的方式显示（动画横向移动）。

- android:gravity：设置文本位置，设置成"center"，文本将居中显示。
- android:linksClickable：设置单击时是否链接，即使设置了 autoLink。
- android:marqueeRepeatLimit：在 ellipsize 设定为 marquee 时，设置重复滚动的次数，设置为 marquee_forever 时表示无限次。
- android:lines：设置文本的行数，设置两行就显示两行，即使第 2 行没有数据。
- android:shadowRadius：设置阴影的半径。设置为 0.1 就变成字体的颜色了，一般设置为 3.0 的效果比较好。
- android:shadowColor：指定文本阴影的颜色，需要与 shadowRadius 一起使用。
- android:singleLine：设置单行显示。
- android:textColorLink：设置文字链接的颜色。
- android:textScaleX：设置文字之间间隔，默认为 1.0f。
- android:textStyle：设置字形 bold（粗体）为 0、italic（斜体）为 1、bolditalic（又粗又斜）为 2，可以设置一个或多个，用"|"隔开。
- android:typeface：设置文本字体，必须是以下常量值之一：normal 为 0、sans 为 1、serif 为 2 和 monospace（等宽字体）为 3。

2.2.2　接收信息实例

本实例通过对 EditText 编辑框的应用，构建一个主要接收用户输入电子邮箱地址和电话号码界面。通过本实例可演示 EditText 编辑框的具体应用。

EditText 控件与 TextView 控件最大的不同就是用户可以对 EditText 控件进行编辑。同时,用户还可以为 EditText 控件设置监听器,用来检测用户的输入是否合法等。

本实例的具体实现步骤如下。

(1) 在 Eclipse 中创建一个 Android 应用项目,命名为 EditText_Test。

(2) 打开 res\layout 目录下的 main.xml 布局文件,在文件中声明对应的编辑框控件。代码为:

```xml
<?xml version = "1.0" encoding = "utf-8"?>
<TableLayout xmlns:android = "http://schemas.android.com/apk/res/android"
    android:layout_width = "fill_parent"
    android:layout_height = "fill_parent"
    android:shrinkColumns = "0,2"
    android:background = "#aabbcc">
<!-- 声明一个 TableLayout -->
    <TableRow
        android:layout_width = "fill_parent"
        android:layout_height = "wrap_content">
<!-- 声明一个 TableRow 控件 -->
        <TextView
            android:id = "@+id/tvEmail"
            android:layout_width = "wrap_content"
            android:layout_height = "wrap_content"
            android:autoLink = "email"
            android:ellipsize = "end"
            android:text = "邮箱地址" />
<!-- 声明一个 TextView 控件 -->
        <EditText
            android:id = "@+id/etEmail"
            android:layout_width = "wrap_content"
            android:layout_height = "wrap_content"
            android:hint = "请输入电子邮件地址"
            android:selectAllOnFocus = "true" />
    </TableRow>
    <TableRow
        android:layout_width = "fill_parent"
        android:layout_height = "wrap_content">
<!-- 声明一个 TableRow -->
        <TextView
            android:id = "@+id/tvPhone"
            android:layout_width = "wrap_content"
            android:layout_height = "wrap_content"
            android:autoLink = "phone"
            android:ellipsize = "middle"
            android:text = "电话号码"/>
<!-- 声明一个 TextView 控件 -->
        <EditText
            android:id = "@+id/etPhone"
            android:layout_width = "wrap_content"
            android:layout_height = "wrap_content"
            android:hint = "请输入电话号码"
            android:maxWidth = "160px"
            android:phoneNumber = "true"
            android:selectAllOnFocus = "true"
            android:singleLine = "true" />
```

```
        </TableRow>
        <EditText
            android:id = "@ + id/etInfo"
            android:layout_width = "wrap_content"
            android:layout_height = "wrap_content"
            android:cursorVisible = "false"
            android:editable = "false"
            android:hint = "此处显示登录信息"
            android:lines = "5"
            android:shadowColor = "@color/shadow"
            android:shadowDx = "2.5"
            android:shadowDy = "2.5"
            android:shadowRadius = "5.0" />
</TableLayout>
```

(3) 在 res\values 目录下创建一个 color.xml 文件，实现颜色资源。代码为：

```
<?xml version = "1.0" encoding = "utf - 8"?>
<resources>
    <color name = "shadow">#fd8d8d</color>
<!-- 声明名为 shadow 的颜色资源 -->
</resources>
```

(4) 打开 src\fs.edittext_test 包下的 MainActivity.java 文件，在文件中实现接收邮箱地址及电话号码的输入信息，并将信息显示在对应的编辑框中。代码为：

```
package fs.edittext_test;
import android.app.Activity;
import android.os.Bundle;
import android.view.KeyEvent;
import android.view.View;
import android.view.View.OnKeyListener;
import android.widget.EditText;
public class MainActivity extends Activity {
    /** 第一次调用 Activity 活动 */
    @Override
    public void onCreate(Bundle savedInstanceState) {
        super.onCreate(savedInstanceState);
        setContentView(R.layout.main);
        EditText etEmail = (EditText) findViewById(R.id.etEmail);
        etEmail.setOnKeyListener(myOnKeyListener);           //为 EditText 控件设置监听器
    }
    //自定义的 OnKeyListner 对象
    private OnKeyListener myOnKeyListener = new OnKeyListener() {
        public boolean onKey(View v, int keyCode, KeyEvent event) {      //重写 onKey 方法
            EditText etInfo = (EditText) findViewById(R.id.etInfo);
            EditText etEmail = (EditText) findViewById(R.id.etEmail);
            etInfo.setText("您输入的邮箱地址为：" + etEmail.getText());
                                                      //设置 EditText 控件的显示内容
            return true;
        }
    };
}
```

运行程序，效果如图 2-8 所示。

图 2-8 编辑框效果图

EditText 控件继承 TextView 控件的属性外，还有自身的一些属性，主要内容如下。
- android:cursorVisible：设置光标是否可见，默认为可见。
- android:lines：通过设置固定的行数来决定 EditText 的高度。
- android:maxLines：设置最大行数。
- android:minLines：设置最小行数。
- android:password：设置文本框中的内容是否显示为密码。
- android:phoneNumber：设置文本框中的内容只能是电话号码。
- android:scrollHorizontally：设置文本框是否可以水平地进行滚动。
- android:selectAllOnFocus：如果文本内容可选中，则当文本框获得焦点时自动选中全部文本内容。
- android:shadowColor：为文本框设置指定颜色的阴影。
- android:shadowDx：为文本框设置阴影的水平偏移，为浮点数。
- android:shadowDy：为文本框设置阴影的垂直偏移，为浮点数。
- android:shadowRadius：为文本框设置阴影的半径，为浮点数。
- android:singleLine：设置文本框为单行模式。
- android:maxLength：设置最大显示长度。

2.2.3 自动搜索实例

本实例利用 AutoCompleteTextView 控件实现当用户输入某些文字时，即自动出现下拉菜单，显示与用户输入文字相关的信息，用户直接单击需要的文字，即可自动填写到文本控件中。通过本实例可以演示 AutoCompleteTextView 控件的具体用法。其具体实现步骤如下。

（1）在 Eclipse 中创建一个 Android 应用项目，命名为 AutoCompleteTextView_test。

(2) 打开 res\layout 目录下的 main.xml 布局文件,在文件中声明自动提示文本框控件。代码为:

```xml
<?xml version = "1.0" encoding = "utf-8"?>
<LinearLayout xmlns:android = "http://schemas.android.com/apk/res/android"
    android:orientation = "vertical"
    android:layout_width = "fill_parent"
    android:layout_height = "fill_parent"
    android:background = "#aabbcc">
    <!-- 当只有一个 EditText 或者 AutoCompleteTextView 的时候,进入画面时是默认得到焦点的。
         要想去除焦点,可以在 auto 之前加一个 0 像素的 layout,并设置它先得到焦点 -->
    <LinearLayout
        android:layout_width = "0px"
        android:layout_height = "0px"
        android:focusable = "true"
        android:focusableInTouchMode = "true"/>
    <!-- 定义一个自动完成文本框,指定输入一个字符后进行提示 -->
    <AutoCompleteTextView
        android:id = "@+id/auto"
        android:layout_width = "fill_parent"
        android:layout_height = "wrap_content"
        android:hint = "请输入文字进行搜索"
        android:completionHint = "最近的 5 条记录"
        android:dropDownHorizontalOffset = "20dp"
        android:completionThreshold = "1"
        android:dropDownHeight = "fill_parent"/>
    <!--
    android:completionHint:设置出现在下拉菜单中的提示标题
    android:completionThreshold:设置用户至少输入多少个字符才会显示提示
    android:dropDownHorizontalOffset:设置下拉菜单于文本框之间的水平偏移。下拉菜单默认
    与文本框左对齐
    android:dropDownVerticalOffset:设置下拉菜单于文本框之间的垂直偏移。下拉菜单默认紧
    跟文本框
    android:dropDownHeight:设置下拉菜单的高度
    android:dropDownWidth:设置下拉菜单的宽度
    -->
    <Button
        android:text = "搜索"
        android:id = "@+id/search"
        android:layout_width = "wrap_content"
        android:layout_height = "wrap_content"/>
</LinearLayout>
```

(3) 打开 src\fs.autocompletetextview_test 包下的 MainActivity.java 文件,在文件中实现当输入关键字时即实现自动提示功能。代码为:

```java
package fs.autocompletetextview_test;
import android.app.Activity;
import android.content.SharedPreferences;
import android.os.Bundle;
import android.view.View;
import android.view.View.OnClickListener;
import android.view.View.OnFocusChangeListener;
import android.widget.ArrayAdapter;
import android.widget.AutoCompleteTextView;
```

```java
import android.widget.Button;
public class MainActivity extends Activity {
    private AutoCompleteTextView autoCompleteTextView;
    @Override
    public void onCreate(Bundle savedInstanceState) {
        super.onCreate(savedInstanceState);
        setContentView(R.layout.main);
        autoCompleteTextView = (AutoCompleteTextView) findViewById(R.id.auto);
        initAutoComplete("history", autoCompleteTextView);
        Button searchButton = (Button) findViewById(R.id.search);
        searchButton.setOnClickListener(new MyOnClickListener());
    }
    private final class MyOnClickListener implements OnClickListener {
        @Override
        public void onClick(View v) {
            saveHistory("history", autoCompleteTextView);
        }
    }
    /**
     * 把指定 AutoCompleteTextView 中的内容保存到 sharedPreference 中指定的字符段
     * 保存在 sharedPreference 中的字段名要操作的 AutoCompleteTextView
     */
    private void saveHistory(String field,
            AutoCompleteTextView autoCompleteTextView) {
        String text = autoCompleteTextView.getText().toString();
        SharedPreferences sp = getSharedPreferences("network_url", 0);
        String longhistory = sp.getString(field, "nothing");
        if (!longhistory.contains(text + ",")) {
            StringBuilder sb = new StringBuilder(longhistory);
            sb.insert(0, text + ",");
            sp.edit().putString("history", sb.toString()).commit();
        }
    }
    /**
     * 初始化 AutoCompleteTextView,最多显示 5 项提示,使 AutoCompleteTextView
     * 在一开始获得焦点时自动提示保存在 sharedPreference 中的字段名
     * 要操作的 AutoCompleteTextView
     */
    private void initAutoComplete(String field,
            AutoCompleteTextView autoCompleteTextView) {
        SharedPreferences sp = getSharedPreferences("network_url", 0);
        String longhistory = sp.getString("history", "nothing");
        String[] histories = longhistory.split(",");
        ArrayAdapter<String> adapter = new ArrayAdapter<String>(this,
                android.R.layout.simple_dropdown_item_1line, histories);
        //只保留最近的 50 条的记录
        if (histories.length > 50) {
            String[] newHistories = new String[50];
            System.arraycopy(histories, 0, newHistories, 0, 50);
            adapter = new ArrayAdapter<String>(this,
                    android.R.layout.simple_dropdown_item_1line, newHistories);
        }
        autoCompleteTextView.setAdapter(adapter);
        autoCompleteTextView
                .setOnFocusChangeListener(new OnFocusChangeListener() {
```

```
            @Override
            public void onFocusChange(View v, boolean hasFocus) {
                AutoCompleteTextView view = (AutoCompleteTextView) v;
                if (hasFocus) {
                    view.showDropDown();
                }
            }
        });
    }
}
```

运行程序,默认效果如图 2-9(a)所示,当在文本框中输入关键字时,效果如图 2-9(b)所示。

(a) 默认界面　　　　　　　　　　(b) 显示自动查找内容

图 2-9　自动提示文本框

2.3　按钮类实例

Android 中提供了普通按钮和图片按钮两种按钮组件。这两种按钮都用于用户界面上生成一个可以单击的按钮。当用户单击按钮时,将会触发一个 onClick 事件,可以通过为按钮添加单击事件监听器指定所要触发的动作,同时,在 Android 中也提供了单选与复选按钮组。下面分别对按钮类控件进行实例分析。

2.3.1　按钮测试实例

Button 按钮是最常见的控件,本实例利用 Button 按钮实现几个按钮的测试,同时向读者演示 Button 控件的具体用法。其具体实现步骤如下。

(1) 在 Eclipse 中创建一个 Android 应用项目,命名为 Button_test。

(2) 打开 res\layout 目录下的 main.xml 布局文件,在文件中声明几个 Button 控件。代码为:

```xml
<LinearLayout xmlns:android="http://schemas.android.com/apk/res/android"
    xmlns:tools="http://schemas.android.com/tools"
    android:layout_width="match_parent"
    android:layout_height="match_parent"
    tools:context=".MainActivity"
    android:orientation="vertical"
    android:background="#aabbcc">
    <TextView
        android:layout_width="wrap_content"
        android:layout_height="wrap_content"
        android:text="@string/hello_world" />
    <Button
        android:id="@+id/button1"
        android:layout_width="match_parent"
        android:layout_height="wrap_content"
        android:text="Button1 测试"/>
    <Button
        android:id="@+id/button2"
        android:layout_width="match_parent"
        android:layout_height="wrap_content"
        android:text="Button2 测试"/>
    <Button
        android:id="@+id/button3"
        android:layout_width="match_parent"
        android:layout_height="wrap_content"
        android:text="Button3 测试"
        android:onClick="clickHandler"/>
</LinearLayout>
```

(3) 打开 src\fs.button_test 包下的 MainActivity.java 文件，在文件中实现单击按钮的触发事件。代码为：

```java
package fs.button_test;
import android.app.Activity;
import android.os.Bundle;
import android.view.Menu;
import android.view.View;
import android.view.View.OnClickListener;
import android.widget.Button;
public class MainActivity extends Activity implements OnClickListener{
    private Button button1 = null;
    private Button button2 = null;
    public void findButton() {
        button1 = (Button)findViewById(R.id.button1);
        button2 = (Button)findViewById(R.id.button2);
    }
    @Override
    protected void onCreate(Bundle savedInstanceState) {
        super.onCreate(savedInstanceState);
        setContentView(R.layout.main);
        findButton();
        button2.setOnClickListener(this);
        button1.setOnClickListener(new OnClickListener() {
            @Override
            public void onClick(View v) {
                //TODO 自动存根法
```

```
      System.out.println("您单击了Button1");
     }
   });
  }
  @Override
  public boolean onCreateOptionsMenu(Menu menu) {
   getMenuInflater().inflate(R.menu.main, menu);
   return true;
  }
  @Override
  public void onClick(View v) {
   //TODO 自动存根法
   switch (v.getId()) {
   case R.id.button2:
     System.out.println("您单击了Button2");
     break;
   default:
     break;
   }
  }
  public void clickHandler(View view) {
   System.out.println("您单击了Button3");
  }
}
```

运行程序,效果如图 2-10 所示。

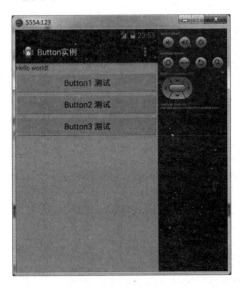

图 2-10 按钮测试

虽然 Button 控件为最基本的控件,但其在 Android 中有其自身的发展,主要内容如下。

- java 代码中通过 btn1 关联次控件:

  android:id = "@ + id/btn1"

- 控件宽度:

android:layout_width = "80px"   //"80dip"或"80dp"
android:layout_width = "wrap_content"

```
android:layout_width = "match_parent"
```

- 控件高度：

```
android:layout_height = "80px"              //"80dip"或"80dp"
android:layout_height = "wrap_content"
android:layout_height = "match_parent"
```

- 控件排布：

```
android:orientation = "horizontal"
android:orientation = "vertical"
```

- 控件间距：

```
android:layout_marginLeft = "5dip"          //距离左边
android:layout_marginRight = "5dip"         //距离右边
android:layout_marginTop = "5dip"           //距离上面
android:layout_marginBottom = "5dip"        //距离下面
```

- 控件显示位置：

```
android:gravity = "center"                  //left,right, top, bottom
android:gravity = "center_horizontal"
android:layout_gravity                      //属性则设置控件本身相对于父控件的显示位置
android:gravity                             //是本元素所有子元素的重力方向
android:layout_gravity = "center_vertical"
android:layout_gravity = "left"
android:layout_gravity = "left|bottom"
```

- TextView 中文本字体：

```
android:text = "@String/text1"              //在 string.xml 中定义 text1 的值
android:textSize = "20sp"
android:textColor = "#ff123456"
android:textStyle = "bold"                  //普通(normal)、斜体(italic)、粗斜体(bold_italic)
```

- 定义控件是否可见：

```
android:visibility = "visible"              //可见
android:visibility = "invisible"            //不可见,但是在布局中占用的位置还在
android:visibility = "gone"                 //不可见,完全从布局中消失
```

- 定义背景图片

```
android:background = "@drawable/img_bg"     //img_bg 为 drawable 下的一张图片
```

- seekbar 控件背景图片及最大值：

```
android:progressDrawable = "@drawable/seekbar_img"
android:thumb = "@drawable/thumb"
android:max = "60"
```

- 在父布局的相对位置：

```
android:layout_alignParentLeft = "true"     //在布局左边
android:layout_alignParentRight = "true"    //在布局右边
android:layout_alignParentTop = "true"      //在布局上面
android:layout_alignParentBottom = "true"   //在布局下面
```

- 在某个控件的相对位置：

```
android:layout_toRightOf = "@id/button1"    //在控件 button1 右边,不仅仅是紧靠着
android:layout_toLeftOf = "@id/button1"     //在控件 button1 左边,不仅仅是紧靠着
android:layout_below = "@id/button1"        //在控件 button1 下面,不仅仅是正下方
android:layout_above = "@id/button1"        //在控件 button1 上面,不仅仅是正上方
```

- 定义和某控件对齐：

```
android:layout_alignTop = "@id/button1"     //和控件 button1 上对齐
android:layout_alignBottom = "@id/button1"  //和控件 button1 下对齐
android:layout_alignLeft = "@id/button1"    //和控件 button1 左对齐
android:layout_alignRight = "@id/button1"   //和控件 button1 右对齐
android:layout_centerHorizontal = "true"    //水平居中
android:layout_centerVertical = "true"
android:layout_centerInParent = "true"
```

- 仅在 LinearLayout 中有效：

```
android:layout_weight = "1"                 //设置控件在一排或一列中所占比例值
```

2.3.2 图片说明实例

本实例通过对 ImageButton 的应用,构建 4 个带图片的小程序。并通过本实例来演示 ImageButton 的具体应用。

在本实例中单击 ImageButton 按钮,可更换按钮图片,即可达到改变背景图片的效果。其具体实现步骤如下。

（1）在 Eclipse 中创建一个 Android 应用项目,命名为 ImageButton_test。
（2）打开 res\layout 目录下的 main.xml 布局文件,在文件中加如下声明。

```xml
<?xml version = "1.0" encoding = "utf-8"?>
<LinearLayout xmlns:android = "http://schemas.android.com/apk/res/android"
    android:orientation = "vertical"
    android:layout_width = "fill_parent"
    android:layout_height = "fill_parent"
    android:background = "#aabbcc">
<ImageButton
    android:id = "@+id/ImageButton01"
    android:layout_width = "wrap_content"
    android:layout_height = "wrap_content"
    android:src = "@drawable/c1"/>
<ImageButton
    android:id = "@+id/ImageButton02"
    android:layout_width = "wrap_content"
    android:layout_height = "wrap_content"
    android:src = "@drawable/c2" />
<ImageButton
    android:id = "@+id/ImageButton03"
    android:layout_width = "wrap_content"
    android:layout_height = "wrap_content"
    android:src = "@drawable/c3" />
<ImageButton
    android:id = "@+id/ImageButton04"
    android:layout_width = "wrap_content"
```

```
android:layout_height = "wrap_content"/>
</LinearLayout>
```

(3) 打开 src\fs.imagebutton_test 包下的 MainActivity.java 文件，在文件中实现当单击某一个图片按钮时，即弹出对应的提示框。代码为：

```java
package fs.imagebutton_test;
import android.app.Activity;
import android.app.AlertDialog;
import android.app.Dialog;
import android.app.AlertDialog.Builder;
import android.content.DialogInterface;
import android.os.Bundle;
import android.view.View;
import android.widget.Button;
import android.widget.ImageButton;
import android.widget.TextView;
public class MainActivity extends Activity {
    /** 第一次调用 Activity 活动. */
    TextView textView;
    ImageButton imageButton1,imageButton2,imageButton3,imageButton4;
    @Override
    public void onCreate(Bundle savedInstanceState) {
        super.onCreate(savedInstanceState);
        setContentView(R.layout.main);
        imageButton1 = (ImageButton)findViewById(R.id.ImageButton01);
        imageButton2 = (ImageButton)findViewById(R.id.ImageButton02);
        imageButton3 = (ImageButton)findViewById(R.id.ImageButton03);
        imageButton4 = (ImageButton)findViewById(R.id.ImageButton04);
        /**
         * 给按钮设置使用的图标,由于 button1、button2、button3、button4
         * 已经在 xml 文件中设置,这里就不再设置了
         */ imageButton4.setImageDrawable(getResources().getDrawable(android.R.drawable.sym_call_incoming));
        //以下分别为每个按钮设置事件监听 setOnClickListener
        imageButton1.setOnClickListener(new Button.OnClickListener(){
            public void onClick(View v) {
                //对话框 Builder 是 AlertDialog 的静态内部类
                Dialog dialog = new AlertDialog.Builder(MainActivity.this)
                //设置对话框的标题
                    .setTitle("提示")
                //设置对话框要显示的消息
                    .setMessage("我真的是 ImageButton1")
                //给对话框设置按钮叫作"确定",并且设置监听器,这种写法也真是有些 BT
                    .setPositiveButton("确定", new DialogInterface.OnClickListener(){
                        public void onClick(DialogInterface dialog, int which) {
                            //单击"确定"按钮之后要执行的操作就写在这里了
                        }
                    }).create();      //创建按钮
                dialog.show();        //显示
            }
        });
```

```java
        imageButton2.setOnClickListener(new Button.OnClickListener(){
            public void onClick(View v) {
                Builder dialog = new AlertDialog.Builder(MainActivity.this);
                dialog.setTitle("提示");
                dialog.setMessage("我是ImageButton2,我要使用ImageButton3的图标");
                dialog.setPositiveButton("确定", new DialogInterface.OnClickListener(){
                    public void onClick(DialogInterface dialog, int which) {
                        //好了,我成功地把Button3的图标掠夺过来
                        imageButton2.setImageDrawable(getResources().getDrawable(R.
                        drawable.c4));
                    }
                }).create();                //创建按钮
                dialog.show();
            }
        });
        imageButton3.setOnClickListener(new Button.OnClickListener(){
            public void onClick(View v) {
                Builder dialog = new AlertDialog.Builder(MainActivity.this);
                dialog.setTitle("提示");
                dialog.setMessage("我是ImageButton3,我要使用系统设置电话图标");
                dialog.setPositiveButton("确定", new DialogInterface.OnClickListener(){
                    public void onClick(DialogInterface dialog, int which) {
                        //把imageButton3的图标设置为系统的打电话图标 imageButton3.
                        setImageDrawable(getResources().getDrawable(android.R.drawable.sym_
                        action_call));
                    }
                }).create();                //创建按钮
                dialog.show();
            }
        });
        imageButton4.setOnClickListener(new Button.OnClickListener(){
            public void onClick(View v) {
                Builder dialog = new AlertDialog.Builder(MainActivity.this);
                dialog.setTitle("提示");
                dialog.setMessage("我没钱买图标,使用的是系统图标");
                dialog.setPositiveButton("确定", new DialogInterface.OnClickListener(){
                    public void onClick(DialogInterface dialog, int which) {
                    }
                }).create();                //创建按钮
                dialog.show();
            }
        });
    }
}
```

运行程序,默认效果如图2-11(a)所示,当单击第3个图片按钮时,效果如图2-11(b)所示。

Android中的ImageButton控件是不能带文字的。本实例为了实现带文字的ImageButton,所以选择将ImageView控件和TextView控件封装在一个LinearLayout里面,整个LinearLayout就是一个按钮,然后对它监听单击等动作。其具体实现步骤如下。

(1) 在Eclipse中创建一个Android应用项目,命名为ImageView_test。

(2) 打开res\layout目录下的main.xml布局文件,在文件中声明一个ImageView控件及一个TextView控件。代码为:

 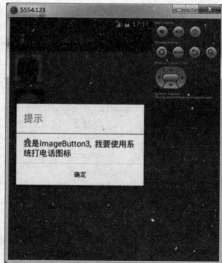

（a）4个图片按钮　　　　　　　　（b）提示框

图 2-11　图片按钮效果

```xml
< LinearLayout xmlns:android = "http://schemas.android.com/apk/res/android"
    android:layout_width = "wrap_content"
    android:layout_height = "wrap_content"
    android:orientation = "vertical"
    android:id = "@ + id/bt"
    android:background = " # aabbcc">
    <! -- 当单击图片时,即改变背景颜色 -->
    < ImageView
        android:id = "@ + id/ib"
        android:layout_width = "wrap_content"
        android:layout_height = "wrap_content"
        android:src = "@drawable/flow"
        android:background = " # 00000000"/>
    < TextView
        android:id = "@ + id/tv"
        android:layout_width = "wrap_content"
        android:layout_height = "wrap_content"
        android:text = "花开花落又一春,日复一日又一年"
        android:paddingLeft = "20px"/>
</LinearLayout >
```

（3）打开 src\fs.imageview_test 包下的 MainActivity.java 文件,在文件中实现当单击界面中的图片时,即改变图片按钮的背景颜色。代码为：

```java
package fs.imageview_test;
import android.app.Activity;
import android.graphics.Color;
import android.os.Bundle;
import android.view.MotionEvent;
import android.view.View;
import android.view.View.OnClickListener;
import android.view.View.OnTouchListener;
import android.widget.LinearLayout;
import android.widget.Toast;
public class MainActivity extends Activity {
    LinearLayout m_ll;
```

```java
/** 第一次调用 Activity 活动 */
@Override
public void onCreate(Bundle savedInstanceState) {
    super.onCreate(savedInstanceState);
    setContentView(R.layout.main);
    m_ll = (LinearLayout)findViewById(R.id.bt);
    m_ll.setClickable(true);
    m_ll.setOnClickListener(ocl);
    m_ll.setOnTouchListener(otl);
}
public OnClickListener ocl = new OnClickListener() {
    @Override
    public void onClick(View v) {
        //TODO 自动存根法
    Toast.makeText(getApplicationContext(), "yes", Toast.LENGTH_SHORT).show();
    }
};
public OnTouchListener otl = new OnTouchListener() {
    @Override
    public boolean onTouch(View v, MotionEvent event) {
        //TODO 自动存根法
        if(event.getAction() == MotionEvent.ACTION_DOWN)
        {
            m_ll.setBackgroundColor(Color.rgb(127,127,127));
        }
        else if(event.getAction() == MotionEvent.ACTION_UP)
        {
            m_ll.setBackgroundColor(Color.TRANSPARENT);
        }
        return false;
    }
};
}
```

运行程序,效果如图 2-12(a)所示,当单击界面中的图片按钮时,效果如图 2-12(b)所示。

(a) 默认界面　　　　　　　　　　　(b) 改变背景颜色

图 2-12　带文字说明的图片按钮

2.3.3 程序开闭实例

ToggleButton 控件是 Android 提供的开关按钮,有选中和未选择两种状态。

本实例通过对 ToggleButton 控件的应用,构建一个关闭开启按钮。通过本实例详细演示了 ToggleButton 控件具体应用。其具体实现步骤如下。

(1) 在 Eclipse 中创建一个 Android 应用项目,命名为 ToggleButton_test。

(2) 打开 res\layout 目录下的 main.xml 布局文件,在文件中声明 ToggleButton 控件。代码为:

```xml
<?xml version = "1.0" encoding = "utf-8"?>
<LinearLayout xmlns:android = "http://schemas.android.com/apk/res/android"
        android:orientation = "vertical"
        android:layout_width = "fill_parent"
        android:background = " #FFF5F5F5"
        android:layout_height = "fill_parent">
<LinearLayout
    android:layout_width = "fill_parent"
    android:layout_height = "wrap_content"
    android:orientation = "horizontal">
    <TextView
        android:textSize = "14.0sp"
        android:id = "@ + id/tvSound"
        android:textColor = "@android:color/black"
        android:layout_width = "wrap_content"
        android:layout_height = "wrap_content"
        android:text = "已关闭" />
    <ToggleButton
        android:id = "@ + id/tglSound"
        android:background = "@drawable/select"
        android:layout_width = "wrap_content"
        android:layout_height = "wrap_content"
        android:checked = "true"
        android:textOn = ""
        android:textOff = ""
        android:text = "" />
</LinearLayout>
</LinearLayout>
```

(3) 在 res 文件夹下新建一个 drawable 文件夹,在文件夹中新建一个 select.xml 文件,用于存放两个图片的资源。代码为:

```xml
<?xml version = "1.0" encoding = "utf-8"?>
<selector
  xmlns:android = "http://schemas.android.com/apk/res/android">
    <item android:state_checked = "true"
        android:drawable = "@drawable/caa" />
    <item android:drawable = "@drawable/baa" />
</selector>
```

(4) 打开 src\fs.togglebutton_test 包下的 MainActivity.java 文件,在文件中实现开关的开启与关闭,并切换到相应的图片。代码为:

```java
package fs.togglebutton_test;
import android.app.Activity;
```

```java
import android.os.Bundle;
import android.view.Window;
import android.widget.CompoundButton;
import android.widget.CompoundButton.OnCheckedChangeListener;
import android.widget.TextView;
import android.widget.ToggleButton;
public class MainActivity extends Activity implements OnCheckedChangeListener{
    private ToggleButton mToggleButton;
    private TextView tvSound;
    @Override
    public void onCreate(Bundle savedInstanceState) {
        super.onCreate(savedInstanceState);
        requestWindowFeature(Window.FEATURE_NO_TITLE);          //隐藏标题栏
        setContentView(R.layout.main);
        initView();                                             //初始化控件方法
    }
    private void initView() {
        mToggleButton = (ToggleButton) findViewById(R.id.tglSound);  //获取控件
        mToggleButton.setOnCheckedChangeListener(this);              //添加监听事件
        tvSound = (TextView) findViewById(R.id.tvSound);
    }
    @Override
    public void onCheckedChanged(CompoundButton buttonView, boolean isChecked) {
        if(isChecked){
            tvSound.setText("已关闭");
        }else{
            tvSound.setText("已开启");
        }
    }
}
```

运行程序,默认效果如图 2-13(a)所示,当单击开关按钮时,即进行图片的切换,效果如图 2-13(b)所示。

(a) 开关关闭　　　　　　　　　　　(b) 开关开启

图 2-13　开关按钮

在Android中，ToggleButton控件除了继承自父类的一些属性外还具有其自身的一些属性，主要内容如下。
- android:textOff：设置当该按钮没有被选中时显示的文本。
- android:textOn：设置当该按钮被选中时显示的文本。

2.3.4 城市选择实例

本实例中，通过对RadioButton控件的应用，构建一个城市选择程序，当选择正确的城市时即弹出对应的Toast提示框，选择错误时，出现对应的Toast提示框。其具体实现步骤如下。

（1）在Eclipse中创建一个Android应用项目，命名为RadioButton_test。

（2）打开res\layout目录下的main.xml布局文件，在文件中定义RadioButton控件。代码为：

```xml
<AbsoluteLayout xmlns:android = "http://schemas.android.com/apk/res/android"
    android:orientation = "vertical"
    android:layout_width = "fill_parent"
    android:layout_height = "fill_parent"
    android:background = "#aabbcc">
  <TextView
        android:text = "哪个城市是最适合度假的?"
        android:textSize = "30px"
        android:layout_height = "wrap_content"
        android:id = "@+id/mytextview"
        android:layout_width = "fill_parent"
        android:textColor = "#FF0000"
        android:layout_x = "0dp"
        android:layout_y = "80dp"/>
  <RadioGroup
        android:orientation = "vertical"
        android:layout_height = "wrap_content"
        android:layout_width = "wrap_content"
        android:id = "@+id/radiogroup"
        android:layout_x = "0dp"
        android:layout_y = "140dp">
    <RadioButton
        android:layout_height = "wrap_content"
        android:text = "杭州"
        android:layout_width = "wrap_content"
        android:id = "@+id/button1"/>
    <RadioButton
        android:layout_height = "wrap_content"
        android:text = "海南"
        android:layout_width = "wrap_content"
        android:id = "@+id/button2"/>
    <RadioButton
        android:layout_height = "wrap_content"
        android:text = "成都"
        android:layout_width = "wrap_content"
        android:id = "@+id/button3"/>
    <RadioButton
        android:layout_height = "wrap_content"
```

```xml
            android:text = "云南"
            android:layout_width = "wrap_content"
            android:id = "@ + id/button4"/>
    </RadioGroup>
</AbsoluteLayout>
```

(3) 打开 src\fs.radiobutton_test 包下的 MainActivity.java 文件,在文件中实现城市的选择,并弹出相应的提示框。代码为:

```java
package com.example.radiobutton_test;
import android.app.Activity;
import android.os.Bundle;
import android.view.Gravity;
import android.widget.RadioButton;
import android.widget.RadioGroup;
import android.widget.Toast;
import android.widget.RadioGroup.OnCheckedChangeListener;
public class MainActivity extends Activity {
    /** 第一次调用 Activity 活动 */
    protected RadioGroup group;
    protected RadioButton radio1,radio2,radio3,radio4;
    @Override
    public void onCreate(Bundle savedInstanceState) {
        super.onCreate(savedInstanceState);
        setContentView(R.layout.main);
        group = (RadioGroup)findViewById(R.id.radiogroup);
        radio1 = (RadioButton)findViewById(R.id.button1);
        radio2 = (RadioButton)findViewById(R.id.button2);
        radio3 = (RadioButton)findViewById(R.id.button3);
        radio4 = (RadioButton)findViewById(R.id.button4);
        group.setOnCheckedChangeListener(new AnswerListener() );
    }
    class AnswerListener implements OnCheckedChangeListener {
        public void onCheckedChanged(RadioGroup group, int checkedId) {
            //TODO 自动存根法
                if (checkedId == radio2.getId())
                {
                    showMessage("正确答案:" + radio2.getText()+"恭喜你,答对了!");
                }
                else if(checkedId == radio1.getId())
                {
                    showMessage("对不起! " + radio1.getText()+"你答错了哦!");
                }
                else if(checkedId == radio3.getId())
                {
                    showMessage("对不起!" + radio3.getText()+"你答错了哦!");
                }
                else
                {
                    showMessage("对不起!" + radio4.getText()+"你答错了哦!");
                }
        }
    }
    public void showMessage(String str)
    {
        Toast toast = Toast.makeText(this, str, Toast.LENGTH_SHORT);
```

```
            toast.setGravity(Gravity.TOP, 0, 420);
            toast.show();
        }
    }
```

运行程序,效果如图 2-14(a)所示,当选择相应的城市时,即弹出对应的 Toast 提示框,效果如图 2-14(b)及图 2-14(c)所示。

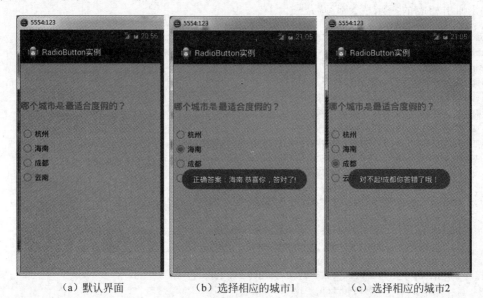

(a) 默认界面　　　　　　(b) 选择相应的城市1　　　　(c) 选择相应的城市2

图 2-14　RadioButton 控件使用

实现单选按钮由两部分组成,也就是 RadioButton 和 RadioGroup 配合使用,RadioButton 和 RadioGroup 的关系如下。

(1) RadioButton 表示单个圆形单选框,而 RadioGroup 是可以容纳多个 RadioButton 的容器。

(2) 每个 RadioGroup 中的 RadioButton 同时只能有一个被选中。

(3) 不同的 RadioGroup 中的 RadioButton 互不相干,即如果组 A 中有一个选中了,组 B 中依然可以有一个被选中。

(4) 大部分场合下,一个 RadioGroup 中至少有两个 RadioButton。

(5) 大部分场合下,一个 RadioGroup 中的 RadioButton 默认会有一个被选中,并建议将它放在 RadioGroup 中的起始位置。

2.3.5　确定选择实例

本实例中,通过对 CheckBox 的 isChecked 属性的应用,构建 CheckBox 的 isChecked 属性控制 Button 按钮的 Enable 的实例。通过本实例演示 CheckBox 复选框的应用。

在本实例的多选按钮选中的情况下,单击"确定"按钮,在多选按钮上显示"已经选择"。其具体实现步骤如下。

(1) 在 Eclipse 中创建一个 Android 应用项目,命名为 isChecked_test。

(2) 打开 res\layout 目录下的 main.xml 布局文件,在文件中声明一个 TextView 控件、一个 CheckBox 控件及一个 Button 控件。代码为:

```xml
<?xml version="1.0" encoding="utf-8"?>
<LinearLayout xmlns:android="http://schemas.android.com/apk/res/android"
    android:orientation="vertical"
    android:layout_width="fill_parent"
    android:layout_height="fill_parent"
    android:background="#aabbcc">
  <TextView
      android:text="初始化"
      android:id="@+id/TextView01"
      android:textSize="28dip"
      android:layout_width="wrap_content"
      android:layout_height="wrap_content"/>
  <CheckBox
      android:text="ischeck"
      android:id="@+id/CheckBox01"
      android:textSize="28dip"
      android:layout_width="wrap_content"
      android:layout_height="wrap_content"/>
  <Button
      android:text="确定"
      android:id="@+id/Button01"
      android:textSize="28dip"
      android:layout_width="wrap_content"
      android:layout_height="wrap_content"/>
</LinearLayout>
```

(3) 打开 src\fs.ischecked_test 包下的 MainActivty.java 文件,在文件中实现复选框的选择,当选择了复选框,并单击"确定"按钮时,即在文本框中显示结果。代码为:

```java
package fs.ischecked_test;
import android.app.Activity;
import android.os.Bundle;
import android.view.View;
import android.view.View.OnClickListener;
import android.widget.Button;
import android.widget.CheckBox;
import android.widget.TextView;
public class MainActivity extends Activity
{
    @Override
    public void onCreate(Bundle savedInstanceState)
    {
        super.onCreate(savedInstanceState);
        setContentView(R.layout.main);
        final TextView tv01 = (TextView)this.findViewById(R.id.TextView01);
        final CheckBox cb = (CheckBox)this.findViewById(R.id.CheckBox01);
        final Button but = (Button)this.findViewById(R.id.Button01);
        cb.setChecked(false);
        but.setEnabled(false);
        cb.setOnClickListener(
            new OnClickListener()
            {
                public void onClick(View v)
                {
```

```
                    if(cb.isChecked())
                    {
                        tv01.setText("");
                        but.setEnabled(true);
                    }
                    else
                    {
                        but.setEnabled(false);
                        tv01.setText("请选择我");
                    }
                }
            }
        );
        but.setOnClickListener
        (
            new OnClickListener()
            {
                public void onClick(View v)
                {
                    if(cb.isChecked())
                    {
                        tv01.setText("已经选择");
                    }
                }
            }
        );
    }
}
```

运行程序,效果如图 2-15(a)所示,当选择复选框时,效果如图 2-15(b)所示,当单击"确定"按钮时,效果如图 2-15(c)所示。

(a) 默认按钮不可见界面　　(b) 选择复选框界面　　(c) 单击"确定"按钮界面

图 2-15　isChecked 属性效果

前面介绍了 RadioButton 控件与 CheckBox 控件,那么它们有什么共同之处与不同之处呢？RadioButton 和 CheckBox 的区别如下。

- 单个 RadioButton 在选中后,通过单击无法变为未选中。
  单个 CheckBox 在选中后,通过单击可以变为未选中。
- 一组 RadioButton,只能同时选中一个。
  一组 CheckBox,能同时选中多个。
- RadioButton 在大部分 UI 框架中默认都以圆形表示。
  CheckBox 在大部分 UI 框架中默认都以矩形表示

而 RadioButton 控件与 CheckBox 控件的共同属性主要如下所示。

- isCheck()：判断是否被选中,如果被选中返回 true,否则返回 false。
- performClick()：通过传入的参数设置控件状态。
- performClick()：调用 OnClickListener 监听器,即模拟一次单击。
- toggle()：置反控件当前的状态。
- setOnCheckedChangeListener(CompoundButton. OnCheckedChangeListener listener)：为控件设置 OnCheckedChangeListener 监听器。

下面通过一个实例来演示 CheckBox 复选框同时选择多个选项的应用,其具体实现步骤如下。

（1）在 Eclipse 中创建一个 Android 应用项目,命名为 CheckBox_test。

（2）打开 res\layout 目录下的 main.xml 布局文件,在文件中声明一个 EditText 控件及 3 个 CheckBox 控件。代码为：

```xml
<?xml version = "1.0" encoding = "utf-8"?>
<LinearLayout xmlns:android = "http://schemas.android.com/apk/res/android"
    android:orientation = "vertical"
    android:layout_width = "fill_parent"
    android:layout_height = "fill_parent"
    android:background = "#ccddee">
<EditText
    android:id = "@ + id/editText1"
    android:layout_width = "fill_parent"
    android:layout_height = "wrap_content"
    android:text = "请选择" />
<CheckBox
    android:id = "@ + id/beijing"
    android:layout_width = "wrap_content"
    android:layout_height = "wrap_content"
    android:text = "天津" />
<CheckBox
    android:id = "@ + id/shanghai"
    android:layout_width = "wrap_content"
    android:layout_height = "wrap_content"
    android:text = "深圳"/>
<CheckBox
    android:id = "@ + id/shenzhen"
    android:layout_width = "wrap_content"
    android:layout_height = "wrap_content"
```

```
            android:text = "汕头"/>
</LinearLayout>
```

(3) 打开 src\fs.checkbox_test 包下的 MainActivity.java 文件，在文件中实现 CheckBox 控件的定义及单击事件的监听并显示结果。代码为：

```java
package fs.checkbox_test;
import android.app.Activity;
import android.os.Bundle;
import android.widget.CheckBox;
import android.widget.CompoundButton;
import android.widget.EditText;
public class MainActivity extends Activity {
    /** 第一次调用 Activity 活动 */
    //对控件对象进行声明
    CheckBox beijing = null;
    CheckBox shanghai = null;
    CheckBox shenzhen = null;
    EditText editText1 = null;
    @Override
    public void onCreate(Bundle savedInstanceState) {
        super.onCreate(savedInstanceState);
        setContentView(R.layout.main);
        //通过控件的 ID 来得到代表控件的对象
        beijing = (CheckBox)findViewById(R.id.beijing);
        shanghai = (CheckBox)findViewById(R.id.shanghai);
        shenzhen = (CheckBox)findViewById(R.id.shenzhen);
        editText1 = (EditText)findViewById(R.id.editText1);
        //给 CheckBox 设置事件监听
        beijing.setOnCheckedChangeListener(new CompoundButton.OnCheckedChangeListener(){
            @Override
            public void onCheckedChanged(CompoundButton buttonView,
                    boolean isChecked) {
                //TODO 自动存根法
                if(isChecked){
                    editText1.setText(buttonView.getText() + "选中");
                }else{
                    editText1.setText(buttonView.getText() + "取消选中");
                }
            }
        });
        shanghai.setOnCheckedChangeListener(new CompoundButton.OnCheckedChangeListener(){
            @Override
            public void onCheckedChanged(CompoundButton buttonView,
                    boolean isChecked) {
                //TODO 自动存根法
                if(isChecked){
                    editText1.setText(buttonView.getText() + "选中");
                }else{
                    editText1.setText(buttonView.getText() + "取消选中");
                }
            }
```

```
                });
                shenzhen.setOnCheckedChangeListener(new CompoundButton.OnCheckedChangeListener(){
                    @Override
                    public void onCheckedChanged(CompoundButton buttonView,
                            boolean isChecked) {
                        //TODO 自动存根法
                        if(isChecked){
                            editText1.setText(buttonView.getText() + "选中");
                        }else{
                            editText1.setText(buttonView.getText() + "取消选中");
                        }
                    }
                });
            }
        }
```

运行程序,效果如图 2-16(a)所示,当选中对应的项时即将结果显示在编辑框中,如图 2-16(b)所示,当取消对应的项时也将结果显示在编辑框中,如图 2-16(c)所示。

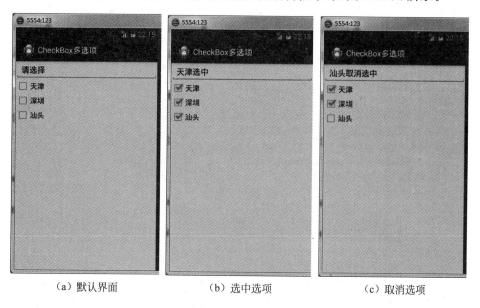

(a)默认界面　　　　　　(b)选中选项　　　　　　(c)取消选项

图 2-16　多项选择

2.3.6　个人性格选择实例

CheckedTextView 控件是在 Android 技术中实现选中的 checked 效果。本实例中,通过对 CheckedTextView 控件的应用,构建一个个人性格特点选择界面。通过本实例可向读者演示 CheckedTextView 控件的具体应用。其具体实现步骤如下。

(1) 在 Eclipse 中创建一个 Android 应用项目,命名为 CheckedTextView_test。

(2) 打开 res\layout 目录下的 main.xml 布局文件,在文件中声明一个 TextView 控件、一个 ListView 控件及 4 个 CheckedTextView 控件。代码为:

```
<?xml version = "1.0" encoding = "utf - 8"?>
< LinearLayout xmlns:android = "http://schemas.android.com/apk/res/android"
```

```xml
    android:orientation = "vertical"
    android:layout_width = "fill_parent"
    android:layout_height = "fill_parent"
    android:background = "#aabbcc">
<TextView
    android:layout_width = "80px"
    android:layout_height = "wrap_content"
    android:text = "TextView多选框"
    android:layout_gravity = "center"
    android:ellipsize = "marquee"
    android:singleLine = "true"
    android:focusable = "true"
    android:marqueeRepeatLimit = "marquee_forever"
    android:focusableInTouchMode = "true"
    android:scrollHorizontally = "true"/>
<ListView
    android:id = "@+id/listView"
    android:layout_width = "fill_parent"
    android:layout_height = "wrap_content"/>
<CheckedTextView
    android:id = "@+id/checkedTextView1"
    android:layout_width = "fill_parent"
    android:layout_height = "wrap_content"
    android:checkMark = "?android:attr/listChoiceIndicatorMultiple"
    android:text = "开朗、活泼、健谈"/>
<CheckedTextView
    android:id = "@+id/checkedTextView2"
    android:layout_width = "fill_parent"
    android:layout_height = "wrap_content"
    android:checkMark = "?android:attr/listChoiceIndicatorMultiple"
    android:text = "多疑、爆躁、吝啬"/>
<CheckedTextView
    android:id = "@+id/checkedTextView3"
    android:layout_width = "fill_parent"
    android:layout_height = "wrap_content"
    android:checkMark = "?android:attr/listChoiceIndicatorMultiple"
    android:text = "喜欢运动、旅游"/>
<CheckedTextView
    android:id = "@+id/checkedTextView4"
    android:layout_width = "fill_parent"
    android:layout_height = "wrap_content"
    android:checkMark = "?android:attr/listChoiceIndicatorMultiple"
    android:text = "讨厌吃蔬菜、喜欢吃肉"/>
</LinearLayout>
```

(3) 打开 src\fs.checkedtextview_test 包下的 MainActivity.java 文件，在文件中实现多项框的选择状态。代码为：

```java
package fs.checkedtextview_test;
import android.app.Activity;
```

```java
import android.os.Bundle;
import android.view.View;
import android.widget.CheckedTextView;
import android.widget.ListView;
public class MainActivity extends Activity {
    private ListView listView;
    private CheckedTextView checkedTextView1,checkedTextView2,checkedTextView3,checkedTextView4;
    /** 第一次调用 Activity 活动 */
    @Override
    public void onCreate(Bundle savedInstanceState) {
        super.onCreate(savedInstanceState);
        setContentView(R.layout.main);
        listView = (ListView)findViewById(R.id.listView);
        checkedTextView1 = (CheckedTextView)findViewById(R.id.checkedTextView1);
        checkedTextView2 = (CheckedTextView)findViewById(R.id.checkedTextView2);
        checkedTextView3 = (CheckedTextView)findViewById(R.id.checkedTextView3);
        checkedTextView4 = (CheckedTextView)findViewById(R.id.checkedTextView4);
        //设置 checkedTextView1 为选中状态
        checkedTextView1.setChecked(true);
//设置 checkedTextView2 的页边距,即距上、下、左、右各 20 像素,默认为未选中状态
        checkedTextView2.setPadding(20, 20, 20, 20);
        //设置 checkedTextView3 为选中状态,并更改其显示图标,使用 Android 系统资源 arrow_down
            _float
        checkedTextView3.setChecked(true); checkedTextView3.setCheckMarkDrawable(android.R.
        drawable.arrow_down_float);
        //设置 checkedTextView4 反转状态,即由默认的未选中反转为选中状态
        checkedTextView4.toggle();
        //单击状态后变更相反,如选中变为未选中,未选中的变为选中
        checkedTextView1.setOnClickListener(new View.OnClickListener() {
            @Override
            public void onClick(View v) {
                //TODO 自动存根法
                checkedTextView1.toggle();
            }
        });
        //单击状态后变更相反,如选中变为未选中,未选中的变为选中
    checkedTextView2.setOnClickListener(new View.OnClickListener() {
            @Override
            public void onClick(View v) {
                //TODO 自动存根法
                checkedTextView2.toggle();
            }
        });
//单击状态后变更相反,即下三角转化为上三角符号
    checkedTextView3.setOnClickListener(new View.OnClickListener() {
            @Override
            public void onClick(View v) {
                //TODO 自动存根法

                checkedTextView3.setCheckMarkDrawable(android.R.drawable.arrow_up_float);
            }
```

```
        });
//单击状态后变更相反,如选中变为未选中,未选中的变为选中
checkedTextView4.setOnClickListener(new View.OnClickListener() {
            @Override
            public void onClick(View v) {
                //TODO 自动存根法
                checkedTextView4.toggle();
            }
        });
        //设置 listView 的模式为 CHOICE_MODE_SINGLE
        listView.setChoiceMode(ListView.CHOICE_MODE_MULTIPLE);
    }
}
```

运行程序,效果如图 2-17(a)所示,当单击界面中的多选框时,即可实现改变选择状态,单击第 3 个右侧的三角符按钮时即可实现改变三角符的上下状态,效果如图 2-17(b)所示。

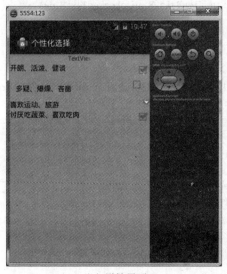
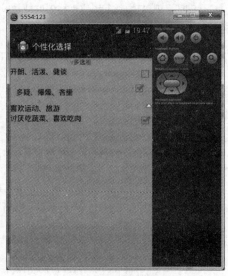

（a）默认界面　　　　　　　　　　　（b）改变选择状态

图 2-17　多项选择

2.4　计时实例

本实例是通过 Chronometer 控件实现一个手机计时器,通过本实例向读者演示 Chronometer 控件的具体用法。

Chronometer 是一个简单的定时器,可以给它一个开始时间,并以此定时,或者如果不给它一个开始时间,它将会使用你的通话开始时间。默认情况下它会显示当前定时器的值的形式为"分:秒"或"MM:SS"。

本实例的具体实现步骤如下。

(1) 在 Eclipse 中创建一个 Android 应用项目,命名为 Chronometer_test。

(2) 打开 res\layout 目录下的 main.xml 布局文件,在文件中声明 4 个 Button 控件和一

个 Chronometer 控件。代码为:

```xml
<?xml version = "1.0" encoding = "utf-8"?>
<LinearLayout xmlns:android = "http://schemas.android.com/apk/res/android"
    android:layout_width = "fill_parent"
    android:layout_height = "fill_parent"
    android:orientation = "vertical"
    android:background = "#aabbcc">
<Chronometer
    android:id = "@+id/myChronometer"
    android:layout_width = "fill_parent"
    android:layout_height = "wrap_content" />
<LinearLayout
    android:layout_width = "fill_parent"
    android:layout_height = "wrap_content"
    android:orientation = "horizontal" >
    <Button
        android:id = "@+id/btn_start"
        android:layout_width = "wrap_content"
        android:layout_height = "wrap_content"
        android:text = "开始" />
    <Button
        android:id = "@+id/btn_stop"
        android:layout_width = "wrap_content"
        android:layout_height = "wrap_content"
        android:text = "停止" />
    <Button
        android:id = "@+id/btn_base"
        android:layout_width = "wrap_content"
        android:layout_height = "wrap_content"
        android:text = "复位" />
    <Button
        android:id = "@+id/btn_format"
        android:layout_width = "wrap_content"
        android:layout_height = "wrap_content"
        android:text = "格式化" />
</LinearLayout>
</LinearLayout>
```

(3) 打开 src\fs.chronometer_test 包下的 MainActivity.java 文件,在文件中实现计时器的计时、停止、复位及格式化。代码为:

```java
package fs.chronometer_test;
import android.app.Activity;
import android.os.Bundle;
import android.os.SystemClock;
import android.os.Vibrator;
import android.view.View;
import android.widget.Button;
import android.widget.Chronometer;
import android.widget.Chronometer.OnChronometerTickListener;
public class MainActivity extends Activity {
    private Vibrator vibrator;
    private Chronometer chronometer;                    //计时组件
    private Button btn_start;
```

```java
        private Button btn_stop;
        private Button btn_base;
        private Button btn_format;
        @Override
        public void onCreate(Bundle savedInstanceState) {
            super.onCreate(savedInstanceState);
            setContentView(R.layout.main);
            vibrator = (Vibrator) getSystemService(VIBRATOR_SERVICE);        //获取振动服务
            chronometer = (Chronometer) findViewById(R.id.myChronometer);
            chronometer
            //给计时组件设置监听对象
            .setOnChronometerTickListener(new OnChronometerTickListenerImpl());
            btn_start = (Button) findViewById(R.id.btn_start);
            btn_stop = (Button) findViewById(R.id.btn_stop);
            btn_base = (Button) findViewById(R.id.btn_base);
            btn_format = (Button) findViewById(R.id.btn_format);
            btn_start.setOnClickListener(new ButtonClickListener());
            btn_stop.setOnClickListener(new ButtonClickListener());
            btn_base.setOnClickListener(new ButtonClickListener());
            btn_format.setOnClickListener(new ButtonClickListener());
        }
    public class OnChronometerTickListenerImpl implements
                                            //计时监听事件,随时随地监听时间的变化
OnChronometerTickListener {
        @Override
        public void onChronometerTick(Chronometer chronometer) {
            String time = chronometer.getText().toString();
            if ("00:05".equals(time)) {                     //判断5秒之后,让手机振动
    //设置振动周期和是否循环振动,如果不想循环振动把0改为-1
                vibrator.vibrate(new long[] { 1000, 10, 100, 10 }, 0);
            }
        }
    }
    public class ButtonClickListener implements View.OnClickListener {
        @Override
        public void onClick(View v) {
            switch (v.getId()) {
            case R.id.btn_start:
                chronometer.start();                //开始计时
                break;
            case R.id.btn_stop:
                chronometer.stop();                 //停止计时
                break;
            case R.id.btn_base:
                chronometer.setBase(SystemClock.elapsedRealtime());        //复位键
                break;
            case R.id.btn_format:
                chronometer.setFormat("显示时间: %s.");              //更改时间显示格式
                break;
            default:
                break;
            }
```

　　　　　　}
　　　　}
}

运行程序,默认效果如图 2-18(a)所示,单击界面中的"开始"按钮,效果如图 2-18(b)所示。当单击界面中的"停止"按钮时,即实现计时停止,当单击界面中的"复位"按钮时即实现重新计时,当单击界面中的"格式化"按钮时即更改时间显示格式。

（a）默认界面　　　　　　　　　　　　（b）开始计时

图 2-18　计时器

Chronometer 控件中有其自身的一些属性及方法,主要内容如下。
- long getBase()：返回当前的时间,由 setBase(long)设置。
- String getFormat()：返回当前字符串格式,此格式是通过 setFormat()实现的。
- void setBase(long base)：设置时间,计数定时器指定的值。
- void setFormat(String format)：设置显示的内容,计时器将会显示这个参数所对应的值。

2.5　条类控件实例

在 Android 的条类控件主要有 ProgressBar(进度条)控件、拖动条(SeekBar)控件及星型等级(RatingBar)控件,下面给予实例介绍。

2.5.1　进度提示实例

本实例中通过对 ProgressBar 的应用,构建一个长方形进度条及圆形进度条程序,通过本程序实现,向读者演示 ProgressBar 的具体应用。

ProgressBar 控件主要用于显示一些操作的进度,应用程序可以修改其长度来表示当前后台操作的完成情况,因为进度条会移动,所以长时间加载某些资源或执行某些耗时的操作时,不会使用户界面失去响应。Android 当中的进度条 ProgressBar 有两种形式,一种为垂直(圆圈),一种为水平(水平线)。

本实例的具体实现步骤如下。

(1) 在 Eclipse 中创建一个 Android 应用项目,命名为 ProgressBar_test。

(2) 打开 res\layout 目录下的 main.xml 布局文件,在文件中声明一个 TextView 控件、一个 Button 控件及两个 ProgressBar 控件。代码为:

```xml
<?xml version = "1.0" encoding = "utf-8"?>
<LinearLayout xmlns:android = "http://schemas.android.com/apk/res/android"
    android:orientation = "vertical"
    android:layout_width = "fill_parent"
    android:layout_height = "fill_parent"
    android:background = "#aabbcc">
    <TextView
        android:layout_width = "fill_parent"
        android:layout_height = "wrap_content"
        android:text = "欢迎来到此界面" />
    <ProgressBar
        android:id = "@+id/rectangleProgressBar"
        style = "?android:attr/progressBarStyleHorizontal"
        android:layout_width = "fill_parent"
        android:layout_height = "wrap_content"
        android:visibility = "gone"/>
    <ProgressBar
        android:id = "@+id/circleProgressBar"
        style = "?android:attr/progressBarStyleLarge"
        android:layout_width = "wrap_content"
        android:layout_height = "wrap_content"
        android:visibility = "gone"/>
    <Button
        android:id = "@+id/button"
        android:text = "显示进度条"
        android:layout_width = "wrap_content"
        android:layout_height = "wrap_content"/>
</LinearLayout>
```

(3) 打开 src\fs.progressbar_test 包下的 MainActivity.java 文件,在文件中当单击界面中的"显示进度条"按钮时,即弹出圆形进度条与长方形进度条。代码为:

```java
package fs.progressbar_test;
import android.app.Activity;
import android.os.Bundle;
import android.os.Handler;
import android.os.Message;
import android.view.View;
import android.widget.Button;
import android.widget.ProgressBar;
public class MainActivity extends Activity {
    private ProgressBar rectangleProgressBar,circleProgressBar;
    private Button mButton;
    protected static final int STOP = 0x10000;
    protected static final int NEXT = 0x10001;
    private int iCount = 0;
    public void onCreate(Bundle savedInstanceState) {
        super.onCreate(savedInstanceState);
```

```java
        setContentView(R.layout.main);
        //查找浏览器的ID
        rectangleProgressBar = (ProgressBar)findViewById(R.id.rectangleProgressBar);
        circleProgressBar = (ProgressBar)findViewById(R.id.circleProgressBar);
        mButton = (Button)findViewById(R.id.button);
        rectangleProgressBar.setIndeterminate(false);
        circleProgressBar.setIndeterminate(false);
        mButton.setOnClickListener(new Button.OnClickListener() {
            public void onClick(View v) {
                rectangleProgressBar.setVisibility(View.VISIBLE);
                circleProgressBar.setVisibility(View.VISIBLE);
                rectangleProgressBar.setMax(100);
                rectangleProgressBar.setProgress(0);
                circleProgressBar.setProgress(0);
                //创建一个线程,每秒步长为5进行增加,到100%时停止
                Thread mThread = new Thread(new Runnable() {
                    public void run() {
                        for(int i = 0 ; i < 20; i++){
                            try{
                                iCount = (i + 1) * 5;
                                Thread.sleep(1000);
                                if(i == 19){
                                    Message msg = new Message();
                                    msg.what = STOP;
                                    mHandler.sendMessage(msg);
                                    break;
                                }else{
                                    Message msg = new Message();
                                    msg.what = NEXT;
                                    mHandler.sendMessage(msg);
                                }
                            }catch (Exception e) {
                                e.printStackTrace();
                            }
                        }
                    }
                });
                mThread.start();
            }
        });
    }
    //定义一个Handler
    private Handler mHandler = new Handler(){
        public void handleMessage(Message msg){
            switch (msg.what) {
            case STOP:
                rectangleProgressBar.setVisibility(View.GONE);
                circleProgressBar.setVisibility(View.GONE);
                Thread.currentThread().interrupt();
                break;
            case NEXT:
                if(!Thread.currentThread().isInterrupted()){
                    rectangleProgressBar.setProgress(iCount);
                    circleProgressBar.setProgress(iCount);
```

```
            }
            break;
        }
    }
};
```

运行程序,效果如图 2-19(a)所示,当单击界面中的"显示进度条"按钮时,效果如图 2-19(b)所示。

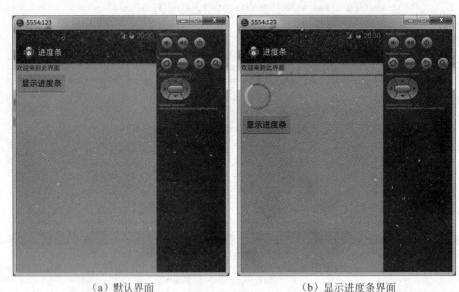

(a)默认界面　　　　　　　　(b)显示进度条界面

图 2-19　进度条

Android 支持几种风格的进度条,通过 style 属性可以为 ProgressBar 指定风格。该属性可支持以下几个属性值。

- @android:style/Widget.ProgressBar.Horizontal:水平进度条。
- @android:style/Widget.ProgressBar.Inverse:普通大小进度条。
- @android:style/Widget.ProgressBar.Large:大进度条。
- @android:style/Widget.ProgressBar.Large.Inverse:普通大进度条。
- @android:style/Widget.ProgressBar.Small:小进度条。
- @android:style/Widget.ProgressBar.Small.Inverse:普通小进度条。

除此之外,ProgressBar 还支持如下所列的属性。

- android:max:设置该进度条的最大值。
- android:progress:设置该进度条的已完成进度值。
- android:progressDrawable:设置该进度条的轨道的绘制形式。
- android:indeterminate:该属性设为 true,设置进度条不精确显示进度。
- android:indeterminateDrawable:设置绘制不显示进度的进度条的 Drawable 对象。
- android:indeterminateDuration:设置不精确显示进度的持续时间

ProgressBar 提供了如下方法来操作进度。

- setProgress(int):设置进度的完成百分比。

- incrementProgressBy(int):设置进度条的进度增加或减少。当参数为正数时进度增加;当参数为负数时进度减少。

2.5.2 音量大小调节实例

本实例中,通过对 SeekBar 的应用,构建一个音量大小调节的程序。通过本实例向读者演示 SeekBar 的具体应用。

SeekBar 是接收用户输入的控件,SeekBar 类似于拖动条,可以直观地显示用户需要的数据,常用于声音调节等场合,SeekBar 不但可以直观地显示数值的大小,而且还可以为其设置标度,类似于显示在屏幕中的一把尺子。

本实例的具体实现步骤如下。

(1) 在 Eclipse 中创建一个 Android 应用项目,命名为 SeekBar_test。

(2) 打开 res\layout 目录下的 main.xml 布局文件,在文件中声明一个 TextView 控件及一个 SeekBar 控件。代码为:

```xml
<?xml version = "1.0" encoding = "utf - 8"?>
<LinearLayout
    xmlns:android = "http://schemas.android.com/apk/res/android"
    android:orientation = "vertical"
    android:layout_width = "fill_parent"
    android:layout_height = "fill_parent"
    android:background = "#aabbcc">
    <TextView
        android:text = "音量大小:0"
        android:textSize = "35dip"
        android:id = "@ + id/TextView01"
        android:layout_width = "wrap_content"
        android:layout_height = "wrap_content">
    </TextView>
    <SeekBar
        android:id = "@ + id/SeekBar01"
        android:layout_width = "fill_parent"
        android:layout_height = "wrap_content">
    </SeekBar>
</LinearLayout>
```

(3) 打开 src\fs.seekbar_test 包下的 MainActivity.java 文件,在文件中实现音量控件,并显示当前值。代码为:

```java
package fs.seekbar_test;
import android.app.Activity;
import android.os.Bundle;
import android.widget.SeekBar;
import android.widget.TextView;
public class MainActivity extends Activity
{
    final static double MAX = 100;                    //SeekBar 的最大值
    SeekBar sb;
    TextView tv;
    @Override
    public void onCreate(Bundle savedInstanceState)
```

```
{
    super.onCreate(savedInstanceState);
    setContentView(R.layout.main);
    sb = (SeekBar)this.findViewById(R.id.SeekBar01);
    tv = (TextView)this.findViewById(R.id.TextView01);
    //普通拖拉条被拉动的处理代码
    sb.setOnSeekBarChangeListener
    (
        new SeekBar.OnSeekBarChangeListener()
        {
            public void onProgressChanged(SeekBar seekBar, int progress,
                    boolean fromUser)
            {
                tv.setText("音量大小:" + (int)sb.getProgress());
            }
            public void onStartTrackingTouch(SeekBar seekBar)
            {
            }
            public void onStopTrackingTouch(SeekBar seekBar) {
                //TODO 自动存根法
            }
        }
    );
}
```

运行程序,效果如图2-20(a)所示,拖动即显示当前的音量,效果如图2-20(b)所示。

(a) 默认界面　　　　　　　　(b) 当前音量

图2-20　控制音量

SeekBar允许用户改变拖动条的滑块外观,改变滑块外观通过如下属性来指定。
android:thumb：指定一个Drawable对象,该对象将作为自定义滑块。
SeekBar中还有几个重要的属性,分别如下。
- android:layout_height="wrap_content"：建议使用wrap_content,否则一定要保证设置的值不小于seekbar图片资源中的最高值。

- android:maxHeight="12px"：说明进度条的最大高度。
- android:minHeight="12px"：说明进度条的最低高度。
- android:paddingLeft="18px"或 android:paddingRight="18px"：解决拖动按钮在最左、最右显示不全的问题，padding 的值一般是 thumb 的一半宽度。
- android:progressDrawable="@drawable/seekbar_style"：设置了此值，就表示使用自定义的进度条样式，在其中可以设置进度条背景图、进度条图、缓冲条图。

2.5.3 等级打分实例

在本实例中，通过对 RatingBar 的应用，构建一个为酒店管理等级进行评分程序，通过本程序向读者演示 RatingBar 控件的具体用法。

RatingBar 是 SeekBar 和 ProgressBar 的一种扩展，用星星表示等级，为评分条控件，默认效果为若干个绿色的星星，如果想将其换成其他自定义图片就要自定义它的 style。

当 RatingBar 使用默认的大小，用户可以单击、拉曳或使用方向键来设置等级。当 RatingBar 使用默认的大小。它有两种样式（小风格用 ratingBarStyleSmall，大风格用 ratingBarStyleIndicator），其中大的只适合指示，不适合于用户交互（用户无法改变）。当使用可以支持用户交互的 RatingBar 时，无论将控件（widgets）放在它的左边还是右边都是不合适的。

本实例的具体实现步骤如下。

(1) 在 Eclipse 中创建一个 Android 应用项目，命名为 RatingBar_test。

(2) 打开 res\layout 目录下的 main.xml 布局文件，在文件中声明一个 TextView 控件、一个 RatingBar 控件及一个 Button 控件。代码为：

```xml
<?xml version="1.0" encoding="UTF-8"?>
<LinearLayout xmlns:android="http://schemas.android.com/apk/res/android"
        android:orientation="vertical"
        android:layout_width="fill_parent"
        android:layout_height="fill_parent"
        android:background="#aabbcc">
<TextView
    android:id="@+id/textView1"
    android:layout_width="wrap_content"
    android:layout_height="wrap_content"
    android:text="酒店星级"
    android:textSize="35dip"
    android:textColor="#000000"/>
<RatingBar
    android:id="@+id/ratingBar1"
    android:layout_width="wrap_content"
    android:layout_height="wrap_content"
    android:numStars="5"
    android:stepSize="0.5" />
<Button
    android:id="@+id/buttn1"
    android:layout_width="wrap_content"
    android:layout_height="wrap_content"
    android:text="评分" />
</LinearLayout>
```

(3) 打开 src\fs.ratingbar_test 包下的 MainActivity.java 文件,在文件中实现当单击界面中的"评分"按钮时,即可实现选择半个星进行评分。代码为:

```java
package fs.ratingbar_test;
import android.support.v4.app.Fragment;
import android.app.Activity;
import android.os.Bundle;
import android.util.Log;
import android.view.LayoutInflater;
import android.view.Menu;
import android.view.MenuItem;
import android.view.View;
import android.view.View.OnClickListener;
import android.view.ViewGroup;
import android.widget.Button;
import android.widget.RatingBar;
import android.widget.RatingBar.OnRatingBarChangeListener;
import android.os.Build;
public class MainActivity extends Activity
{
    //定义控件
    RatingBar ratingBar;
    Button button1;
    public static String TAG = "MainActivity";
    @Override
    protected void onCreate(Bundle savedInstanceState)
    {
        super.onCreate(savedInstanceState);
        setContentView(R.layout.main);
        ratingBar = (RatingBar)findViewById(R.id.ratingBar1);
        ratingBar.setOnRatingBarChangeListener(new RatingBarChangeListener());
        button1 = (Button)findViewById(R.id.buttn1);
        button1.setOnClickListener(new ClickListener());
    }
    class RatingBarChangeListener implements OnRatingBarChangeListener
    {
        @Override
        public void onRatingChanged(RatingBar ratingBar, float rating,
                boolean fromUser)
        {
            Log.i(TAG, "当前评分 = " + rating);
            System.out.println("当前评分 = " + rating);
        }
    }
    class ClickListener implements OnClickListener
    {
        @Override
        public void onClick(View v)
        {                                              //在当前加 0.5 分
            ratingBar.setRating(ratingBar.getRating() + 0.5f);
        }
    }
}
```

```
    @Override
    public boolean onCreateOptionsMenu(Menu menu)
    {
        getMenuInflater().inflate(R.menu.main, menu);
        return true;
    }
}
```

运行程序，效果如图 2-21 所示。

图 2-21　等级评分

在 Android 中，RatingBar 控件有其自身的属性，主要内容如下。
- android:isIndicator：设置该星级评分条是否允许用户改变（true 为不允许修改）。
- android:numStars：设置该星级评分条总共有多少个星级。
- android:rating：设置该星级评分条默认的星级。
- android:stepSize：设置每次最少需要改变多少个星级。

2.6　手机图片查看实例

本实例通过 ImageView 控件，实现手机的简单图片查看器。通过本实例的实现向读者演示 ImageView 控件的具体用法。

ImageView 类可以加载各种来源的图片（如资源或图片库），需要计算图像的尺寸，方便它可以在其他布局中使用，并提供例如缩放和着色（渲染）各种显示选项。

本实例的具体实现步骤如下。

(1) 在 Eclipse 中创建一个 Android 应用项目，命名为 ImageView。

(2) 打开 res\layout 目录下的 main.xml 布局文件，在文件中声明 ImageView 控件及 Button 控件。代码为：

```
<?xml version = "1.0" encoding = "utf-8"?>
<LinearLayout xmlns:android = "http://schemas.android.com/apk/res/android"
```

```xml
        android:layout_width = "fill_parent"
        android:layout_height = "fill_parent"
        android:orientation = "vertical"
        android:background = "#aabbcc">
    <!-- 声明了一个垂直分布的线性布局 -->
    <ImageView
        android:id = "@+id/iv"
        android:layout_width = "wrap_content"
        android:layout_height = "wrap_content"
        android:layout_gravity = "center_horizontal"
        android:src = "@drawable/a01" />
    <!-- 声明了 ImageView 控件 -->
    <Button
        android:id = "@+id/previous"
        android:layout_width = "match_parent"
        android:layout_height = "wrap_content"
        android:text = "上一张" />
    <LinearLayout
        android:layout_width = "fill_parent"
        android:layout_height = "wrap_content"
        android:layout_gravity = "center_horizontal"
        android:orientation = "horizontal" >
        <Button
            android:id = "@+id/alpha_plus"
            android:layout_width = "match_parent"
            android:layout_height = "wrap_content"
            android:text = "透明度增加" />
    </LinearLayout>
    <Button
        android:id = "@+id/alpha_minus"
        android:layout_width = "match_parent"
        android:layout_height = "wrap_content"
        android:text = "透明度增小" />
    <Button
        android:id = "@+id/next"
        android:layout_width = "match_parent"
        android:layout_height = "wrap_content"
        android:text = "下一张" />
</LinearLayout>
```

（3）打开 src\fs.imageview 包下的 MainActivity.java 文件，在文件中实现当单击界面中的"下一张"按钮时，即查看下一张图片，当单击界面中的"透明度增加"时，即增加图像的透明度。代码为：

```java
package fs.imageview;
import android.app.Activity;
import android.os.Bundle;
import android.view.View;
import android.widget.Button;
import android.widget.ImageView;
public class MainActivity extends Activity {
    ImageView iv;                        //ImageView 对象引用
        Button btnNext;                  //Button 对象引用
        Button btnPrevious;              //Button 对象引用
        Button btnAlphaPlus;             //Button 对象引用
```

```java
        Button btnAlphaMinus;                              //Button 对象引用
        int currImgId = 0;                                 //记录当前 ImageView 显示的图片 id
        int alpha = 255;                                   //记录 ImageView 的透明度
        int [] imgId = {                                   //ImageView 显示的图片数组
            R.drawable.a01,
            R.drawable.a02,
            R.drawable.a03,
            R.drawable.a04,
            R.drawable.a05
        };
        @Override
        public void onCreate(Bundle savedInstanceState) {   //重写 onCreate 方法
            super.onCreate(savedInstanceState);
            setContentView(R.layout.main);
            iv = (ImageView)findViewById(R.id.iv);          //获得 ImageView 对象引用
            btnNext = (Button)findViewById(R.id.next);      //获得 Button 对象引用
            btnPrevious = (Button)findViewById(R.id.previous);  //获得 Button 对象引用
            //获得 ImageView 对象引用
            btnAlphaPlus = (Button)findViewById(R.id.alpha_plus);
            btnAlphaMinus = (Button)findViewById(R.id.alpha_minus);     //获得 ImageView 对象引用
            btnNext.setOnClickListener(myListener);         //为 Button 对象设置 OnClickListener 监听器
            btnPrevious.setOnClickListener(myListener);     //为 Button 对象设置 OnClickListener 监听器
            btnAlphaPlus.setOnClickListener(myListener);    //为 Button 对象设置 OnClickListener 监听器
            btnAlphaMinus.setOnClickListener(myListener);   //为 Button 对象设置 OnClickListener 监听器
        }
        private View.OnClickListener myListener = new View.OnClickListener(){
                                                            //自定义的 OnClickListener 监听器
            @Override
            public void onClick(View v) {                   //判断单击下的是哪个 Button
                if(v == btnNext){                           //下一张图片按钮被单击
                    currImgId = (currImgId + 1) % imgId.length;
                    iv.setImageResource(imgId[currImgId]);  //设置 ImageView 的显示图片
                }
                else if(v == btnPrevious){                  //上一张图片按钮被单击
                    currImgId = (currImgId - 1 + imgId.length) % imgId.length;
                    iv.setImageResource(imgId[currImgId]);  //设置 ImageView 的显示图片
                }
                else if(v == btnAlphaPlus){                 //增加透明度按钮被单击
                    alpha -= 25;
                    if(alpha < 0){
                        alpha = 0;
                    }
                    iv.setAlpha(alpha);                     //设置 ImageView 的透明度
                }
                else if(v == btnAlphaMinus){                //减少透明度按钮被单击
                    alpha += 25;
                    if(alpha > 255){
                        alpha = 255;
                    }
                    iv.setAlpha(alpha);                     //设置 ImageView 的透明度
                }
            }
        };
}
```

运行程序，默认效果如图 2-22(a)所示，当单击界面中的"透明度增加"按钮时，效果如图 2-22(b)所示。

　　　　　（a）默认界面　　　　　　　　　　（b）增加图像的透明度

图 2-22　图片查看器

在 Android 中，ImageView 控件有其自身的常用属性，主要内容如下。
- android:adjustViewBounds：用于设置 ImageView 是否调整自己的边界来保持所显示图片的长宽比。
- android:maxHeight：设置 ImageView 的最大高度，需要设置 android:adjustViewBounds 属性值为 true，否则不起使用。
- android:maxWidth：设置 ImageView 的最大宽度，需要设置 android:adjustViewBounds 属性值为 true，否则不起作用。
- android:scaleType：用于设置所显示的图片怎样缩放或移动以适应 ImageView 的大小，其属性值可以是 matrix(使用 matrix 方式进行缩放)、fitXY(对图片横向、纵向独立缩放，使得该图片完全适应于该 ImageView，图片的纵横比可能会改变)、fitStart(保持纵横比缩放图片，直到该图片能完全显示在 ImageView 中，缩放完全显示在 ImageView 的左上角)、fitCenter(保持纵横比缩放图片，直到该图片能完全显示在 ImageView 中，缩放完成后该图片放在 ImageView 的中央)、fitEnd(保持纵横比缩放图片，直到该图片能完全显示在 ImageView 中，缩放完成后该图片放在 ImageView 的右下角)、center(把图片放在 ImageView 的中间，但不进行任何缩放)、centerCrop(保持纵横比缩放图片，以使得图片能完全覆盖 ImageView)或 centerInside(保持纵横比缩放图片，以使得 ImageView 能完全显示该图片)。
- android:src：用于设置 ImageView 所显示的 Drawable 对象的 ID，例如，设置显示保存在 res/drawable 目录下的名称为 a04.jpg 的图片，可以将属性值设置为 android:src="@drawable/a04"。
- android:tint：用于为图片着色，其属性值可以是♯rgb、♯argb、♯rrggbb 或♯aarrggbb 表示的颜色值。

同时，ImageView 类中还有一些成员方法比较常用，主要如下。

- setAlpha(int alpha)：设置 ImageView 的透明度。
- setImageBitmap(Bitmap bm)：设置 ImageView 所显示的内容为指定的 Bitmap 对象。
- setImageDrawable(Drawable drawable)：设置 ImageView 所显示的内容为指定的 Drawable 对象。
- setImageResource(int resId)：设置 ImageView 所显示的内容为指定 id 的资源。
- setImageURI(Uri uri)：设置 ImageView 所显示的内容为指定 Uri。
- setSelected(boolean selected)：设置 ImageView 的选中状态。

2.7 色彩选择实例

本实例通过 Spinner 控件，构建一个实现选择自己的色彩程序。通过本实例向读者演示 Spinner 控件的具体使用。

Spinner 是一个列表选择框，会在用户选择后，展示一个列表供用户进行选择。Spinner 是 ViewGroup 的间接子类，它和其他的 Android 控件一样，数据需要使用 Adapter 进行封装。

本实例的具体实现步骤如下。

(1) 在 Eclipse 中创建一个 Android 应用项目，命名为 Spinner_test。

(2) 打开 res\layout 目录下的 strings.xml 文件，直接通过资源文件配置。代码为：

```xml
<?xml version = "1.0" encoding = "utf-8"?>
<resources>
    <string name = "app_name">下拉列表</string>
    <string name = "action_settings">Settings</string>
    <string name = "hello_world">Hello world!</string>
    <string name = "color">选择色彩</string>
    <string-array name = "colors">
        <item>黑色 | Black</item>
        <item>蓝色 | Blue</item>
        <item>棕色 | Brown</item>
        <item>绿色 | Green</item>
        <item>灰色 | Grey</item>
        <item>粉色 | Pink</item>
        <item>紫色 | Purple</item>
        <item>红色 | Red</item>
        <item>白色 | White</item>
        <item>黄色 | Yellow</item>
    </string-array>
</resources>
```

(3) 打开 res\layout 目录下的 main.xml 布局文件，在文件中声明 TextView 控件及 Spinner 控件。代码为：

```xml
<?xml version = "1.0" encoding = "utf-8"?>
<LinearLayout xmlns:android = "http://schemas.android.com/apk/res/android"
    android:layout_width = "fill_parent"
    android:layout_height = "fill_parent"
    android:orientation = "vertical"
    android:background = " # aabbcc">
    <TextView
        android:layout_width = "fill_parent"
```

```xml
        android:layout_height = "wrap_content"
        android:layout_marginTop = "10dip"
        android:text = "@string/color" />
    <Spinner
        android:id = "@+id/spinner"
        android:layout_width = "fill_parent"
        android:layout_height = "wrap_content"
        android:prompt = "@string/color" />
</LinearLayout>
```

(4) 打开 src\fs.spinner_test1 包下的 MainActivity.java 文件，在文件中利用下拉列表框对色彩进行选择。代码为：

```java
package fs.spinner_test1;
import android.app.Activity;
import android.os.Bundle;
import android.view.View;
import android.widget.AdapterView;
import android.widget.ArrayAdapter;
import android.widget.Spinner;
import android.widget.Toast;
public class MainActivity extends Activity {
    private Spinner spinner;
    @Override
    public void onCreate(Bundle savedInstanceState) {
        super.onCreate(savedInstanceState);
        setContentView(R.layout.main);
        spinner = (Spinner) findViewById(R.id.spinner);
        //准备一个数组适配器
        //R.array.colors : 直接从 strings.xml 去数据
        //android.R.layout.simple_spinner_item : 设置 Spinner 样式(无下拉列表时)
        ArrayAdapter<CharSequence> arrayAdapter = ArrayAdapter
                .createFromResource(this, R.array.colors, android.R.layout.simple_spinner_item);
        //设置下拉列表样式
        arrayAdapter.setDropDownViewResource(android.R.layout.simple_spinner_dropdown_item);
        //为下拉列表设置适配器
        spinner.setAdapter(arrayAdapter);
        //为下拉列表绑定事件监听器
        spinner.setOnItemSelectedListener(new AdapterView.OnItemSelectedListener() {
            @Override
            public void onItemSelected(AdapterView<?> parent, View view,
                    int position, long id) {
                Toast.makeText(
                        MainActivity.this,
                        "选择的色彩: "
                                + parent.getItemAtPosition(position).toString(),
                        Toast.LENGTH_LONG).show();
            }
            @Override
            public void onNothingSelected(AdapterView<?> parent) {
            }
        });
    }
}
```

运行程序，默认效果如图 2-23(a)所示，当单击下拉列表框右侧的三角按钮时，即弹出对应的下拉选项，效果如图 2-23(b)所示。

（a）默认界面　　　　　　　　　　（b）下拉选项

图 2-23　色彩选择

Android 中的 Spinner 也有其自身相应的 getter、setter 方法，主要内容如下。

- android:spinnerMode：列表显示的模式，有两个选择，分别为弹出列表(dialog)和下拉列表(dropdown)，如果不特别设置，默认为下拉列表。
- android:entries：使用资源配置数据源。
- android:prompt：对当前下拉列表设置标题，仅在 dialog 模式下有效。传递一个"@string/name"资源，需要在资源文件中定义。

作为一个列表选择控件，Spinner 具有一些选中选项可以触发的事件，但它本身没有定义这些事件，均继承自间接父类 AdapterView。Spinner 支持的几个常用事件有以下几个。

- AdapterView.OnItemCLickListener：列表项被单击时触发。
- AdapterView.OnItemLongClickListener：列表项被长按时触发。
- AdapterView.OnItemSelectedListener：列表项被选择时触发。
- PS：因为适配器可以设置各种不同的样式，有选择、单选和多选，所以 OnItemCLickListener 和 OnItemSelectedListener 是适用于不同场景的。

对于 Spinner 展示的数据源，一般使用两种方式设定数据。

- 通过 XML 资源文件设置，这种方式比较死板，但是如果仅仅需要展示固定的、简单的数据，这种方式还是可以考虑的，比较直观。
- 使用 Adapter 接口设置，这是最常见的方式，动态、灵活，可以设定各种样式以及数据来源。

2.8　手机模拟时钟实例

Android UI 设计模拟 AnalogClock 和数字 DigitalClock，此控件的功能实现非常简单，主要为了获取系统时间动态更新 TextView 控件，显示系统时间。通过本实例向读者演示

AnalogClock 控件与 DigitalClock 的具体用法。

本实例首先显示的是 AnalogClock 模拟时钟,模拟时钟随系统时间改变而更新,然后在 AnalogClock 下边显示的是 DigitalClock 数字时钟,同样随系统时间改变而更新,并动态将时间显示在 TextView 控件中。

本实例的具体实现步骤如下。

(1) 在 Eclipse 中创建一个 Android 应用项目,命名为 Clock_test。

(2) 打开 res\layout 目录下的 main.xml 布局文件,在文件中声明一个 AnalogClock 控件、一个 DigitalClock 控件及一个 TextView 控件。代码为:

```xml
<?xml version = "1.0" encoding = "utf-8"?>
<LinearLayout xmlns:android = "http://schemas.android.com/apk/res/android"
        android:orientation = "vertical"
        android:layout_width = "fill_parent"
        android:layout_height = "fill_parent"
        android:background = "#000000" >
    <AnalogClock
        android:id = "@+id/AnalogClock01"
        android:layout_width = "wrap_content"
        android:layout_height = "wrap_content"
        android:layout_marginLeft = "90dip"
        android:layout_marginTop = "30dip">
    </AnalogClock>
    <DigitalClock
        android:text = "@+id/DigitalClock01"
        android:textSize = "25dip"
        android:id = "@+id/DigitalClock01"
        android:layout_width = "wrap_content"
        android:layout_height = "wrap_content"
        android:layout_marginTop = "20dip"
        android:layout_marginLeft = "120dip">
    </DigitalClock>
    <TextView
        android:id = "@+id/TextView01"
        android:textSize = "25dip"
        android:layout_width = "fill_parent"
        android:layout_height = "wrap_content"
        android:textColor = "#FFFFFF"
        android:gravity = "center"
        android:layout_marginTop = "20dip">
    </TextView>
</LinearLayout>
```

(3) 打开 src\fs.clock_test 包下的 MainActivity.java 文件,在文件中实现模拟时钟,并把系统的时间动态更新显示在文本框中。代码为:

```java
package fs.clock_test;
import java.util.Calendar;
import android.app.Activity;
import android.os.Bundle;
import android.os.Handler;
import android.os.Message;
import android.widget.TextView;
public class MainActivity extends Activity
```

```java
{
    public int mHour;
    public int mMinute;
    public int mSecond;
    Handler hd = new Handler()
    {
        @Override
        public void handleMessage(Message msg)
        {
            switch(msg.what)
            {
                case 0:
                    long time = System.currentTimeMillis();        //得到系统时间
                    final Calendar c = Calendar.getInstance();     //得到 Calendar 引用
                    c.setTimeInMillis(time);                       //把系统时间设置到 Calendar 中
                    mHour = c.get(Calendar.HOUR);                  //得到系统的小时数
                    mMinute = c.get(Calendar.MINUTE);              //得到当前时间中的分数
                    mSecond = c.get(Calendar.SECOND);

                    TextView tv = (TextView)findViewById(R.id.TextView01);
                    tv.setText("现在时间是: " + String.valueOf(mHour) + ":" + String.
                    valueOf(mMinute) + ":" + String.valueOf(mSecond));
            }
        }
    };
    @Override
    public void onCreate(Bundle savedInstanceState)
    {
        super.onCreate(savedInstanceState);
        setContentView(R.layout.main);
        Thread_test thread = new Thread_test(this);
        thread.start();
    }
}
```

(4) 在 src\fs.clock_test 目录下创建一个 Thread_test.java 文件，用于实现发送消息的线程。代码为：

```java
package fs.clock_test;
public class Thread_test extends Thread
{
    boolean flag = true;
    MainActivity activity;
    public Thread_test(MainActivity activity)
    {
        this.activity = activity;
    }
    @Override
    public void run()
    {
        while(flag)
        {
            try
            {
                Thread_test.sleep(1000);
            }
```

```
            catch(Exception e)
            {
                e.printStackTrace();
            }
            activity.hd.sendEmptyMessage(0);
        }
    }
}
```

运行程序,效果如图 2-24 所示。

图 2-24 模拟时钟

AnalogClock 和 DigitalClock 控件它们都负责显示时钟,所不同的是 AnalogClock 控件显示模拟时钟,且只显示时针和分针,而 DigitalClock 显示数字时钟,可精确到秒。

2.9 记录购书时间实例

本实例通过 DatePicker 控件及 TimePicker 控件实现一个记录购书的具体实现程序,通过本程序向读者演示 DatePicker 控件及 TimePicker 控件的具体用法。

在 Android 中,DatePicker 控件的主要功能是向用户提供包含了年月日的日期数据并允许用户对其进行选择。而 TimePicker 控件主要功能是向用户显示一天中的时间(可以为 24 小时制,也可以为 AM/PM 制),并允许用户进行选择。

本实例的具体实现步骤如下。

(1) 在 Eclipse 中创建一个 Android 应用项目,命名为 Picker_test。

(2) 打开 res\layout 目录下的 main.xml 布局文件,在文件中声明一个 TextView 控件、一个 EditText 控件、一个 DatePicker 控件及一个 TimePicker 控件。代码为:

```
<?xml version = "1.0" encoding = "utf - 8"?>
< LinearLayout xmlns:android = "http://schemas.android.com/apk/res/android"
    android:layout_width = "fill_parent"
    android:layout_height = "fill_parent"
```

```xml
        android:orientation = "vertical"
        android:background = "@drawable/bj2">
    <TextView
        android:id = "@ + id/textView1"
        android:layout_width = "wrap_content"
        android:layout_height = "wrap_content"
        android:textSize = "26dip"
        android:text = "请选择购买本书的具体时间" />
    <DatePicker
        android:id = "@ + id/datePicker"
        android:layout_width = "wrap_content"
        android:layout_height = "wrap_content"
        android:layout_gravity = "center_horizontal" />
    <TimePicker
        android:id = "@ + id/timePicker"
        android:layout_width = "wrap_content"
        android:layout_height = "wrap_content"
        android:layout_gravity = "center_horizontal" />
    <EditText
        android:id = "@ + id/show"
        android:layout_width = "wrap_content"
        android:layout_height = "wrap_content"
        android:cursorVisible = "false"
        android:editable = "false" />
</LinearLayout>
```

(3) 打开 src/fs.picker_test 包下的 MainActivity.java 文件，在文件中实现时间的设置与选择，设置购书的具体实现。代码为：

```java
package fs.picker_test;
import java.util.Calendar;
import android.app.Activity;
import android.os.Bundle;
import android.widget.DatePicker;
import android.widget.DatePicker.OnDateChangedListener;
import android.widget.EditText;
import android.widget.TimePicker;
import android.widget.TimePicker.OnTimeChangedListener;
public class MainActivity extends Activity {
    //记录当前的时间
    private int year;
    private int month;
    private int day;
    private int hour;
    private int minute;
    @Override
    protected void onCreate(Bundle savedInstanceState) {
        //TODO 自动存根法
        super.onCreate(savedInstanceState);
        setContentView(R.layout.main);
        DatePicker date = (DatePicker) findViewById(R.id.datePicker);
        TimePicker time = (TimePicker) findViewById(R.id.timePicker);
        //获取当前的年月日、小时、分钟
        Calendar ca = Calendar.getInstance();
        year = ca.get(Calendar.YEAR);
        month = ca.get(Calendar.MONTH);
```

```
            day = ca.get(Calendar.DAY_OF_MONTH);
            hour = ca.get(Calendar.HOUR);
            minute = ca.get(Calendar.MINUTE);
            //初始化 DatePicker
            date.init(year, month, day, new OnDateChangedListener() {
                @Override
        public void onDateChanged(DatePicker arg0, int year, int month, int day) {
                    MainActivity.this.year = year;
                    MainActivity.this.month = month;
                    MainActivity.this.day = day;
                    //显示当前日期和时间
                    showDate(year, month, day, hour, minute);
                }
            });
            //为TimerPicker指定事件监听器
            time.setOnTimeChangedListener(new OnTimeChangedListener() {
                @Override
            public void onTimeChanged(TimePicker arg0, int hour, int minute) {
                    MainActivity.this.hour = hour;
                    MainActivity.this.minute = minute;
                }
            });
        }
        protected void showDate(int year2, int month2, int day2, int hour2, int minute2) {
            EditText text = (EditText) findViewById(R.id.show);
            text.setText("您的购买时间为：" + year2 + "年" + month2 + "月" + day2 + "日" +
                hour2 + "时" + minute2 + "分");
        }
    }
```

运行程序，默认效果如图 2-25 所示。

图 2-25　购书时间

通过以上实例可知，在 Android 中，如果要捕获用户修改日期选择控件中数据的事件，需要为 DatePicker 添加 onDateChangedListener 监听器。DatePicker 控件的主要成员方法主要内容如下：

- getDayOfMonth()：获取日期天数。
- getMonth()：获取日期月份。
- getYear()：获取日期年份。
- init(int year,int monthOfYear,int dayOfMonth,DatePicker.OnDateChangedListener onDateChangedListener)：初始化 DatePicker 控件的属性，参数 onDateChangedListener 为监听器对象，负责监听日期数据的变化。
- setEnabled(boolean enabled)：根据传入的参数设置日期选择控件是否可用。
- updateDate(int year,int monthOfYear,int dayOfMonth)：根据传入的参数更新日期选择控件的各个属性值。

如果要捕获用户修改时间数据的事件，即需要为 TimePicker 添加 OnTimeChangedListener 监听器。TimePicker 类的主要成员方法内容如下。

- getCurrentHour()：获取时间选择控件的当前小时，返回 Integer 对象。
- getCurrentMinute()：获取时间选择控件的当前分钟，返回 Integer 对象。
- is24HourView()：判断时间选择控件是否为 24 小时制。
- setCurrentHour(Integer currentHour)：设置时间选择控件的当前小时，传入 Integer 对象。
- setCurrentMinute(Integer currentMinute)：设置时间选择控件的当前分钟，传入 Integer 对象。
- setEnabled(boolean enabled)：根据传入的参数设置时间选择控件是否可用。
- setIs24HourView(boolean is24HourView)：根据传入的参数设置时间选择控件是否可用。
- setOnTimeChangedListener(TimePicker.OnTimeChangedListener onTimeChangedListener)：为时间选择控件添加 OnTimeChangedListener 监听器。

第 3 章

Android 深入开发实例

在第 2 章已经对 Android 的界面开发所需要的布局和控件通过实例进行介绍,本章主要在其的基础上对 Android 深入一层进行开发。

3.1 Android 视图实例

本节主要对 Android 的列表视图、网格视图、画廊视图、滚动视图和多页视图等进行介绍。

3.1.1 左右浏览影片实例

本实例通过 HorizontalScrollView 控件实现一个左右浏览影片封面程序,通过本实例向读者演示 HorizontalScrollView 控件的具体用法。

HorizontalScrollView 控件是一种水平滚动视图,是用于布局的容器,可以放置让用户滚动条查看的视图层次结构,允许视图结构比手机的屏幕大。

本实例的具体实现步骤如下:

(1) 在 Eclipse 中创建一个 Android 应用项目,命名为 HorizontalScrollView_test。

(2) 打开 res\layout 目录下的 main.xml 布局文件,在文件中声明一个 TextView 控件及一个 HorizontalScrollView 控件。代码为:

```
<RelativeLayout xmlns:android = "http://schemas.android.com/apk/res/android"
    xmlns:tools = "http://schemas.android.com/tools"
    android:layout_width = "match_parent"
    android:layout_height = "match_parent"
    tools:context = ".MainActivity"
    android:background = "#aabbcc">
    <TextView
        android:layout_width = "wrap_content"
        android:layout_height = "wrap_content"
        android:layout_centerHorizontal = "true"
        android:layout_centerVertical = "true"
        android:text = "浏览电影封面" />
    <HorizontalScrollView
        android:layout_width = "match_parent"
        android:layout_height = "wrap_content" >
```

```xml
<LinearLayout
    android:orientation = "horizontal"
    android:id = "@ + id/myhorizon"
    android:layout_width = "wrap_content"
    android:layout_height = "wrap_content" />
    </HorizontalScrollView>
</RelativeLayout>
```

（3）打开 src\fs.horizontalscrollview_test 包下的 MainActivity.java 文件，在文件中实现水平浏览电影的封面。代码为：

```java
package fs.horizontalscrollview_test;
import android.os.Bundle;
import android.app.Activity;
import android.view.Gravity;
import android.view.Menu;
import android.view.View;
import android.view.ViewGroup.LayoutParams;
import android.widget.ImageView;
import android.widget.LinearLayout;
public class MainActivity extends Activity {
    private LinearLayout myhorizonLayout;
    @Override
    protected void onCreate(Bundle savedInstanceState) {
        super.onCreate(savedInstanceState);
        setContentView(R.layout.main);
        myhorizonLayout = (LinearLayout) findViewById(R.id.myhorizon);
        //图片资源放置在 drawable 文件夹中
        int[] imageIDs = {
                R.drawable.gd1, R.drawable.gd2, R.drawable.gd3,
                R.drawable.gd
        };
        for(Integer id:imageIDs){
            myhorizonLayout.addView(insertImage(id));
        }
    }
    private View insertImage(Integer id) {
        LinearLayout layout = new LinearLayout(getApplicationContext());
        layout.setLayoutParams(new LayoutParams(320,320));
        layout.setGravity(Gravity.CENTER);
        ImageView imageView = new ImageView(getApplicationContext());
        imageView.setLayoutParams(new LayoutParams(300,300));
        imageView.setBackgroundResource(id);
        layout.addView(imageView);
        return layout;
    }
    @Override
    public boolean onCreateOptionsMenu(Menu menu) {
        getMenuInflater().inflate(R.menu.main, menu);
        return true;
    }
}
```

运行程序，效果如图 3-1(a)所示，左右拖动封面图片效果如图 3-1(b)所示。

在 Android 中，HorizontalScrollView 控件是一种布局方式，其子项被滚动查看时是整体

 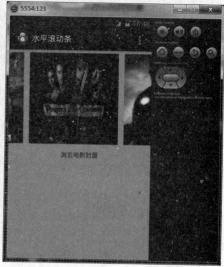

（a）封面图片1　　　　　　　　（b）封面图片2

图 3-1　封面图片的左右浏览

移动的，并且子项本身可以是一个复杂层次结构的布局管理器。一个常见的应用是子项在水平方向中，用户可以滚动显示顶层的水平排列的子项（items）。而且 HorizontalScrollView 只支持水平方向的滚动显示。

3.1.2　单击显示控件实例

本实例通过 ScrollView 控件，构建一个单击界面中的按钮即弹出另一个文本框控件及按钮控件。通过本实例向读者演示 ScrollView 控件的具体用法。

ScrollView 就是一个用于为普通组件添加滚动条的组件。ScrollView 中最多只能包含一个组件，而 ScrollView 控件的作用就是为该组件添加垂直滚动条。

本实例的具体实现步骤如下。

（1）在 Eclipse 中创建一个 Android 应用项目，命名为 ScrollView_test。

（2）打开 res\layout 目录下的 main.xml 布局文件，在文件中声明一个 ScrollView 控件、一个 LinearLayout 布局、一个 TextView 控件及一个 Button 控件。代码为：

```xml
<?xml version = "1.0" encoding = "utf - 8"?>
<ScrollView xmlns:android = "http://schemas.android.com/apk/res/android"
    android:id = "@ + id/ScrollView"
    android:layout_width = "fill_parent"
    android:layout_height = "wrap_content"
    android:scrollbars = "vertical"
    android:background = " # aabbcc">
    <LinearLayout
        android:id = "@ + id/LinearLayout"
        android:orientation = "vertical"
        android:layout_width = "fill_parent"
        android:layout_height = "wrap_content">
        <TextView
            android:id = "@ + id/TestView"
            android:layout_width = "fill_parent"
```

```xml
            android:layout_height = "wrap_content"
            android:text = "文本框控件 0" />
        <Button
            android:id = "@ + id/Button"
            android:text = "按钮控件 0"
            android:layout_width = "fill_parent"
            android:layout_height = "wrap_content"></Button>
    </LinearLayout>
</ScrollView>
```

(3) 打开 src\fs.scrollview_test 包下的 MainActivity.java 文件,在文件中实现单击界面中的"按钮控件 0"即列出另一个按钮控件及文本框控件。代码为:

```java
package fs.scrollview_test;
import android.app.Activity;
import android.os.Bundle;
import android.os.Handler;
import android.view.KeyEvent;
import android.view.View;
import android.widget.Button;
import android.widget.LinearLayout;
import android.widget.ScrollView;
import android.widget.TextView;
public class MainActivity extends Activity {
    /** 第一次调用 Activity 活动 */
    private LinearLayout mLayout;
    private ScrollView sView;
    private final Handler mHandler = new Handler();
    @Override
    public void onCreate(Bundle savedInstanceState) {
        super.onCreate(savedInstanceState);
        setContentView(R.layout.main);
        //创建一个线性布局
        mLayout = (LinearLayout) this.findViewById(R.id.LinearLayout);
        //创建一个 ScrollView 对象
        sView = (ScrollView) this.findViewById(R.id.ScrollView);
        Button mBtn = (Button) this.findViewById(R.id.Button);
        mBtn.setOnClickListener(mClickListener);              //添加单击事件监听
    }
    public boolean onKeyDown(int keyCode, KeyEvent event){
        Button b = (Button) this.getCurrentFocus();
        int count = mLayout.getChildCount();
        Button bm = (Button) mLayout.getChildAt(count - 1);
        if(keyCode == KeyEvent.KEYCODE_DPAD_UP && b.getId() == R.id.Button){
            bm.requestFocus();
            return true;
        }else if(keyCode == KeyEvent.KEYCODE_DPAD_DOWN && b.getId() == bm.getId()){
            this.findViewById(R.id.Button).requestFocus();
            return true;
        }
        return false;
    }
    //Button 事件监听,当单击第一个按钮时增加一个 Button 和一个 TextvIew
    private Button.OnClickListener mClickListener = new Button.OnClickListener() {
        private int index = 1;
        @Override
```

```java
        public void onClick(View v) {
            TextView tView = new TextView(MainActivity.this);          //定义一个 TextView
            tView.setText("文件框控件" + index);       //设置 TextView 的文本信息
            //设置线性布局的属性
            LinearLayout.LayoutParams params = new LinearLayout.LayoutParams(
                    LinearLayout.LayoutParams.FILL_PARENT,
                    LinearLayout.LayoutParams.WRAP_CONTENT);
            mLayout.addView(tView, params);             //添加一个 TextView 控件
            Button button = new Button(MainActivity.this);               //定义一个 Button
            button.setText("控件控件" + index);        //设置 Button 的文本信息
            button.setId(index++);
            mLayout.addView(button, params);            //添加一个 Button 控件
            mHandler.post(mScrollToButton);             //传递一个消息进行滚动
        }
    };
    private Runnable mScrollToButton = new Runnable() {
        @Override
        public void run() {
            int off = mLayout.getMeasuredHeight() - sView.getHeight();
            if (off > 0) {
                sView.scrollTo(0, off);                 //改变滚动条的位置
            }
        }
    };
}
```

运行程序，默认效果如图 3-2(a)所示，当多次单击界面中的"按钮控件 0"时，效果如图 3-2(b)所示。

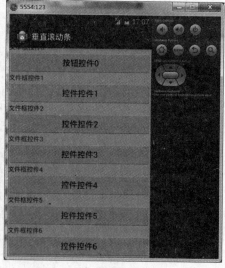

(a) 默认界面　　　　　　　　(b) 显示多个控件

图 3-2　垂直滚动条

ScrollView 是 ViewGroup 的派生类，ViewGroup 是 View 的派生类。ScrollView 控件具有如下自身特点。

- 屏幕大小总是有限制的，对移动设备来说更是如此。当有很多内容需要显示的时候，一屏显示不完时，就需要使用滚动的方式了。

- ScrollView 只能包含一个直接子 view,这是因为 ScrollView 是 FrameLayout 的派生类,通常情况下,这个直接子 view 是一个 LinearLayout,在直接子 view(例如 LinearLayout)中,可以再包含其他对象。
- ScrollView 只支持垂直滚动。
- HorizontalScrollView 除只支持水平滚动外,其他都和 ScrollView 一样。
- 如果要屏幕支持垂直滚动和水平滚动,那么就要让 HorizontalScrollView 作为 ScrollView 的直接子 view,或者让 ScrollView 作为 HorizontalScrollView 的直接子 view。

3.1.3 选择条目实例

本实例通过 ListView 控件,构建一个选择条目的程序。通过本程序向读者演示 ListView 控件的用法。

ListView 控件是手机系统中使用非常广泛的一组组件,它以垂直列表的形式显示所有列表项。创建 ListView 有两种方式。

- 直接使用 ListView 进行创建。
- 让 Activity 继承 ListActivity。

本实例的具体实现步骤如下。

(1) 在 Eclipse 中创建一个 Android 应用项目,命名为 ListView_test。

(2) 打开 res\layout 目录下的 main.xml 布局文件,在文件中定义一个 ListView 控件。代码为:

```xml
<?xml version = "1.0" encoding = "utf-8"?>
<LinearLayout xmlns:android = "http://schemas.android.com/apk/res/android"
    xmlns:tools = "http://schemas.android.com/tools"
    android:layout_width = "fill_parent"
    android:layout_height = "fill_parent"
    android:orientation = "vertical"
    tools:context = ".ListTestActivity"
    android:background = "#aabbcc">
    <ListView
        android:id = "@+id/listView"
        android:layout_width = "match_parent"
        android:layout_height = "wrap_content" >
    </ListView>
</LinearLayout>
```

(3) 在 res\layout 目录下创建一个 item.xml 文件,用于实现条目及图片的显示,在文件中声明两个 TextView 控件和一个 ImageView 控件。代码为:

```xml
<?xml version = "1.0" encoding = "utf-8"?>
<LinearLayout xmlns:android = "http://schemas.android.com/apk/res/android"
    android:layout_width = "fill_parent"
    android:layout_height = "fill_parent"
    android:orientation = "horizontal"
    android:background = "#ccddee">
    <ImageView
        android:id = "@+id/img"
        android:layout_width = "71dp"
```

```xml
            android:layout_height = "71dp"
            android:src = "@drawable/ic_launcher" />
    <LinearLayout
        android:layout_width = "wrap_content"
        android:layout_height = "wrap_content"
        android:orientation = "vertical" >
        <TextView
            android:id = "@+id/title"
            android:layout_width = "242dp"
            android:layout_height = "wrap_content"
            android:text = "222"
            android:textAppearance = "?android:attr/textAppearanceMedium" />
        <TextView
            android:id = "@+id/info"
            android:layout_width = "match_parent"
            android:layout_height = "47dp"
            android:text = "11" />
    </LinearLayout>
</LinearLayout>
```

(4) 打开 src\fs.listview_test 包下的 MainActivity.java 文件，在文件中利用 ListView 实现显示图片及条目，并当选中某个条目时，即显示相应的 ID 信息。代码为：

```java
package fs.listview_test;
import java.util.ArrayList;
import java.util.List;
import android.annotation.SuppressLint;
import android.app.Activity;
import android.content.Context;
import android.os.Bundle;
import android.view.ContextMenu;
import android.view.LayoutInflater;
import android.view.MenuItem;
import android.view.View;
import android.view.ViewGroup;
import android.view.ContextMenu.ContextMenuInfo;
import android.view.View.OnCreateContextMenuListener;
import android.widget.AdapterView;
import android.widget.BaseAdapter;
import android.widget.ImageView;
import android.widget.ListView;
import android.widget.TextView;
import android.widget.Toast;
import android.widget.AdapterView.OnItemClickListener;
@SuppressLint("ParserError")
public class MainActivity extends Activity {
    ListView listView;                              //声明一个 ListView 对象
    private List<info> mlistInfo = new ArrayList<info>();
                                                    //声明一个 List,动态存储要显示的信息
    @Override
    public void onCreate(Bundle savedInstanceState) {
        super.onCreate(savedInstanceState);
        setContentView(R.layout.main);
        listView = (ListView)this.findViewById(R.id.listView);    //将 ListView 与布局对象关联
        setInfo();                                  //给信息赋值函数,用来测试
        listView.setAdapter(new ListViewAdapter(mlistInfo));
```

```java
        //处理Item的单击事件
        listView.setOnItemClickListener(new OnItemClickListener(){
    public void onItemClick(AdapterView<?> parent, View view,int position, long id) {
            info getObject = mlistInfo.get(position);        //通过position获取所单击的对象
                int infoId = getObject.getId();              //获取信息id
                String infoTitle = getObject.getTitle();              //获取信息标题
                String infoDetails = getObject.getDetails();          //获取信息详情
                //Toast显示测试
Toast.makeText(MainActivity.this, "信息ID:" + infoId,Toast.LENGTH_SHORT).show();
            }
        });
        //长按菜单显示
        listView.setOnCreateContextMenuListener(new OnCreateContextMenuListener() {
            public void onCreateContextMenu(ContextMenu conMenu, View view , ContextMenuInfo info) {
                conMenu.setHeaderTitle("菜单");
                conMenu.add(0, 0, 0, "条目一");
                conMenu.add(0, 1, 1, "条目二");
                conMenu.add(0, 2, 2, "条目三");
            }
        });
    }
    //长按菜单处理函数
    public boolean onContextItemSelected(MenuItem aItem) {
        AdapterView.AdapterContextMenuInfo info = (AdapterView.AdapterContextMenuInfo)aItem.getMenuInfo();
        switch (aItem.getItemId()) {
            case 0:
Toast.makeText(MainActivity.this, "你单击了条目一",Toast.LENGTH_SHORT).show();
                return true;
            case 1:
Toast.makeText(MainActivity.this, "你单击了条目二",Toast.LENGTH_SHORT).show();
                return true;
            case 2:
Toast.makeText(MainActivity.this, "你单击了条目三",Toast.LENGTH_SHORT).show();
                return true;
        }
        return false;
    }
    public class ListViewAdapter extends BaseAdapter {
        View[] itemViews;
        public ListViewAdapter(List<info> mlistInfo) {
            //TODO 自动存根法
            itemViews = new View[mlistInfo.size()];
            for(int i = 0;i<mlistInfo.size();i++){
                info getInfo = (info)mlistInfo.get(i);            //获取第i个对象
                //调用makeItemView,实例化一个Item
                itemViews[i] = makeItemView(
                        getInfo.getTitle(), getInfo.getDetails(),getInfo.getAvatar()
                        );
            }
        }
        public int getCount() {
            return itemViews.length;
        }
```

```java
        public View getItem(int position) {
            return itemViews[position];
        }
        public long getItemId(int position) {
            return position;
        }
        //绘制 Item 的函数
        private View makeItemView(String strTitle, String strText, int resId) {
            LayoutInflater inflater = (LayoutInflater) MainActivity.this
                    .getSystemService(Context.LAYOUT_INFLATER_SERVICE);
            //使用 View 的对象 itemView 与 R.layout.item 关联
            View itemView = inflater.inflate(R.layout.item, null);
            //通过 findViewById()方法实例 R.layout.item 内各组件
            TextView title = (TextView) itemView.findViewById(R.id.title);
            title.setText(strTitle);                //填入相应的值
            TextView text = (TextView) itemView.findViewById(R.id.info);
            text.setText(strText);
            ImageView image = (ImageView) itemView.findViewById(R.id.img);
            image.setImageResource(resId);
            return itemView;
        }
        public View getView(int position, View convertView, ViewGroup parent) {
            if (convertView == null)
                return itemViews[position];
            return convertView;
        }
    }
    public void setInfo(){
        mlistInfo.clear();
        int i = 0;
        while(i<10){
            info information = new info();
            information.setId(1000 + i);
            information.setTitle("标题" + i);
            information.setDetails("详细信息" + i);
            information.setAvatar(R.drawable.food);
            mlistInfo.add(information);         //将新的 info 对象加入到信息列表中
            i++;
        }
    }
}
```

(5) 在 src\fs.listview_test 包下新建一个 info.java 类,主要用于实现条目的 id 信息。代码为:

```java
package fs.listview_test;
public class info {
    private int id;                         //信息 id
    private String title;                   //信息标题
    private String details;                 //详细信息
    private int avatar;                     //图片 id
    //信息 id 处理函数
    public void setId(int id) {
        this.id = id;
    }
    public int getId() {
```

```
            return id;
    }
    //标题
    public void setTitle(String title) {
        this.title = title;
    }
    public String getTitle() {
        return title;
    }
    //详细信息
    public void setDetails(String info) {
        this.details = info;
    }
    public String getDetails() {
        return details;
    }
    //图片
    public void setAvatar(int avatar) {
        this.avatar = avatar;
    }
    public int getAvatar() {
        return avatar;
    }
}
```

运行程序,效果如图 3-3 所示。

图 3-3　条目选择

3.1.4　显示文本列表实例

本实例利用 ListView 控件,构建一个不弹出对话框的列表选项的选择程序。通过本实例向读者演示怎样利用 ListView 显示文本列表功能。

本实例的具体实现步骤如下。

(1) 在 Eclipse 中创建一个 Android 应用项目,命名为 ListView_Text。

(2) 打开 res\layout 目录下的 main.xml 布局文件,在文件中声明一个 ListView 控件。代码为:

```xml
<?xml version = "1.0" encoding = "utf-8"?>
<LinearLayout xmlns:android = "http://schemas.android.com/apk/res/android"
    android:orientation = "vertical"
    android:layout_width = "fill_parent"
    android:layout_height = "fill_parent"
    android:background = "#aabbcc">
    <ListView
        android:layout_height = "wrap_content"
        android:id = "@+id/listView1"
        android:layout_width = "match_parent"></ListView>
</LinearLayout>
```

(3) 打开 src\.fs.listview_text 包下的 MainActivity.java 文件,在该文件中实现文本框显示。代码为:

```java
package fs.listview_text;
import java.util.ArrayList;
import android.app.Activity;
import android.os.Bundle;
import android.util.Log;
import android.view.View;
import android.widget.AdapterView;
import android.widget.ArrayAdapter;
import android.widget.ListView;
import android.widget.AdapterView.OnItemClickListener;
public class MainActivity extends Activity {
    private ListView listView1;
    private ArrayList cityList = new ArrayList();
    @Override
    public void onCreate(Bundle savedInstanceState) {
        super.onCreate(savedInstanceState);
        setContentView(R.layout.main);
        listView1 = (ListView) this.findViewById(R.id.listView1);
        cityList.add("深圳");
        cityList.add("香港");
        cityList.add("珠海");
        cityList.add("澳门");
        ArrayAdapter arrayAdapterRef = new ArrayAdapter(this,
                android.R.layout.simple_list_item_1, cityList);
        listView1.setAdapter(arrayAdapterRef);
        listView1.setOnItemClickListener(new OnItemClickListener() {
            public void onItemClick(AdapterView<?> arg0, View arg1, int arg2,
                    long arg3) {
                Log.v("listView1.setOnItemClickListener", ""
                        + cityList.get(arg2));
            }
        });
    }
}
```

运行程序,效果如图 3-4 所示。单击列表中的条目,在 LogCat 中打印出当前选中条目的文本内容为:

```
10 - 25 18:49:48.900: V/listView1.setOnItemClickListener(1492): 澳门
10 - 25 18:49:49.370: V/listView1.setOnItemClickListener(1492): 澳门
10 - 25 18:47:23.880: V/listView1.setOnItemClickListener(1492): 珠海
10 - 25 18:50:43.310: V/listView1.setOnItemClickListener(1492): 深圳
```

图 3-4 ListView 显示文本框

3.1.5 实现多选条目实例

本实例通过 ListView 控件,构建一个利用 CheckBox 控件选择同时选择多个条目程序。通过本实例向读者演示在 ListView 中使用 CheckBox 的功能。

在 ListView 控件中除了可以显示文本列表功能外,还可以在 ListView 控件中使用多选 CheckedBox 控件。

本实例的具体实现步骤如下。

(1) 在 Eclipse 中创建一个 Android 应用项目,命名为 ListView_CheckBox。

(2) 打开 res\layout 目录下的 main.xml 布局文件,在文件中声明一个 ListView 控件及 3 个 Button 控件。代码为:

```
<?xml version = "1.0" encoding = "utf - 8"?>
<LinearLayout xmlns:android = "http://schemas.android.com/apk/res/android"
    android:orientation = "vertical"
    android:layout_width = "fill_parent"
    android:layout_height = "fill_parent"
    android:background = "#ccddee">
    <ListView
        android:layout_weight = "1"
        android:id = "@ + id/listView1"
        android:layout_height = "wrap_content"
        android:layout_width = "match_parent"/>
    <LinearLayout
        android:gravity = "center"
        android:orientation = "horizontal"
        android:layout_width = "fill_parent"
```

```
            android:layout_height = "wrap_content">
        <Button
            android:text = "全选"
            android:id = "@ + id/button1"
            android:layout_width = "wrap_content"
            android:layout_height = "wrap_content"/>
        <Button
            android:text = "反选"
            android:id = "@ + id/button2"
            android:layout_width = "wrap_content"
            android:layout_height = "wrap_content"/>
        <Button
            android:text = "取值"
            android:id = "@ + id/button3"
            android:layout_width = "wrap_content"
            android:layout_height = "wrap_content"/>
    </LinearLayout>
</LinearLayout>
```

(3) 打开 src\fs.listview_checkbox 包下的 MainActivity.java 文件，在文件中实现在 ListView 控件中使用多选 checkedbox 控件，并当单击界面中的"全选"按钮时即选择所有项，单击"反选"按钮时即实现取消或选择所有项，单击"取值"按钮时即在 LogCat 中输出当前条目信息。代码为：

```java
package fs.listview_checkbox;
import java.util.ArrayList;
import java.util.List;
import android.app.Activity;
import android.os.Bundle;
import android.util.Log;
import android.util.SparseBooleanArray;
import android.view.View;
import android.view.View.OnClickListener;
import android.widget.AdapterView;
import android.widget.ArrayAdapter;
import android.widget.Button;
import android.widget.ListView;
import android.widget.TextView;
import android.widget.AdapterView.OnItemClickListener;
public class MainActivity extends Activity {
    private List cityList = new ArrayList();
    private ListView listView;
    private Button button1;                    //全选
    private Button button2;                    //反选
    private Button button3;                    //取值
    @Override
    public void onCreate(Bundle savedInstanceState) {
        super.onCreate(savedInstanceState);
        setContentView(R.layout.main);
        listView = (ListView) this.findViewById(R.id.listView1);
        button1 = (Button) this.findViewById(R.id.button1);
        button2 = (Button) this.findViewById(R.id.button2);
        button3 = (Button) this.findViewById(R.id.button3);
        final boolean[] isCheckedArray = new boolean[8];
        isCheckedArray[0] = false;
```

```java
        isCheckedArray[1] = true;                    //默认值为true
        isCheckedArray[2] = false;
        isCheckedArray[3] = true;                    //默认值为true
        isCheckedArray[4] = false;
        isCheckedArray[5] = true;                    //默认值为true
        isCheckedArray[6] = false;
        isCheckedArray[7] = true;                    //默认值为true
        cityList.add("accp1");
        cityList.add("accp2");
        cityList.add("accp3");
        cityList.add("accp4");
        cityList.add("accp5");
        cityList.add("accp6");
        cityList.add("accp7");
        cityList.add("accp8");
        ArrayAdapter adapter = new ArrayAdapter(this,
                android.R.layout.simple_list_item_multiple_choice, cityList);
        listView.setChoiceMode(ListView.CHOICE_MODE_MULTIPLE);
        listView.setAdapter(adapter);
        listView.setOnItemClickListener(new OnItemClickListener() {
            public void onItemClick(AdapterView<?> arg0, View arg1, int arg2,
                    long arg3) {
                Log.v("------------", "" + ((TextView) arg1).getText());
            }
        });
        //赋初始值
        for (int i = 0; i < isCheckedArray.length; i++) {
            listView.setItemChecked(i, isCheckedArray[i]);
        }
        //全选
        button1.setOnClickListener(new OnClickListener() {
            public void onClick(View arg0) {
                Log.v("单击了全选", "单击了全选");
                for (int i = 0; i < isCheckedArray.length; i++) {
                    listView.setItemChecked(i, true);
                }
            }
        });
        //反选
        button2.setOnClickListener(new OnClickListener() {
            public void onClick(View arg0) {
                Log.v("单击了反选", "单击了反选");
                SparseBooleanArray sparseBooleanArrayRef = listView
                        .getCheckedItemPositions();
                for (int i = 0; i < sparseBooleanArrayRef.size(); i++) {
                    if (sparseBooleanArrayRef.get(i) == true) {
                        listView.setItemChecked(i, false);
                    } else {
                        listView.setItemChecked(i, true);
                    }
                }
            }
        });
        //取值
        button3.setOnClickListener(new OnClickListener() {
```

```
            public void onClick(View arg0) {
                Log.v("单击了取值", "单击了取值");
                SparseBooleanArray sparseBooleanArrayRef = listView
                        .getCheckedItemPositions();
                for (int i = 0; i < sparseBooleanArrayRef.size(); i++) {
                    if (sparseBooleanArrayRef.get(i) == true) {
                        Log.v("值为：", "" + listView.getAdapter().getItemId(i)
                                + " " + listView.getAdapter().getItem(i));
                    }
                }
            }
        });
    }
}
```

运行程序，默认效果如图3-5(a)所示，单击"反选"按钮时，效果如图3-5(b)所示，单击"全选"按钮时，效果如图3-5(c)所示，单击"取值"按钮时，即LogCat输出信息如下：

```
10-25 20:09:17.620: V/单击了取值(1705): 单击了取值
10-25 20:09:17.620: V/值为: (1705): 0 accp1
10-25 20:09:17.630: V/值为: (1705): 1 accp2
10-25 20:09:17.630: V/值为: (1705): 2 accp3
10-25 20:09:17.630: V/值为: (1705): 3 accp4
10-25 20:09:17.630: V/值为: (1705): 4 accp5
10-25 20:09:17.630: V/值为: (1705): 5 accp6
10-25 20:09:17.630: V/值为: (1705): 6 accp7
10-25 20:09:17.630: V/值为: (1705): 7 accp8
```

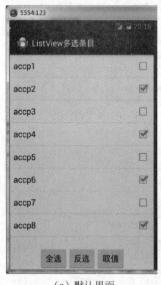
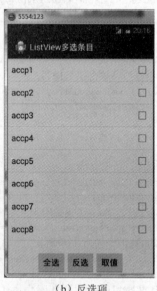
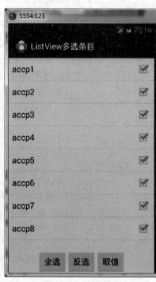

(a) 默认界面　　　　　　　　(b) 反选项　　　　　　　　(c) 全选项

图 3-5　选择多个项目

3.1.6　实现单选条目实例

本实例通过ListView控件，构建一个利用单选按钮实现单选条目程序，通过本实例演示了在ListView控件中使用单选radioButton控件的功能。

本实例的具体实现步骤如下。

(1) 在 Eclipse 中创建一个 Android 应用项目,命名为 ListView_radio。

(2) 打开 res\layout 目录下的 main.xml 布局文件,在文件中声明一个 ListView 控件。代码为:

```xml
<?xml version = "1.0" encoding = "utf-8"?>
<LinearLayout xmlns:android = "http://schemas.android.com/apk/res/android"
    android:orientation = "vertical"
    android:layout_width = "fill_parent"
    android:layout_height = "fill_parent"
    android:background = "#aabbcc">
    <ListView
        android:layout_weight = "1"
        android:id = "@+id/listView1"
        android:layout_height = "wrap_content"
        android:layout_width = "match_parent"/>
</LinearLayout>
```

(3) 打开 src\fs.listview_radio 包下的 MainActivity.java 文件,在文件中实现单选条目功能。代码为:

```java
package fs.listview_radio;
import java.util.ArrayList;
import java.util.List;
import android.app.Activity;
import android.os.Bundle;
import android.util.Log;
import android.view.View;
import android.widget.AdapterView;
import android.widget.ArrayAdapter;
import android.widget.ListView;
import android.widget.TextView;
import android.widget.AdapterView.OnItemClickListener;
public class MainActivity extends Activity {
    private List cityList = new ArrayList();
    private ListView listView;
    @Override
    public void onCreate(Bundle savedInstanceState) {
        super.onCreate(savedInstanceState);
        setContentView(R.layout.main);
        listView = (ListView) this.findViewById(R.id.listView1);
        final boolean[] isCheckedArray = new boolean[8];
        isCheckedArray[0] = false;
        isCheckedArray[1] = true;
        isCheckedArray[2] = false;
        isCheckedArray[3] = false;
        isCheckedArray[4] = false;
        isCheckedArray[5] = false;
        isCheckedArray[6] = false;
        cityList.add("条目 1");
        cityList.add("条目 2");
        cityList.add("条目 3");
        cityList.add("条目 4");
        cityList.add("条目 5");
```

```
            cityList.add("条目6");
            cityList.add("条目7");
            ArrayAdapter adapter = new ArrayAdapter(this,
                    android.R.layout.simple_list_item_single_choice, cityList);
            listView.setChoiceMode(ListView.CHOICE_MODE_SINGLE);
            listView.setAdapter(adapter);
            listView.setOnItemClickListener(new OnItemClickListener() {
                public void onItemClick(AdapterView<?> arg0, View arg1, int arg2,
                        long arg3) {
                    Log.v(" ", "" + ((TextView) arg1).getText());
                }
            });
            //设默认选中状态
            for (int i = 0; i < isCheckedArray.length; i++) {
                listView.setItemChecked(i, isCheckedArray[i]);
            }
        }
    }
```

运行程序,效果如图3-6所示。单击对应的条目,即在LogCat打印日志。代码为:

10 - 28 11:29:12.570: V/ (1067): 条目6
10 - 28 11:29:13.740: V/ (1067): 条目7

图3-6 单选条目

3.1.7 自定义ListView控件实例

在前面几小节中都通过实例介绍了利用ListView控件实现文本框功能、图片功能、多选功能及单选功能。本实例效果是实现一个ListView,ListView里面有标题、内容和图片,并加入单击和长按响应。

本实例的具体实现步骤如下:

(1) 在Eclipse中创建一个Android应用项目,命名为Userinfo_ListView。

(2) 打开res\layout目录下的main.xml文件,在文件中声明:

```xml
<?xml version="1.0" encoding="utf-8"?>
<LinearLayout
    android:id="@+id/LinearLayout01"
    android:layout_width="fill_parent"
    android:layout_height="fill_parent"
    xmlns:android="http://schemas.android.com/apk/res/android"
    android:background="#aabbcc">
<ListView
    android:layout_width="wrap_content"
    android:layout_height="wrap_content"
    android:id="@+id/ListView01"/>
</LinearLayout>
```

(3) 在 res\layout 目录下新建一个 items.xml 文件，用于定义 ListView 每个条目的 Layout，用 RelativeLayout 实现。代码为：

```xml
<?xml version="1.0" encoding="utf-8"?>
<RelativeLayout
    android:id="@+id/RelativeLayout01"
    android:layout_width="fill_parent"
    xmlns:android="http://schemas.android.com/apk/res/android"
    android:layout_height="wrap_content"
    android:paddingBottom="4dip"
    android:paddingLeft="12dip"
    android:paddingRight="12dip">
<ImageView
    android:paddingTop="12dip"
    android:layout_alignParentRight="true"
    android:layout_width="wrap_content"
    android:layout_height="wrap_content"
    android:id="@+id/ItemImage"/>
<TextView
    android:text="TextView01"
    android:layout_height="wrap_content"
    android:textSize="20dip"
    android:layout_width="fill_parent"
    android:id="@+id/ItemTitle" />
<TextView
    android:text="TextView02"
    android:layout_height="wrap_content"
    android:layout_width="fill_parent"
    android:layout_below="@+id/ItemTitle"
    android:id="@+id/ItemText"/>
</RelativeLayout>
```

(4) 打开 src\fs.userinfo_listview 包下的 MainActivity.java 文件，在文件中实现一个 ListView，ListView 里面有标题、内容和图片，并加入单击和长按响应。代码为：

```java
package fs.userinfo_listview;
import java.util.ArrayList;
import java.util.HashMap;
import android.app.Activity;
import android.os.Bundle;
import android.view.ContextMenu;
import android.view.MenuItem;
import android.view.View;
```

```java
import android.view.ContextMenu.ContextMenuInfo;
import android.view.View.OnCreateContextMenuListener;
import android.widget.AdapterView;
import android.widget.ListView;
import android.widget.SimpleAdapter;
import android.widget.AdapterView.OnItemClickListener;
public class MainActivity extends Activity {
    @Override
    public void onCreate(Bundle savedInstanceState) {
        super.onCreate(savedInstanceState);
        setContentView(R.layout.main);
        //绑定 Layout 里面的 ListView
        ListView list = (ListView) findViewById(R.id.ListView01);
        //生成动态数组,加入数据
        ArrayList<HashMap<String, Object>> listItem = new ArrayList<HashMap<String, Object>>();
        for(int i = 0;i < 10;i++)
        {
            HashMap<String, Object> map = new HashMap<String, Object>();
            map.put("ItemImage", R.drawable.bt1);    //图像资源的 id
            map.put("ItemTitle", "Level " + i);
            map.put("ItemText", "Finished in 1 Min 54 Secs, 70 Moves! ");
            listItem.add(map);
        }
        //生成适配器的 Item 和动态数组对应的元素
        SimpleAdapter listItemAdapter = new SimpleAdapter(this,listItem,            //数据源
            R.layout.items,                      //ListItem 的 XML 实现
            //动态数组与 ImageItem 对应的子项
            new String[] {"ItemImage","ItemTitle", "ItemText"},
            //ImageItem 的 XML 文件里面的一个 ImageView,两个 TextView id
            new int[] {R.id.ItemImage,R.id.ItemTitle,R.id.ItemText}
        );
        //添加并且显示
        list.setAdapter(listItemAdapter);
        //添加单击
        list.setOnItemClickListener(new OnItemClickListener() {
            @Override
            public void onItemClick(AdapterView<?> arg0, View arg1, int arg2,
                    long arg3) {
                setTitle("单击第" + arg2 + "个项目");
            }
        });
        //添加长按单击
        list.setOnCreateContextMenuListener(new OnCreateContextMenuListener() {
            @Override
            public void onCreateContextMenu(ContextMenu menu, View v,ContextMenuInfo menuInfo)
{
                menu.setHeaderTitle("长按菜单-ContextMenu");
                menu.add(0, 0, 0, "弹出长按菜单 0");
                menu.add(0, 1, 0, "弹出长按菜单 1");
            }
        });
    }
    //长按菜单响应函数
    @Override
```

```
public boolean onContextItemSelected(MenuItem item) {
    setTitle("单击了长按菜单里面的第" + item.getItemId() + "个项目");
    return super.onContextItemSelected(item);
}
```
}

运行程序,单击项目,效果如图3-7(a)所示,在界面中长按鼠标,弹出对应的菜单项,效果如图3-7(b)所示。

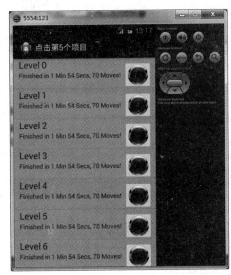 

(a)单击项目　　　　　　　　　　　(b)长按菜单

图3-7　自定义ListView

3.1.8　制作相片集实例

当今大部分的手机都具有照相功能,而且像素越来越高,但是其对相片的管理还是停留在单个相片的管理上,很少有实现相片集管理的功能。Android的Gallery控件是个很不错的看图控件,大大减轻了开发者对于看图功能的开发,而且效果也比较美观。

本实例应用Gallery控件,制作一个相片集程序,通过本程序演示了Gallery控件的具体用法。本实例的具体实现步骤如下。

(1)在Eclipse中创建一个Android应用项目,命名为Gallery_test。

(2)打开res\layout目录下的main.xml布局文件,在文件中声明一个Gallery控件。代码为:

```
<?xml version = "1.0" encoding = "utf - 8"?>
<LinearLayout xmlns:android = "http://schemas.android.com/apk/res/android"
    android:orientation = "vertical"
    android:layout_width = "fill_parent"
    android:layout_height = "fill_parent">
<Gallery
    android:id = "@ + id/gallery"
    android:layout_height = "fill_parent"
    android:layout_width = "fill_parent"/>
</LinearLayout>
```

(3) 打开 src\fs.gallery_test 包下的 MainActivity.java 文件，在文件中实现制作相片集。代码为：

```java
package fs.gallery_test;
import java.lang.reflect.Field;
import java.util.ArrayList;
import android.app.Activity;
import android.content.Context;
import android.graphics.Bitmap;
import android.graphics.BitmapFactory;
import android.os.Bundle;
import android.view.View;
import android.view.ViewGroup;
import android.widget.AdapterView;
import android.widget.BaseAdapter;
import android.widget.Gallery;
import android.widget.ImageView;
import android.widget.AdapterView.OnItemClickListener;
public class MainActivity extends Activity {
    //第一次调用 Activity 活动
    private Gallery mGallery;
    @Override
    public void onCreate(Bundle savedInstanceState) {
        super.onCreate(savedInstanceState);
        setContentView(R.layout.main);
        mGallery = (Gallery)findViewById(R.id.gallery);
        try {
            mGallery.setAdapter(new ImageAdapter(this));
        } catch (IllegalArgumentException e) {
            //TODO 自动存根法
            e.printStackTrace();
        } catch (IllegalAccessException e) {
            //TODO 自动存根法
            e.printStackTrace();
        }
        mGallery.setOnItemClickListener(new OnItemClickListener() {
            public void onItemClick(AdapterView parent, View v, int position, long id) {
                MainActivity.this.setTitle(String.valueOf(position));
            }
        });
    }
    private class ImageAdapter extends BaseAdapter{
        private Context mContext;
        private ArrayList<Integer> imgList = new ArrayList<Integer>();
        private ArrayList<Object> imgSizes = new ArrayList<Object>();
        public ImageAdapter(Context c) throws IllegalArgumentException, IllegalAccessException{
            mContext = c;
            //用反射机制来获取资源中的图片 id 和尺寸
            Field[] fields = R.drawable.class.getDeclaredFields();
            for (Field field : fields)
            {
                if (!"gud".equals(field.getName()))    //除了 icon 之外的图片
                {
                    int index = field.getInt(R.drawable.class);
                    //保存图片 id
                    imgList.add(index);
                    //保存图片大小
                    int size[] = new int[2];
```

```java
                Bitmap bmImg = BitmapFactory.decodeResource(getResources(),index);
                size[0] = bmImg.getWidth();size[1] = bmImg.getHeight();
                imgSizes.add(size);
            }
        }
    }
    @Override
    public int getCount() {
        //TODO 自动存根法
        return imgList.size();
    }
    @Override
    public Object getItem(int position) {
        //TODO 自动存根法
        return position;
    }
    @Override
    public long getItemId(int position) {
        //TODO 自动存根法
        return position;
    }
    @Override
    public View getView(int position, View convertView, ViewGroup parent) {
        //TODO 自动存根法
        ImageView i = new ImageView(mContext);
        //从 imgList 取得图片 id
        i.setImageResource(imgList.get(position).intValue());
        i.setScaleType(ImageView.ScaleType.FIT_XY);
        //从 imgSizes 取得图片大小
        int size[] = new int[2];
        size = (int[]) imgSizes.get(position);
        i.setLayoutParams(new Gallery.LayoutParams(size[0], size[1]));
        return i;
    }
};
}
```

运行程序,效果如图 3-8 所示。

图 3-8　相片集

3.1.9 手机图片查看器实例

本实例通过 ImageSwitcher 控件,构建一个手机图片查看器。通过本实例演示了 ImageSwitcher 控件的具体用法。

ImageSwitcher 控件用于实现类似于 Windows 操作系统下的"Windows 照片查看器"中的上一张、下一张切换图片的功能。在使用 ImageSwitcher 时,必须实现 ViewSwitcher.ViewFactory 接口,并通过 makeView()方法来创建用于显示图片的 ImageView。makeView()方法将返回一个显示图片的 ImageView。在使用图片切换器时,还有一个方法非常重要,那就是 setImageResource()方法,该方法用于指定要在 ImageSwitcher 中显示的图片资源。

本实例的具体实现步骤如下。

(1) 在 Eclipse 中创建一个 Android 应用项目,命名为 ImageSwitcher_test。

(2) 打开 res\layout 目录下的 main.xml 布局文件,在文件中声明一个 ImageSwitcher 控件及一个 Gallery 控件。代码为:

```xml
<RelativeLayout xmlns:android = "http://schemas.android.com/apk/res/android"
    xmlns:tools = "http://schemas.android.com/tools"
    android:layout_width = "match_parent"
    android:layout_height = "match_parent"
    android:paddingBottom = "@dimen/activity_vertical_margin"
    android:paddingLeft = "@dimen/activity_horizontal_margin"
    android:paddingRight = "@dimen/activity_horizontal_margin"
    android:paddingTop = "@dimen/activity_vertical_margin"
    tools:context = ".MainActivity"
    android:background = "#aabbcc">
    <ImageSwitcher
        android:id = "@+id/switcher"
        android:layout_width = "fill_parent"
        android:layout_height = "fill_parent"
        android:layout_alignParentTop = "true"
        android:layout_alignParentLeft = "true"/>
    <Gallery
        android:id = "@+id/gallery"
        android:background = "#55000000"
        android:layout_width = "fill_parent"
        android:layout_height = "60dp"
        android:layout_alignParentBottom = "true"
        android:layout_alignParentLeft = "true"
        android:gravity = "center_vertical"
        android:spacing = "16dp"/>
</RelativeLayout>
```

(3) 打开 src\fs.imageswitcher_test 包下的 MainActivity.java 文件,在文件中实现图片查看器,当单击界面中的某一幅图片时,即该幅图片显示在屏幕上。代码为:

```java
package fs.imageswitcher_test;
import android.app.Activity;
import android.content.Context;
import android.os.Bundle;
import android.view.View;
import android.view.ViewGroup;
```

```java
import android.view.ViewGroup.LayoutParams;
import android.view.Window;
import android.view.animation.AnimationUtils;
import android.widget.AdapterView;
import android.widget.AdapterView.OnItemSelectedListener;
import android.widget.BaseAdapter;
import android.widget.Gallery;
import android.widget.ImageSwitcher;
import android.widget.ImageView;
import android.widget.ViewSwitcher.ViewFactory;
public class MainActivity extends Activity implements ViewFactory{
    //第一次调用 Activity 活动
    private ImageSwitcher imageSwitcher;
    private Gallery gallery;
    //图片集合
    private Integer[] Images = { R.drawable.gud1, R.drawable.gud2,
            R.drawable.gud3, R.drawable.gud4
    };
    @Override
    protected void onCreate(Bundle savedInstanceState) {
        super.onCreate(savedInstanceState);
        requestWindowFeature(Window.FEATURE_NO_TITLE);
        setContentView(R.layout.main);
        imageSwitcher = (ImageSwitcher) findViewById(R.id.switcher);
        //为它指定一个 ViewFactory,也就是定义它是如何把内容显示出来的,实现 ViewFactory
        //接口并覆盖对应的 makeView 方法
        imageSwitcher.setFactory(this);
        //添加动画效果
        imageSwitcher.setInAnimation(AnimationUtils.loadAnimation(this,
                android.R.anim.fade_in));
        imageSwitcher.setOutAnimation(AnimationUtils.loadAnimation(this,
                android.R.anim.fade_out));
        gallery = (Gallery) findViewById(R.id.gallery);
        //添加适配器
        gallery.setAdapter(new ImageAdapter(this));
        //设置监听器
        gallery.setOnItemSelectedListener(new onItemSelectedListener());
    }
    //重写 makeView()方法
    public View makeView() {
        ImageView imageView = new ImageView(this);
        imageView.setBackgroundColor(0xFF000000);
        //设置填充方式
        imageView.setScaleType(ImageView.ScaleType.FIT_XY);
        imageView.setLayoutParams(new ImageSwitcher.LayoutParams(
                LayoutParams.MATCH_PARENT, LayoutParams.MATCH_PARENT));
        return imageView;
    }
    //适配器
    public class ImageAdapter extends BaseAdapter {
        private Context mContext;
        public ImageAdapter(Context c) {
```

```java
            mContext = c;
        }
        public int getCount() {
            return Images.length;
        }
        public Object getItem(int position) {
            return position;
        }
        public long getItemId(int position) {
            return position;
        }
        public View getView(int position, View convertView, ViewGroup parent) {
            ImageView imageView = new ImageView(mContext);
            imageView.setImageResource(Images[position]);
            imageView.setAdjustViewBounds(true);
            imageView.setLayoutParams(new Gallery.LayoutParams(
                    LayoutParams.WRAP_CONTENT, LayoutParams.WRAP_CONTENT));
            imageView.setBackgroundResource(R.drawable.gud5);
            return imageView;
        }
    }
    private class onItemSelectedListener implements OnItemSelectedListener{
        public void onItemSelected(AdapterView<?> arg0, View arg1, int arg2,
                long arg3) {
            imageSwitcher.setImageResource(Images[arg2]);
        }
        public void onNothingSelected(AdapterView<?> arg0) {
        }

    }
}
```

运行程序,效果如图 3-9(a)所示,当单击某一幅图片时,效果如图 3-9(b)所示。

(a) 显示图片1　　　　　　　　　　(b) 显示图片2

图 3-9　图片查看器

3.1.10 九宫布局实例

GridView(网格视图)是按照行列的方式来显示内容的,一般用于显示图片等内容,例如实现九宫格图,用 GridView 是首选,也是最简单的。主要用于设置 Adapter。

本实例利用 GridView 控件,构建一个九宫布局程序,通过本实例演示了 GridView 控件的具体用法。其实现步骤如下。

(1) 在 Eclipse 中创建一个 Android 应用项目,命名为 GridView_test。

(2) 打开 res\layout 目录下的 main.xml 文件,用于存放 GridView 控件。代码为:

```xml
<?xml version = "1.0" encoding = "utf-8"?>
<!--
android:numColumns = "auto_fit",          //GridView 的列数设置为自动
android:columnWidth = "90dp",             //每列的宽度,也就是 Item 的宽度
android:stretchMode = "columnWidth",      //缩放与列宽大小同步
android:verticalSpacing = "10dp",
                                          //两行之间的边距,如行一(NO.0~NO.2)与行二(NO.3~NO.5)间距为10dp
android:horizontalSpacing = "10dp",       //两列之间的边距
-->
<GridView xmlns:android = "http://schemas.android.com/apk/res/android"
    android:id = "@ + id/gridview"
    android:layout_width = "fill_parent"
    android:layout_height = "fill_parent"
    android:numColumns = "auto_fit"
    android:verticalSpacing = "10dp"
    android:horizontalSpacing = "10dp"
    android:columnWidth = "90dp"
    android:stretchMode = "columnWidth"
    android:gravity = "center"
    android:background = "#aabbcc"/>
```

(3) 在 res\layout 目录下新建一个 item_n.xml,用于存放显示控件。代码为:

```xml
<?xml version = "1.0" encoding = "utf-8"?>
<RelativeLayout
    xmlns:android = "http://schemas.android.com/apk/res/android"
    android:layout_height = "wrap_content"
    android:paddingBottom = "4dip"
    android:layout_width = "fill_parent"
    android:background = "#aabbcc">
    <ImageView
        android:layout_height = "wrap_content"
        android:layout_width = "wrap_content"
        android:layout_centerHorizontal = "true"
        android:id = "@ + id/itemImage"/>
    <TextView
        android:layout_width = "wrap_content"
        android:layout_below = "@ + id/itemImage"
        android:layout_height = "wrap_content"
        android:text = "TextView01"
        android:layout_centerHorizontal = "true"
        android:id = "@ + id/itemText"/>
</RelativeLayout>
```

（4）在 src\fs.gridview_test 包下新建 3 个文件，分别命名为 TestActivity1.java、TestActivity2.java 和 TestActivity3.java，用于实现跳转类，代码分别如下。

TestActivity1.java 文件的代码为：

```java
package fs.gridview_test;
import android.app.Activity;
import android.os.Bundle;
public class TestActivity1 extends Activity {
    @Override
    public void onCreate(Bundle savedInstanceState) {
        super.onCreate(savedInstanceState);
        //setContentView(R.layout.main);
    }
}
```

TestActivity2.java 文件的代码为：

```java
package fs.gridview_test;
import android.app.Activity;
import android.os.Bundle;
public class TestActivity2 extends Activity {
    @Override
    public void onCreate(Bundle savedInstanceState) {
        super.onCreate(savedInstanceState);
        //setContentView(R.layout.main);
    }
}
```

TestActivity3.java 文件的代码为：

```java
package fs.gridview_test;
import android.app.Activity;
import android.os.Bundle;
public class TestActivity3 extends Activity {
    @Override
    public void onCreate(Bundle savedInstanceState) {
        super.onCreate(savedInstanceState);
        //setContentView(R.layout.main);
    }
}
```

（5）打开 src\fs.gridview_test 包下的 MainActivity.java 文件，在该文件中实现九宫布局效果。代码为：

```java
package fs.gridview_test;
import java.util.ArrayList;
import java.util.HashMap;
import android.app.Activity;
import android.content.Intent;
import android.os.Bundle;
import android.view.View;
import android.widget.AdapterView;
import android.widget.GridView;
import android.widget.SimpleAdapter;
import android.widget.Toast;
import android.widget.AdapterView.OnItemClickListener;
```

```java
public class MainActivity extends Activity {
    private String texts[] = null;
    private int images[] = null;
    public void onCreate(Bundle savedInstanceState) {
        super.onCreate(savedInstanceState);
        setContentView(R.layout.main);
        images = new int[]{R.drawable.c1, R.drawable.c2,
                R.drawable.d1, R.drawable.d2,
                R.drawable.d3, R.drawable.d4,
                R.drawable.d5, R.drawable.face};
        texts = new String[]{ "宫式布局1", "宫式布局2",
                "宫式布局3", "宫式布局4", "宫式布局5",
                "宫式布局6", "宫式布局7", "宫式布局8"};
        GridView gridview = (GridView) findViewById(R.id.gridview);
        ArrayList<HashMap<String, Object>> lstImageItem = new ArrayList<HashMap<String, Object>>();
        for (int i = 0; i < 8; i++) {
            HashMap<String, Object> map = new HashMap<String, Object>();
            map.put("itemImage", images[i]);
            map.put("itemText", texts[i]);
            lstImageItem.add(map);
        }
        SimpleAdapter saImageItems = new SimpleAdapter(this,
                lstImageItem,                                //数据源
                R.layout.item_n,                             //显示布局
                new String[] { "itemImage", "itemText" },
                new int[] { R.id.itemImage, R.id.itemText });
        gridview.setAdapter(saImageItems);
        gridview.setOnItemClickListener(new ItemClickListener());
    }
    class ItemClickListener implements OnItemClickListener {
        /**
         * 单击项时触发事件
         * @param parent 发生单击动作的 AdapterView
         * @param view 在 AdapterView 中被单击的视图(它是由 adapter 提供的一个视图).
         * @param position 视图在 adapter 中的位置.
         * @param rowid 被单击元素的行 id.
         */
        public void onItemClick(AdapterView<?> parent, View view, int position, long rowid) {
            HashMap<String, Object> item = (HashMap<String, Object>) parent.
                    getItemAtPosition(position);
            //获取数据源的属性值
            String itemText = (String)item.get("itemText");
            Object object = item.get("itemImage");
            Toast.makeText(MainActivity.this, itemText, Toast.LENGTH_LONG).show();
            //根据图片进行相应的跳转
            switch (images[position]) {
            case R.drawable.c1:
                startActivity(new Intent(MainActivity.this, TestActivity1.class));   //启动另一个 Activity
                finish();                                    //结束此 Activity,可回收
                break;
            case R.drawable.c2:
                startActivity(new Intent(MainActivity.this, TestActivity2.class));
                finish();
                break;
```

```java
            case R.drawable.d1:
                startActivity(new Intent(MainActivity.this, TestActivity3.class));
                finish();
                break;
            }

        }
    }
}
```

(6) 打开 res\layout 目录下的 strings.xml 文件,用于实现变量赋值。代码为:

```xml
<?xml version = "1.0" encoding = "utf - 8"?>
<resources>
    <string name = "app_name">九宫布局</string>
    <string name = "action_settings">Settings</string>
    <string name = "hello_world">Hello world!</string>
    <string name = "test_name1">跳转到 TestActivity1</string>
    <string name = "test_name2">跳转到 TestActivity2</string>
    <string name = "test_name3">跳转到 TestActivity3</string>
</resources>
```

(7) 打开 AndroidManifest.xml 文件,用于声明变量。代码为:

```xml
<?xml version = "1.0" encoding = "utf - 8"?>
<manifest xmlns:android = "http://schemas.android.com/apk/res/android"
    package = "fs.gridview_test"
    android:versionCode = "1"
    android:versionName = "1.0" >
    <uses - sdk
        android:minSdkVersion = "8"
        android:targetSdkVersion = "18" />
    <application
        android:allowBackup = "true"
        android:icon = "@drawable/ic_launcher"
        android:label = "@string/app_name"
        android:theme = "@style/AppTheme" >
        <activity
            android:name = "fs.gridview_test.MainActivity"
            android:label = "@string/app_name" >
            <intent - filter>
                <action android:name = "android.intent.action.MAIN" />
                <category android:name = "android.intent.category.LAUNCHER" />
            </intent - filter>
        </activity>
        <activity android:name = ".TestActivity1" android:label = "@string/test_name1"/>
        <activity android:name = ".TestActivity2" android:label = "@string/test_name2"/>
        <activity android:name = ".TestActivity3" android:label = "@string/test_name3"/>
    </application>
</manifest>
```

运行程序,效果如图 3-10(a)所示,单击对应的宫布局时,即显示对应的提示,如图 3-10(b)所示。

（a）默认九宫布局　　　　　　　　（b）选择布局的提示

图 3-10　九宫布局

3.1.11　带图片文字的 ListView 实例

SimpleAdapter 的扩展性最好，可以定义各种各样的布局出来，可以放上 ImageView（图片），还可以放上 Button（按钮）和 CheckBox（复选框）等。下面实例直接继承了 ListActivity，ListActivity 和普通的 Activity 没有太大的差别，不同就是对显示 ListView 做了许多优化，方便显示而已。

本实例的具体实现步骤如下。

（1）在 Eclipse 中创建一个 Android 应用项目，命名为 SimpleAdapter_test。

（2）打开 res\layout 目录下的 main.xml 布局文件，在文件中声明一个 ListView 控件。代码为：

```
<LinearLayout xmlns:android = "http://schemas.android.com/apk/res/android"
    xmlns:tools = "http://schemas.android.com/tools"
    android:layout_width = "match_parent"
    android:layout_height = "match_parent"
    android:orientation = "horizontal"
    tools:context = ".MainActivity"
    android:background = " # aabbcc" >
    <ListView
        android:id = "@ + id/lt1"
        android:layout_width = "match_parent"
        android:layout_height = "wrap_content" >
    </ListView>
</LinearLayout>
```

（3）在 res\layout 目录下新建一个 listview_n.xml 布局文件，在文件中声明一个 ImageView 控件和两个 TextView 控件。代码为：

```
<?xml version = "1.0" encoding = "utf - 8"?>
<LinearLayout xmlns:android = "http://schemas.android.com/apk/res/android"
```

```xml
    android:layout_width = "match_parent"
    android:layout_height = "match_parent"
    android:orientation = "horizontal"
    android:background = "#aabbcc" >
    <ImageView
        android:id = "@+id/head"
        android:layout_width = "wrap_content"
        android:layout_height = "wrap_content"
        android:paddingLeft = "10dp" />
    <LinearLayout
        android:layout_width = "match_parent"
        android:layout_height = "wrap_content"
        android:orientation = "vertical" >
        <TextView
            android:id = "@+id/name"
            android:layout_width = "wrap_content"
            android:layout_height = "wrap_content"
            android:textSize = "20dp"
            android:textColor = "#f0f"
            android:paddingLeft = "10dp"/>
        <TextView
            android:id = "@+id/desc"
            android:layout_width = "wrap_content"
            android:layout_height = "wrap_content"
            android:textSize = "14dp"
            android:paddingLeft = "10dp"/>
    </LinearLayout>
</LinearLayout>
```

（4）打开 src\fs.simpleadapter_test 包下的 MainActivity.java 文件，在文件中实现带图文 ListView 的效果。代码为：

```java
package fs.simpleadapter_test;
import java.util.ArrayList;
import java.util.HashMap;
import java.util.List;
import java.util.Map;
import android.app.Activity;
import android.os.Bundle;
import android.view.Menu;
import android.widget.ListView;
import android.widget.SimpleAdapter;
public class MainActivity extends Activity {
private String[] name = {"卡通长毛狗","叮咚猫","杀生丸","福尔摩斯","大力士"};
    private String[] desc = {"吉祥如意","机灵活泼,通人性","妖颜银发,额生月印","一个才华横溢的侦探形象","浑身惊人的力气"};
    private int[] imageids = { R.drawable.kt2, R.drawable.kt1,
            R.drawable.po1, R.drawable.po3, R.drawable.po4};
    private ListView lt1;
    @Override
    protected void onCreate(Bundle savedInstanceState) {
        super.onCreate(savedInstanceState);
        setContentView(R.layout.main);
        List<Map<String, Object>> listems = new ArrayList<Map<String, Object>>();
```

```java
        for (int i = 0; i < name.length; i++) {
            Map<String, Object> listem = new HashMap<String, Object>();
            listem.put("head", imageids[i]);
            listem.put("name", name[i]);
            listem.put("desc", desc[i]);
            listems.add(listem);
        }
        /* SimpleAdapter 的参数说明
         * 第 1 个参数 表示访问整个 android 应用程序接口,基本上所有的组件都需要
         * 第 2 个参数表示生成一个 Map(String ,Object)列表选项
         * 第 3 个参数表示界面布局的 id 表示该文件作为列表项的组件
         * 第 4 个参数表示该 Map 对象的哪些 key 对应 value 来生成列表项
         * 第 5 个参数表示来填充的组件 Map 对象 key 对应的资源,与依次填充组件 顺序有对应
         关系
         * 注意的是 Map 对象可以 key 可以找不到 但组件的必须要有资源填充,因为找不到 key 也
         会返回 null,其实就相当于给了一个 null 资源
         * 下面的程序中如果 new String[] { "name", "head", "desc","name" } new int[] {R.id.
         name,R.id.head,R.id.desc,R.id.head}
         * 这个 head 的组件会被 name 资源覆盖
         * */
        SimpleAdapter simplead = new SimpleAdapter(this, listems,
                R.layout.listview_n, new String[] { "name", "head", "desc" },
                new int[] {R.id.name,R.id.head,R.id.desc});
        lt1 = (ListView)findViewById(R.id.lt1);
        lt1.setAdapter(simplead);
    }
    @Override
    public boolean onCreateOptionsMenu(Menu menu) {
        getMenuInflater().inflate(R.menu.main, menu);
        return true;
    }
}
```

运行程序,效果如图 3-11 所示。

图 3-11　带图文的 ListView 效果

3.1.12 手机浏览网页实例

网页浏览视图(WebView)类似于常用的浏览器,在Android手机中内置了一款高性能WebKit内核浏览器,WebView组件就是由WebKit封装而来的,可以用它来显示一个Web页面。通过WebView控件可以直接访问网页,或者把输入的HTML字符串显示出来,功能比较强大,有以下几个优点。

- 功能强大,支持CSS,Java script等HTML语言,这样页面就能更漂亮。
- 能够对浏览器控件进行非常详细的设置,例如字体大小、背景色和滚动条样式等。
- 能够捕捉到所有浏览器操作,例如点击URL、打开或关闭URL。
- 能够很好地融入布局。
- 甚至WebView还能和JS进行交互。

本实例通过WebView控件,构建一个在安卓中浏览网页的程序。通过本实例演示了WebView控件的具体用法。

本实例的具体实现步骤如下。

(1) 在Eclipse中创建一个Android应用项目,命名为WebView_test。

(2) 打开res\layout目录下的main.xml布局文件,在文件中声明一个ScrollView控件,在控件中定义一个LinearLayout布局及3个WebView控件。代码为:

```xml
<?xml version = "1.0" encoding = "utf-8"?>
<ScrollView
    xmlns:android = "http://schemas.android.com/apk/res/android"
    android:layout_width = "match_parent"
    android:layout_height = "wrap_content"
    android:orientation = "vertical"
    android:background = "#aabbcc">
    <LinearLayout
        android:orientation = "vertical"
        android:layout_width = "match_parent"
        android:layout_height = "wrap_content"
        android:background = "#aabbcc">
        <WebView
            android:id = "@+id/wv1"
            android:layout_height = "wrap_content"
            android:layout_width = "match_parent"/>
        <WebView
            android:id = "@+id/wv2"
            android:layout_height = "wrap_content"
            android:layout_width = "match_parent"/>
        <WebView
            android:id = "@+id/wv3"
            android:layout_height = "wrap_content"
            android:layout_width = "match_parent"/>
    </LinearLayout>
</ScrollView>
```

(3) 打开src\fs.webview_test包下的MainActivity.java文件,在文件中利用WebView打开对应的网页。代码为:

```java
package fs.webview_test;
import android.app.Activity;
```

```java
import android.os.Bundle;
import android.webkit.WebSettings;
import android.webkit.WebView;
import android.webkit.WebViewClient;
public class MainActivity extends Activity {
    @Override
    public void onCreate(Bundle icicle) {
        super.onCreate(icicle);
        setContentView(R.layout.main);
        final String mimeType = "text/html";
        final String encoding = "utf-8";
        WebView wv;
        wv = (WebView) findViewById(R.id.wv1);
        wv.loadDataWithBaseURL("http://www.google.com","<a href='http://www.baidu.com'>百度搜索</a>", mimeType, encoding, "");
        wv = (WebView) findViewById(R.id.wv2);
        wv.loadDataWithBaseURL("http://www.google.com","<a href='www.cnblogs.com'>博客园</a>", mimeType, encoding, "");
        //出现乱码,因此建议一般情况下不要使用此方法
        wv = (WebView) findViewById(R.id.wv3);
        wv.loadData("<a href='x'>http://ent.sina.com.cn/</a>", mimeType, encoding);
    }
}
```

运行程序,默认界面如图 3-12 所示。

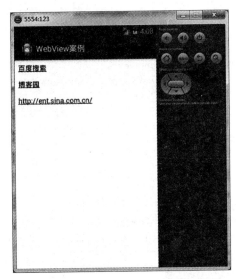

图 3-12　WebView 的使用

3.1.13　手势滑动实例

ViewPager 用于实现多页面的切换效果,该类在于 Google 的兼容包中,所以在引用时记住在 BuilldPath 中加入"android-support-v4.jar"。

本实例利用 ViewPager 控件,构建一个多页面切换程序,通过本实例演示了 ViewPager 的用法。本实例的具体实现步骤如下:

(1) 在 Eclipse 中创建一个 Android 应用项目,命名为 ViewPage_test。

(2) 打开 res\layout 目录下的 main.xml 布局文件,在文件中定义一个 ViewPager 控件及

一个 PagerTitleStrip 控件。代码为：

```xml
<?xml version = "1.0" encoding = "utf-8"?>
<LinearLayout xmlns:android = "http://schemas.android.com/apk/res/android"
    android:layout_width = "fill_parent"
    android:layout_height = "fill_parent"
    android:orientation = "vertical"
    android:background = "#aabbcc">
    <android.support.v4.view.ViewPager
        android:id = "@+id/viewpager"
        android:layout_width = "wrap_content"
        android:layout_height = "wrap_content"
        android:layout_gravity = "center" >
    <android.support.v4.view.PagerTitleStrip
        android:id = "@+id/pagertitle"
        android:layout_width = "wrap_content"
        android:layout_height = "wrap_content"
        android:layout_gravity = "top" />
    </android.support.v4.view.ViewPager>
</LinearLayout>
```

(3) 打开 src\fs.viewpage_test 包下的 MainActivity.java 文件，用于实现图像的翻页浏览。代码为：

```java
package fs.viewpage_test;
import java.util.ArrayList;
import android.os.Bundle;
import android.app.Activity;
import android.graphics.drawable.Drawable;
import android.support.v4.view.PagerAdapter;
import android.support.v4.view.PagerTitleStrip;
import android.support.v4.view.ViewPager;
import android.view.LayoutInflater;
import android.view.Menu;
import android.view.View;
import android.widget.ImageView;
import android.widget.LinearLayout;
public class MainActivity extends Activity {
    /** 第一次调用 Activity 活动。*/
    private ViewPager mViewPager;
    private PagerTitleStrip mPagerTitleStrip;
    private int[] pics = { R.drawable.g1, R.drawable.g2, R.drawable.g4 };
    final ArrayList<View> views = new ArrayList<View>();
    @Override
    public void onCreate(Bundle savedInstanceState) {
        super.onCreate(savedInstanceState);
        setContentView(R.layout.main);
        mViewPager = (ViewPager) findViewById(R.id.viewpager);
        mPagerTitleStrip = (PagerTitleStrip) findViewById(R.id.pagertitle);
        LinearLayout.LayoutParams mParams = new LinearLayout.LayoutParams(
                LinearLayout.LayoutParams.WRAP_CONTENT,
                LinearLayout.LayoutParams.WRAP_CONTENT);
        //将要分页显示的 View 装入数组中
        for (int i = 0; i < pics.length; i++) {
            ImageView iv = new ImageView(this);
            iv.setLayoutParams(mParams);
            iv.setImageResource(pics[i]);
            views.add(iv);
```

```
        }
    //每个页面的 Title 数据
    final ArrayList<String> titles = new ArrayList<String>();
    titles.add("tab1");
    titles.add("tab2");
    titles.add("tab3");
    //填充 ViewPager 的数据适配器
    PagerAdapter mPagerAdapter = new PagerAdapter() {
            @Override
            public boolean isViewFromObject(View arg0, Object arg1) {
                return arg0 == arg1;
            }
            @Override
            public int getCount() {
                return views.size();
            }
            @Override
            public void destroyItem(View container, int position, Object object) {
                ((ViewPager) container).removeView(views.get(position));
            }
            @Override
            public CharSequence getPageTitle(int position) {
                return titles.get(position);
            }
            @Override
            public Object instantiateItem(View container, int position) {
                ((ViewPager) container).addView(views.get(position));
                return views.get(position);
            }
    };
    mViewPager.setAdapter(mPagerAdapter);
    }
    @Override
    public boolean onCreateOptionsMenu(Menu menu) {
        getMenuInflater().inflate(R.menu.main, menu);
        return true;
    }
}
```

运行程序,效果如图 3-13 所示。

图 3-13　图片的翻页浏览

3.1.14 多个标签栏实例

Android 程序中，Tab 标签窗口是一种常用的 UI 界面元素。它的实现主要利用了 TabHost 类。

TabHost 是一个标签窗口的容器。一个 TabHost 对象包含两个子元素对象。
- 一个对象是 Tab 标签集合（TabWidget），用户单击它们来选择一个特定的标签；
- 一个是 FrameLayout 对象，展示当前页的内容。

子元素通常是通过容器对象来控制，而不是直接设置子元素的值。下面实现几个 Tab 实例。

第一个实例，利用 TabActivity 实现多个标签栏。本实例的具体实现步骤如下。

（1）在 Eclipse 中创建一个 Android 应用项目，命名为 TabActivity_test。

（2）打开 res\layout 目录下的 main.xml 布局文件，在文件中声明 3 个 TextView 控件。代码为：

```xml
<?xml version = "1.0" encoding = "utf - 8"?>
<FrameLayout xmlns:android = "http://schemas.android.com/apk/res/android"
    android:layout_width = "match_parent"
    android:layout_height = "match_parent"
    android:background = "#ff0000">
    <TextView
        android:id = "@ + id/view1"
        android:layout_width = "match_parent"
        android:layout_height = "match_parent"
        android:background = "#770000ff"
        android:text = "标签 1" />
    <TextView
        android:id = "@ + id/view2"
        android:layout_width = "match_parent"
        android:layout_height = "match_parent"
        android:background = "#7f00"
        android:text = "标签 2" />
    <TextView
        android:id = "@ + id/view3"
        android:layout_width = "match_parent"
        android:layout_height = "match_parent"
        android:background = "#7700ff00"
        android:text = "标签 3" />
</FrameLayout>
```

（3）打开 src\fs.tableactivity_test 包下的 MainActivity.java 文件，在文件中实现 3 个标签的切换。代码为：

```java
package fs.tabactivity_test;
import android.os.Bundle;
import android.view.LayoutInflater;
import android.widget.TabHost;
import android.app.TabActivity;
@SuppressWarnings("deprecation")
public class MainActivity extends TabActivity
{
```

```
@Override
protected void onCreate(Bundle savedInstanceState)
{
    super.onCreate(savedInstanceState);
    //得到 TabActivity 中的 TabHost 对象
    TabHost tabHost = getTabHost();
    //内容：采用布局文件中的布局
    LayoutInflater.from(this).inflate(R.layout.main,
            tabHost.getTabContentView(), true);
    //加上标签
    //参数设置：新增的 TabSpec 的标签,标签中显示的字样
    //setContent 设置内容对应的 View 资源标号
    tabHost.addTab(tabHost.newTabSpec("tab1")
            .setIndicator("tab1 indicator").setContent(R.id.view1));
    tabHost.addTab(tabHost.newTabSpec("tab3").setIndicator("tab2")
            .setContent(R.id.view2));
    tabHost.addTab(tabHost.newTabSpec("tab3").setIndicator("tab3")
            .setContent(R.id.view3));
}
```

运行程序,标签 TAB1 的界面如图 3-14(a)所示,标签 TAB2 的界面如图 3-14(b)所示。

 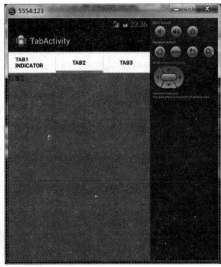

（a）TAB1的界面　　　　　（b）TAB2的界面

图 3-14　TabActivity 实现多个标签

第二个实例,使用 TabHost.TabContentFactory 实现多个标签栏。本实例的具体实现步骤如下。

（1）在 Eclipse 中创建一个 Android 应用项目,命名为 TabHost_test。

（2）打开 src\fs.tabhost_test 包下的 MainActivity.java 文件,在该文件中利用 TabHost 控件实现多个标签栏。代码为:

```
package fs.tabhost_test;
import android.os.Bundle;
import android.view.LayoutInflater;
import android.view.View;
```

```java
import android.widget.TabHost;
import android.widget.TextView;
import android.app.TabActivity;
@SuppressWarnings("deprecation")
public class MainActivity extends TabActivity implements
        TabHost.TabContentFactory
{
    @Override
    protected void onCreate(Bundle savedInstanceState)
    {
        super.onCreate(savedInstanceState);
        TabHost tabHost = getTabHost();
        //不再需要载入布局文件,如果此句不注释掉会导致 content 的重叠
        //LayoutInflater.from(this).inflate(R.layout.activity_hello_tab,
        //tabHost.getTabContentView(), true);
        //setContent 中传递 this
        tabHost.addTab(tabHost.newTabSpec("tab1")
                .setIndicator("tab1 indicator").setContent(this));
        tabHost.addTab(tabHost.newTabSpec("tab2").setIndicator("tab2")
                .setContent(this));
        tabHost.addTab(tabHost.newTabSpec("tab3").setIndicator("tab3")
                .setContent(this));
    }
    //setContent 的参数设为 this 时,从这个方法得到每一个 Tab 的内容(此次不用布局文件,用的话
      会重叠)
    @Override
    public View createTabContent(String tag)
    {
        //参数:这个方法会接收到被选择的 tag 的标签
        final TextView tv = new TextView(this);
        tv.setText("tag 标签的内容 " + tag);
        return tv;
    }
}
```

运行程序,效果如图 3-15 所示。

图 3-15　TabHost 实现多个标签栏

第三个实例，不继承 TabActivity。前面两个程序例子中都是继承了 TabActivity 类，如果不继承它，需要自己写 TabHost 的布局，其中包含了两个必要的子元素：TabWidget 和 FrameLayout，其 id 都是固定值。本实例的具体实现步骤如下。

(1) 在 Eclipse 中创建一个 Android 应用项目，命名为 UnTabActivity_test。

(2) 打开 res\layout 目录下的 main.xml 布局文件，在文件中定义所需要的布局。代码为：

```xml
<?xml version = "1.0" encoding = "utf-8"?>
<LinearLayout xmlns:android = "http://schemas.android.com/apk/res/android"
    android:layout_width = "match_parent"
    android:layout_height = "match_parent"
    android:orientation = "vertical"
    android:background = "#aabbcc">
    <!-- TabHost 必须包含一个 TabWidget 和一个 FrameLayout -->
    <TabHost
        android:id = "@+id/myTabHost"
        android:layout_width = "match_parent"
        android:layout_height = "match_parent" >
        <LinearLayout
            android:layout_width = "fill_parent"
            android:layout_height = "fill_parent"
            android:orientation = "vertical" >
            <!-- TabWidget 的 id 属性必须为 @android:id/tabs -->
            <TabWidget
                android:id = "@android:id/tabs"
                android:layout_width = "match_parent"
                android:layout_height = "wrap_content"
                android:layout_weight = "0"
                android:orientation = "horizontal" />
            <!-- FrameLayout 的 id 属性必须为 @android:id/tabcontent -->
            <FrameLayout
                android:id = "@android:id/tabcontent"
                android:layout_width = "match_parent"
                android:layout_height = "match_parent"
                android:layout_weight = "0" >
                <TextView
                    android:id = "@+id/view1"
                    android:layout_width = "match_parent"
                    android:layout_height = "match_parent"
                    android:text = "Tab1 的内容" />
                <TextView
                    android:id = "@+id/view2"
                    android:layout_width = "match_parent"
                    android:layout_height = "match_parent"
                    android:text = "Tab2 的内容" />
                <TextView
                    android:id = "@+id/view3"
                    android:layout_width = "match_parent"
                    android:layout_height = "match_parent"
                    android:text = "Tab3 的内容" />
            </FrameLayout>
        </LinearLayout>
    </TabHost>
</LinearLayout>
```

（3）打开 src\fs.untabactivity_test 包下的 MainActivity.java 文件，在文件中实现多个标签栏。代码为：

```java
package fs.untabactivity_test;
import android.os.Bundle;
import android.app.Activity;
import android.view.Menu;
import android.widget.TabHost;
public class MainActivity extends Activity
{
    @Override
    protected void onCreate(Bundle savedInstanceState)
    {
        super.onCreate(savedInstanceState);
        setContentView(R.layout.main);
        TabHost tabHost = (TabHost) findViewById(R.id.myTabHost);
        //如果不是继承 TabActivity,则必须在得到 tabHost 之后,添加标签之前调用 tabHost.setup()
        tabHost.setup();
        //这里 content 的设置采用了布局文件中的 view
        tabHost.addTab(tabHost.newTabSpec("tab1")
                .setIndicator("tab1 indicator").setContent(R.id.view1));
        tabHost.addTab(tabHost.newTabSpec("tab3").setIndicator("tab2")
                .setContent(R.id.view2));
        tabHost.addTab(tabHost.newTabSpec("tab3").setIndicator("tab3")
                .setContent(R.id.view3));
    }
}
```

运行程序，效果如图 3-16 所示。

图 3-16 不继承 TabActivity 实现多个标签

第四个实例：利用 Scrolling Tab 实现多个标签。当标签太多时，需要把标签设置到一个 ScrollView 中进行滚动。本实例的具体实现步骤如下。

（1）在 Eclipse 中创建一个 Android 应用项目，命名为 Scrolling_test。

（2）打开 res\layout 目录下的 main.xml 布局文件，在文件中定义一个 TabHost 容器、一个 HorizontalScrollView 控件和一个 TabWidget 控件。代码为：

```xml
<?xml version = "1.0" encoding = "utf-8"?>
<LinearLayout xmlns:android = "http://schemas.android.com/apk/res/android"
    android:layout_width = "match_parent"
    android:layout_height = "match_parent"
    android:orientation = "vertical"
    android:background = "#f0f0">
    <TabHost
        android:id = "@+id/myTabHost"
        android:layout_width = "match_parent"
        android:layout_height = "match_parent" >
        <LinearLayout
            android:layout_width = "match_parent"
            android:layout_height = "match_parent"
            android:orientation = "vertical"
            android:padding = "5dp" >
            <HorizontalScrollView
                android:layout_width = "match_parent"
                android:layout_height = "wrap_content"
                android:scrollbars = "none" >
                <TabWidget
                    android:id = "@android:id/tabs"
                    android:layout_width = "wrap_content"
                    android:layout_height = "wrap_content" />
            </HorizontalScrollView>
            <FrameLayout
                android:id = "@android:id/tabcontent"
                android:layout_width = "match_parent"
                android:layout_height = "match_parent"
                android:padding = "5dp" />
        </LinearLayout>
    </TabHost>
</LinearLayout>
```

（3）打开 src\fs.scrolling_test 包下的 MainActivity.java 文件，在文件中实现多个标签，并可实现滚动。代码为：

```java
package fs.scrolling_test;
import android.os.Bundle;
import android.app.Activity;
import android.view.View;
import android.widget.TabHost;
import android.widget.TextView;
public class MainActivity extends Activity implements TabHost.TabContentFactory
{
    @Override
    protected void onCreate(Bundle savedInstanceState)
    {
        super.onCreate(savedInstanceState);
        setContentView(R.layout.main);
        //从布局中获取 TabHost 并建立
        TabHost tabHost = (TabHost) findViewById(R.id.myTabHost);
        tabHost.setup();
        //加上 30 个标签
        for (int i = 1; i <= 30; i++)
```

```
            {
                String name = "Tab " + i;
                tabHost.addTab(tabHost.newTabSpec(name).setIndicator(name)
                    .setContent(this));
            }
        }
        @Override
        public View createTabContent(String tag)
        {
            final TextView tv = new TextView(this);
            tv.setText("tag 标签的内容 " + tag);
            return tv;
        }
    }
```

运行程序,效果如图3-17(a)所示,水平方向拖动标签即可实现标签的水平滚动,效果如图3-17(b)所示。

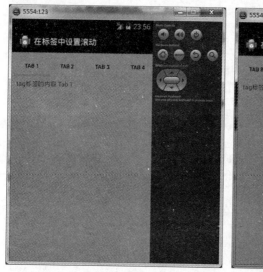

(a)默认界面　　　　　　　　　　(b)水平滚动标签

图 3-17 在标签中设置滚动

3.2 温馨的提示实例

Toast 和 Notification 是 Android 系统为用户提供的轻量级的信息提醒机制。这种方式不会打断用户当前的操作,也不会获取到焦点,非常方便。

Toast 向用户提供比较快速的即时消息,当 Toast 被显示时,虽然其悬浮于应用程序的最上方,但是 Toast 从不获得焦点。因为设计 Toast 时就是为了让其在提示有用信息时尽量不显眼。Toast 应用于提示用户某项设置成功等。

Notification 是另外一种消息提示方式,Notification 位于手机的状态栏(Status Bar),状态栏位于手机屏幕的最上层,通常显示电池电量、信号强度等信息,在 Android 手机中,用手指按下状态栏并往下拉可以打开状态栏查看系统的提示消息。

3.2.1 "通知单"实例

可以将 Toast 理解成为一种通知,也就是在操作 Android 之后 Android 系统反馈给人们的信息,或者数据。本实例的样式就可以理解为不同形式的通知了,实现 5 种样式的"通知单"。

本实例的具体实现步骤如下。

(1) 在 Eclipse 中创建一个 Android 应用项目,命名为 Toast_test。

(2) 打开 res\layout 目录下的 main.xml 布局文件,在文件中声明 5 个 Button 控件。代码为:

```xml
<?xml version = "1.0" encoding = "utf-8"?>
<LinearLayout xmlns:android = "http://schemas.android.com/apk/res/android"
    android:orientation = "vertical"
    android:layout_width = "fill_parent"
    android:layout_height = "fill_parent"
    android:padding = "5dip"
    android:gravity = "center"
    android:background = "#77ffff00">
<Button
    android:layout_height = "wrap_content"
    android:layout_width = "fill_parent"
    android:id = "@+id/btnSimpleToast"
    android:text = "默认"/>
<Button
    android:layout_height = "wrap_content"
    android:layout_width = "fill_parent"
    android:text = "自定义显示位置"
    android:id = "@+id/btnSimpleToastWithCustomPosition"/>
<Button
    android:layout_height = "wrap_content"
    android:layout_width = "fill_parent"
    android:id = "@+id/btnSimpleToastWithImage"
    android:text = "带图片"/>
<Button
    android:layout_height = "wrap_content"
    android:layout_width = "fill_parent"
    android:text = "完全自定义"
    android:id = "@+id/btnCustomToast"/>
<Button
    android:layout_height = "wrap_content"
    android:layout_width = "fill_parent"
    android:text = "其他线程"
    android:id = "@+id/btnRunToastFromOtherThread"/>
</LinearLayout>
```

(3) 在 res\layout 目录下新建一个 cutom.xml 自定义布局文件,在文件中声明两个 TextView 控件及一个 ImageView 控件。代码为:

```xml
<?xml version = "1.0" encoding = "utf-8"?>
<LinearLayout
    xmlns:android = "http://schemas.android.com/apk/res/android"
    android:layout_height = "wrap_content"
    android:layout_width = "wrap_content"
    android:orientation = "vertical"
```

```xml
            android:id = "@ + id/llToast"
            android:background = "#0000ff">
    <TextView
            android:layout_height = "wrap_content"
            android:layout_margin = "1dip"
            android:textColor = "#ffffffff"
            android:layout_width = "fill_parent"
            android:gravity = "center"
            android:background = "#bb000000"
            android:id = "@ + id/tvTitleToast" />
    <LinearLayout
            android:layout_height = "wrap_content"
            android:orientation = "vertical"
            android:id = "@ + id/llToastContent"
            android:layout_marginLeft = "1dip"
            android:layout_marginRight = "1dip"
            android:layout_marginBottom = "1dip"
            android:layout_width = "wrap_content"
            android:padding = "15dip"
            android:background = "#44000000" >
        <ImageView
            android:layout_height = "wrap_content"
            android:layout_gravity = "center"
            android:layout_width = "wrap_content"
            android:id = "@ + id/tvImageToast" />
        <TextView
            android:layout_height = "wrap_content"
            android:paddingRight = "10dip"
            android:paddingLeft = "10dip"
            android:layout_width = "wrap_content"
            android:gravity = "center"
            android:textColor = "#ff000000"
            android:id = "@ + id/tvTextToast" />
    </LinearLayout>
</LinearLayout>
```

（4）打开 src\fs.Toast_test 包下的 MainActivity.java 文件，在文件中实现当单击界面中的各个按钮时，即显示相应的 Toast"通知单"。代码为：

```java
package fs.toast_test;
import android.app.Activity;
import android.os.Bundle;
import android.os.Handler;
import android.view.Gravity;
import android.view.LayoutInflater;
import android.view.View;
import android.view.ViewGroup;
import android.view.View.OnClickListener;
import android.widget.ImageView;
import android.widget.LinearLayout;
import android.widget.TextView;
import android.widget.Toast;
public class MainActivity extends Activity implements OnClickListener {
    Handler handler = new Handler();
    @Override
    public void onCreate(Bundle savedInstanceState) {
        super.onCreate(savedInstanceState);
        setContentView(R.layout.main);
```

```java
        findViewById(R.id.btnSimpleToast).setOnClickListener(this); findViewById(R.id.
        btnSimpleToastWithCustomPosition).setOnClickListener(this);
        findViewById(R.id.btnSimpleToastWithImage).setOnClickListener(this);
        findViewById(R.id.btnCustomToast).setOnClickListener(this);
        findViewById(R.id.btnRunToastFromOtherThread).setOnClickListener(this);
    }
    public void showToast() {
        handler.post(new Runnable() {
            @Override
            public void run() {
                Toast.makeText(getApplicationContext(), "我来自其他线程!", Toast.LENGTH_
                SHORT).show();
            }
        });
    }
    @Override
    public void onClick(View v) {
        Toast toast = null;
        switch (v.getId()) {
        case R.id.btnSimpleToast:
            Toast.makeText(getApplicationContext(), "默认 Toast 样式", Toast.LENGTH_SHORT).
            show();
            break;
            case R.id.btnSimpleToastWithCustomPosition:
                toast = Toast.makeText(getApplicationContext(),"自定义位置 Toast", Toast.
                LENGTH_LONG);
                toast.setGravity(Gravity.CENTER, 0, 0);
                toast.show();
                break;
            case R.id.btnSimpleToastWithImage:
                toast = Toast.makeText(getApplicationContext(), "带图片的 Toast", Toast.
                LENGTH_LONG);
                toast.setGravity(Gravity.CENTER, 0, 0);
                LinearLayout toastView = (LinearLayout) toast.getView();
                ImageView imageCodeProject = new ImageView(getApplicationContext());
                imageCodeProject.setImageResource(R.drawable.kt1);
                toastView.addView(imageCodeProject, 0);
                toast.show();
                break;
            case R.id.btnCustomToast:
                LayoutInflater inflater = getLayoutInflater();
                View layout = inflater.inflate(R.layout.custom, (ViewGroup) findViewById(R.id.llToast));
                ImageView image = (ImageView) layout.findViewById(R.id.tvImageToast);
                image.setImageResource(R.drawable.kt2);
                TextView title = (TextView) layout.findViewById(R.id.tvTitleToast);
                title.setText("Attention");
                TextView text = (TextView) layout.findViewById(R.id.tvTextToast);
                text.setText("完全自定义 Toast");
                toast = new Toast(getApplicationContext());
                toast.setGravity(Gravity.RIGHT | Gravity.TOP, 12, 40);
                toast.setDuration(Toast.LENGTH_LONG);
                toast.setView(layout);
                toast.show();
                break;
            case R.id.btnRunToastFromOtherThread:
                new Thread(new Runnable() {
                    public void run() {
                        showToast();
```

```
                })).start();
                break;
            }
        }
    }
```

运行程序,默认效果如图 3-18(a)所示,单击界面中的"默认"按钮,即显示对应的 Toast "通知单",如图 3-18(b)所示。单击界面中的"自定义显示位置"按钮,即显示对应的 Toast "通知单",如图 3-18(c)所示。单击界面中的"带图片"按钮,即显示对应的 Toast "通知单",如图 3-18(d)所示。单击界面中的"完全自定义"按钮,即显示对应的 Toast "通知单",如图 3-18(e)所示。单击界面中的"其他线程"按钮,即显示对应的 Toast "通知单",如图 3-18(f)所示。

(a) 默认界面

(b) 默认Toast样式通知

(c) 自定义位置Toast样式通知

(d) 带图片Toast样式通知

(e) 完全自定义Toast样式通知

(f) 其他线程Toast样式通知

图 3-18 5 种类型的"通知单"

3.2.2 手机消息提醒实例

本实例利用 Notification 在状态栏上显示通知,通过本实例演示了 Notification 控件的具体用法。

Android 也提供了用于处理显示通知栏中的信息类,即 Notification 和 NotificationManager。其中,Notification 代表具有全局效果的通知,而 NotificationManager 则是用于发送 Notification 通知的系统服务。

使用 Notification 和 NotificationManager 类发送和显示通知也比较简单,大致可分 4 个步骤实现。

- 调用 getSystemService()方法获取系统的 NotificationManager 服务。
- 创建一个 Notification 对象,并为其设置各种属性。
- 为 Notification 对象设置事件信息。
- 通过 NotificationManager 类的 notify()方法发送 Notification 通知。

本实例的具体实现步骤如下。

(1) 在 Eclipse 中创建一个 Android 应用项目,命名为 Notification_test。

(2) 打开 res\layout 目录下的 main.xml 主布局文件,在文件中声明一个 Button 控件及一个 TextView 控件。代码为:

```xml
<?xml version = "1.0" encoding = "utf-8"?>
<LinearLayout xmlns:android = "http://schemas.android.com/apk/res/android"
    android:orientation = "vertical"
    android:layout_width = "fill_parent"
    android:layout_height = "fill_parent"
    android:background = "#f0f0">
<TextView
    android:layout_width = "fill_parent"
    android:layout_height = "wrap_content"
    android:gravity = "center"
    android:textColor = "#EEE"
    android:textStyle = "bold"
    android:textSize = "25sp"
    android:text = "NotificationDemo 实例" />
<Button
    android:id = "@+id/btnSend"
    android:text = "send notification"
    android:layout_width = "wrap_content"
    android:layout_height = "wrap_content"
    android:layout_gravity = "center"/>
</LinearLayout>
```

(3) 在 res\layout 目录下新建一个 second.xml 次布局文件,在文件中声明一个 TextView 控件及一个 Button 控件。代码为:

```xml
<?xml version = "1.0" encoding = "utf-8"?>
<LinearLayout
    xmlns:android = "http://schemas.android.com/apk/res/android"
    android:orientation = "vertical"
    android:layout_width = "fill_parent"
    android:layout_height = "fill_parent"
```

```xml
            android:background="#ffffff00">
    <TextView
        android:layout_width="fill_parent"
        android:layout_height="wrap_content"
        android:gravity="center"
        android:textColor="#EEE"
        android:textStyle="bold"
        android:textSize="25sp"
        android:text="显示通知界面" />
    <Button
        android:id="@+id/btnCancel"
        android:text="cancel notification"
        android:layout_width="wrap_content"
        android:layout_height="wrap_content"
        android:layout_gravity="center" />
</LinearLayout>
```

(4) 打开 src\fs.notification_test 包下的 MainActivity.java 文件，在文件中第一次调用 Activity 活动，发送 Broadcast 广播。代码为：

```java
package fs.notification_test;
import android.app.Activity;
import android.content.Intent;
import android.os.Bundle;
import android.view.View;
import android.widget.Button;
public class MainActivity extends Activity {
    private Button btnSend;
    //定义 BroadcastReceiver 的 action
    private static final String NotificationDemo_Action = "com.andyidea.notification.NotificationDemo_Action";
    /** 第一次调用 Activity 活动 */
    @Override
    public void onCreate(Bundle savedInstanceState) {
        super.onCreate(savedInstanceState);
        setContentView(R.layout.main);
        btnSend = (Button)findViewById(R.id.btnSend);
        btnSend.setOnClickListener(new View.OnClickListener() {
            @Override
            public void onClick(View v) {
                Intent intent = new Intent();
                intent.setAction(NotificationDemo_Action);
                sendBroadcast(intent);
            }
        });
    }
}
```

(5) 在 src\fs.notification_test 包下新建一个 SecondActivity.java 文件，在文件中实现 Notification 通知发送及在状态栏上显示通知。代码为：

```java
package fs.notification_test;
import android.app.Activity;
import android.app.Notification;
import android.app.NotificationManager;
import android.app.PendingIntent;
```

```java
import android.content.Intent;
import android.os.Bundle;
import android.view.View;
import android.widget.Button;
public class SecondActivity extends Activity {
    private Button btnCancel;
    //声明 Notification
    private Notification notification;
    //声明 NotificationManager
    private NotificationManager mNotification;
    //标识 Notification 的 ID
    private static final int ID = 1;
    @Override
    protected void onCreate(Bundle savedInstanceState) {
        super.onCreate(savedInstanceState);
        setContentView(R.layout.second);
        btnCancel = (Button)findViewById(R.id.btnCancel);
        //获得 NotificationManager 的实例
        String service = NOTIFICATION_SERVICE;
        mNotification = (NotificationManager)getSystemService(service);
        //获得 Notification 的实例
        notification = new Notification();
        //设置该图标会在状态栏显示
        int icon = notification.icon = android.R.drawable.stat_sys_phone_call;
        //设置提示信息
        String tickerText = "Test Notification";
        //设置显示时间
        long when = System.currentTimeMillis();
        notification.icon = icon;
        notification.tickerText = tickerText;
        notification.when = when;
        Intent intent = new Intent(this, MainActivity.class);
        PendingIntent pi = PendingIntent.getActivity(this, 0, intent, 0);
        notification.setLatestEventInfo(this, "消息", "SMS Android", pi);
        mNotification.notify(ID, notification);
        btnCancel.setOnClickListener(new View.OnClickListener() {
            @Override
            public void onClick(View v) {
                mNotification.cancel(ID);           //取消通知
            }
        });
    }
}
```

(6) 在 src\fs.notification_test 包下新建一个 NotificationReceiver.java 文件,在该文件中实现显示通知。代码为:

```java
package fs.notification_test;
import android.content.BroadcastReceiver;
import android.content.Context;
import android.content.Intent;
public class NotificationReceiver extends BroadcastReceiver {
    @Override
    public void onReceive(Context context, Intent intent) {
        //实例化 Intent
        Intent i = new Intent();
```

```
            //在新任务中启动 Activity
            i.setFlags(Intent.FLAG_ACTIVITY_NEW_TASK);
            //设置 Intent 启动的组件名称
            i.setClass(context, SecondActivity.class);
            //启动 Activity,显示通知
            context.startActivity(i);
        }
    }
```

运行程序,效果如图 3-19 所示。

图 3-19　Notification 通知

3.3　友好界面实例

　　控件 Menu 的功能是为用户提供一个友好的界面显示效果。菜单在 Android 中的使用非常重要,几乎所有重要的选项和设置都在菜单中进行处理,所以掌握菜单的使用是开发操作方便软件的基础。本节主要介绍 Android 中几种常用的菜单。

3.3.1　选项菜单实例

　　本实例在 Menu 对象中添加菜单项 MenuItem 和子菜单 SubMenu 实现创建多个选项菜单。通过本实例演示了在 Android 中创建选项菜单的具体方法。
　　本实例的具体实现步骤如下。
　　(1) 在 Eclipse 中创建一个 Android 应用项目,命名为 Options_menu。
　　(2) 打开 res\layout 目录下的 main.xml 布局文件,文件代码为:

```
<RelativeLayout xmlns:android = "http://schemas.android.com/apk/res/android"
    xmlns:tools = "http://schemas.android.com/tools"
    android:layout_width = "match_parent"
    android:layout_height = "match_parent"
    android:paddingBottom = "@dimen/activity_vertical_margin"
```

```xml
        android:paddingLeft = "@dimen/activity_horizontal_margin"
        android:paddingRight = "@dimen/activity_horizontal_margin"
        android:paddingTop = "@dimen/activity_vertical_margin"
        tools:context = ".MainActivity"
        android:background = "#77ffff00" >
        <TextView
            android:layout_width = "wrap_content"
            android:layout_height = "wrap_content"
            android:text = "@string/hello_world" />
</RelativeLayout>
```

（3）打开 src\fs.options_menu 包下的 MainActivity.java 文件，在文件中实现选项菜单的创建。代码为：

```java
package fs.options_menu;
import android.app.Activity;
import android.os.Bundle;
import android.util.Log;
import android.view.Menu;
import android.view.MenuItem;
public class MainActivity extends Activity {
    private final static int MyMENU1_ID = 1;
    private final static int MyMENU2_ID = 2;
    private final static int MyMENU3_ID = 3;
    private final static int MyMENU4_ID = 4;
    private final static int MyMENU5_ID = 5;
    private final static int MyMENU6_ID = 6;
    private final static int MyMENU7_ID = 7;
    @Override
    public void onCreate(Bundle savedInstanceState) {
        super.onCreate(savedInstanceState);
        setContentView(R.layout.main);
    }
    @Override
    public boolean onCreateOptionsMenu(Menu menu) {
        Log.v("onCreateOptionsMenu", "执行了 onCreateOptionsMenu");
        menu.add(0, MyMENU1_ID, 1, "菜单 1").setIcon(R.drawable.ic_launcher);
        menu.add(0, MyMENU2_ID, 2, "菜单 2").setIcon(R.drawable.ic_launcher);
        menu.add(0, MyMENU3_ID, 3, "菜单 3").setIcon(R.drawable.ic_launcher);
        menu.add(0, MyMENU4_ID, 4, "菜单 4").setIcon(R.drawable.ic_launcher);
        menu.add(0, MyMENU5_ID, 5, "菜单 5").setIcon(R.drawable.ic_launcher);
        menu.add(0, MyMENU6_ID, 6, "菜单 6").setIcon(R.drawable.ic_launcher);
        menu.add(0, MyMENU7_ID, 7, "菜单 7").setIcon(R.drawable.ic_launcher);
        return super.onCreateOptionsMenu(menu);
    }
    @Override
    public boolean onPrepareOptionsMenu(Menu menu) {
        Log.v("onPrepareOptionsMenu", "执行了 onPrepareOptionsMenu");
        return super.onPrepareOptionsMenu(menu);
    }
    @Override
    public boolean onOptionsItemSelected(MenuItem item) {
        switch (item.getItemId()) {
        case MyMENU1_ID:
            Log.v("菜单事件", "单击菜单 1");
            break;
```

```
                case MyMENU2_ID:
                    Log.v("菜单事件","单击菜单 2");
                    break;
                case MyMENU3_ID:
                    Log.v("菜单事件","单击菜单 3");
                    break;
                case MyMENU4_ID:
                    Log.v("菜单事件","单击菜单 4");
                    break;
                case MyMENU5_ID:
                    Log.v("菜单事件","单击菜单 5");
                    break;
                case MyMENU6_ID:
                    Log.v("菜单事件","单击菜单 6");
                    break;
                case MyMENU7_ID:
                    Log.v("菜单事件","单击菜单 7");
                    break;
        }
        return super.onOptionsItemSelected(item);
    }
}
```

运行程序,默认效果如图 3-20(a)所示,单击模拟中的"▇"按钮弹出菜单,效果如图 3-20(b)所示,并且在 LogCat 中看到打印出来的日志顺序如下。

```
11-04 16:10:04.730: V/onCreateOptionsMenu(1308): 执行了 onCreateOptionsMenu
11-04 16:10:04.880: V/onPrepareOptionsMenu(1308): 执行了 onPrepareOptionsMenu
11-04 16:14:13.560: V/菜单事件(1308): 单击菜单 5
11-04 16:14:14.220: I/Choreographer(361): Skipped 38 frames! The application may be doing too much work on its main thread.
```

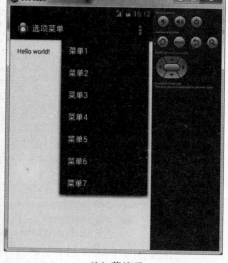

(a) 默认界面　　　　　　　　(b) 菜单项

图 3-20　创建选项菜单

从上面例子可知,在 Android 中通过回调方法来创建菜单并处理菜单项按下的事件,这些回调方法主要如下。

(1) onCreateOptionsMenu(Menu menu)：初始化选项菜单，该方法只在第一次显示菜单时调用，如果需要每次显示菜单时更新菜单项，则需要重写 onPrepareOptionsMenu(Menu)方法。

(2) public boolean onOptionsItemSelected(MenuItem item)：当选项菜单中某个选项被选中时调用该方法，默认是一个返回 false 的空实现。

(3) public void onOptionsMenu(Menu menu)：当选项菜单关闭时(或由于用户按下了返回键或是选择了某个菜单项)调用该方法。

(4) public boolean onPrepareOptionsMenu(Menu menu)：为程序准备选项菜单，每次选项菜单显示前会调用该方法。可以通过该方法设置某些菜单项可用或不可用或修改菜单项的内容。重写该方法时需要返回 true，否则选项菜单将不会显示。

Menu 中常用的方法主要如下。

- add：向 Menu 添加一个菜单项，返回 MenuItem 对象。
- addSubMenu：向 Menu 添加一个子菜单，返回 SubMenu 对象。
- clear：移除菜单中所有的子项。
- close：如果菜单正在显示，则关闭菜单。
- findItem：返回指定 id 的 MenuItem 对象。
- removeGroup：如果指定 id 的组不为空，则从菜单中移除该组。
- removeItem：移除指定 id 的 MenuItem。
- size：返回 Menu 中菜单项的个数。

在 MenuItem 中常用的成员方法主要如下。

- setAlphabeticShortcut(char alphaChar)：设置 MenuItem 的字母快捷键。
- MenuItem setNumericShortcut (char numericChar)：设置 MenuItem 的数字快捷键。
- MenuItem setIcon(Drawable icon)：设置 MenuItem 的图标。
- MenuItem setIntent(Intent intent)：为 MenuItem 绑定 Intent 对象，当被选中时，将会调用 startActivity 方法处理相应的 Intent。
- setOnMenuItemClickListener(MenuItem. OnMenuItemClickListener menuItemClickListener)：为 MenuItem 设置自定义的监听器，一般情况下，使用回调方法 onOptionsItemSelected 会更有效率。
- setShortcut(char numericChar, char alphaChar)：为 MenuItem 设置数字快捷键和字母快捷键，当按下快捷键或按住 Alt 的同时按下快捷键时将会触发 MenuItem 的选中事件。
- setTitle：为 MenuItem 设置标题。
- setTitleCondensed(CharSequence title)：设置 MenuItem 的缩略标题，当 MenuItem 不能显示全部的标题时，将显示缩略标题。

SubMenu 继承自 Menu，每个 SubMenu 实例代表一个子菜单。SubMenu 中常用的方法主要如下。

- setHeaderIcon：设置子菜单的标题图标。
- setHeaderTitle：设置子菜单的标题。
- setIcon：设置子菜单在父菜单中显示的图标。
- setHeaderView：设置指定的 View 对象为子菜单图标。

3.3.2 在菜单中添加单、多选功能实例

本实例用于实现在菜单的子菜单中添加多选菜单和单选功能。通过本实例进一步加深对菜单项用法的认识。

本实例的具体实现步骤如下。

(1) 在 Eclipse 中创建一个 Android 应用项目，命名为 Radio_check_menu。

(2) 打开 res\layout 目录下的 main.xml 布局文件，在文件中声明一个 ScrollView 控件和一个 EditText 控件。代码为：

```xml
<?xml version = "1.0" encoding = "utf-8"?>
<LinearLayout android:id = "@+id/LinearLayout01"
    android:layout_width = "fill_parent"
    android:layout_height = "fill_parent"
    android:orientation = "vertical"
    xmlns:android = "http://schemas.android.com/apk/res/android"
    android:background = "#770000ff">
    <ScrollView
        android:id = "@+id/scrollView"
        android:layout_width = "fill_parent"
        android:layout_height = "fill_parent">
        <EditText
            android:id = "@+id/editText"
            android:layout_width = "fill_parent"
            android:layout_height = "fill_parent"
            android:editable = "false"
            android:cursorVisible = "false"
            android:text = "您的选择为\n">
        </EditText>
    </ScrollView>
</LinearLayout>
```

(3) 打开 src\fs.radio_check_menu 包下的 MainActivity.java 文件，在文件中实现在子菜单中添加多选和单选功能，并将选择的结果显示在 EditText 框中。代码为：

```java
package fs.radio_check_menu;
import android.app.Activity;
import android.os.Bundle;
import android.view.Menu;
import android.view.MenuItem;
import android.view.SubMenu;
import android.view.MenuItem.OnMenuItemClickListener;
import android.widget.EditText;
public class MainActivity extends Activity {
    private final int MENU_GENDER_MALE = 0;
    private final int MENU_GENDER_FEMALE = 1;
    private final int MENU_HOBBY1 = 2;
    private final int MENU_HOBBY2 = 3;
    private final int MENU_HOBBY3 = 4;
    private final int MENU_OK = 5;
```

```java
        private final int MENU_GENDER = 6;
        private final int MENU_HOBBY = 7;
        private final int GENDER_GROUP = 0;
        private final int HOBBY_GROUP = 1;
        private final int MAIN_GROUP = 2;
        MenuItem[] hoddyMenuItems = new MenuItem[3];       //爱好菜单项组
        MenuItem maleMenuItem = null;                       //男性菜单项
        @Override
        public void onCreate(Bundle savedInstanceState) {
            super.onCreate(savedInstanceState);
            setContentView(R.layout.main);
        }
        /**
         * 初始化选项菜单
         */
        @Override
        public boolean onCreateOptionsMenu(Menu menu) {
            //单选菜单选项
            SubMenu genderMenu = menu.addSubMenu(MAIN_GROUP, MENU_GENDER, 0, "性别");
            genderMenu.setIcon(R.drawable.b9);
            genderMenu.setHeaderIcon(R.drawable.b9);
            maleMenuItem = genderMenu.add(GENDER_GROUP, MENU_GENDER_MALE, 0, "男");
            maleMenuItem.setChecked(true);
            genderMenu.add(GENDER_GROUP, MENU_GENDER_FEMALE, 0, "女");
            //设置菜单项为单选菜单项,互斥的
            genderMenu.setGroupCheckable(GENDER_GROUP, true, true);
            //复选菜单选项
            SubMenu hobbyMenu = menu.addSubMenu(MAIN_GROUP,MENU_HOBBY,0,"爱好");
            hobbyMenu.setIcon(R.drawable.fb1);
            hobbyMenu.setHeaderIcon(R.drawable.fb1);
hoddyMenuItems[0] = hobbyMenu.add(HOBBY_GROUP, MENU_HOBBY1, 0, "游泳");
hoddyMenuItems[1] = hobbyMenu.add(HOBBY_GROUP, MENU_HOBBY2, 0, "唱歌");
hoddyMenuItems[2] = hobbyMenu.add(HOBBY_GROUP, MENU_HOBBY3, 0, "编程");
            //设置菜单项为复选菜单项
            hoddyMenuItems[0].setCheckable(true);
            hoddyMenuItems[1].setCheckable(true);
            hoddyMenuItems[2].setCheckable(true);
            //确定菜单项
            MenuItem ok = menu.add(MAIN_GROUP,MENU_OK,0,"确定");
            ok.setOnMenuItemClickListener(new OnMenuItemClickListener(){
                public boolean onMenuItemClick(MenuItem item) {
                    appendStateStr();
                    return true;
                }
            });
            //给确定菜单项添加快捷键
            ok.setAlphabeticShortcut('o');              //设置字符快捷键
            //ok.setNumericShortcut('1');               //设置数字快捷键
            //ok.setShortcut('a', '2');
                        //同时设置两种快捷键。注意:同时设置多次时只有最后一个设置起作用
```

```java
        return true;                                    //记得返回true,否则无效
    }
    /**
     * 在菜单选项中某个选项被选中时调用该事件
     */
    @Override
    public boolean onOptionsItemSelected(MenuItem item) {
        switch (item.getItemId()) {
        case MENU_GENDER_MALE:
         case MENU_GENDER_FEMALE:
            item.setChecked(true);
            appendStateStr();
            break;
        case MENU_HOBBY1:
        case MENU_HOBBY2:
        case MENU_HOBBY3:
            item.setChecked(!item.isChecked());
            appendStateStr();
            break;
        }
        return true;
    }
    public void appendStateStr() {
        String result = "您选择的性别为:";
        if (maleMenuItem.isChecked()) {
            result = result + "男";
        } else {
            result = result + "女";
        }
        String hobbyStr = "";
        for (MenuItem hoddy : hoddyMenuItems) {
            if (hoddy.isChecked()) {
                hobbyStr = hobbyStr + hoddy.getTitle() + "、";
            }
        }
        if (hobbyStr.length() > 0) {
            result = result + ",您的爱好为:"
                    + hobbyStr.substring(0, hobbyStr.length() - 1) + ".\n";
        } else {
            result = result + ".\n";
        }
        EditText et = (EditText) MainActivity.this.findViewById(R.id.editText);
        et.append(result);
    }
}
```

运行程序,默认效果如图3-21(a)所示,单击界面中右侧的"MENU"按钮时,效果如图3-21(b)所示,单击界面中右上角的"■"按钮,效果如图3-21(c)所示,当选择"性别"项时,即弹出单选项,效果如图3-21(d)所示,当选择"爱好"项时,即弹出单选项,效果如图3-21(e)所示,效果如图3-21(f)为所选的显示结果。

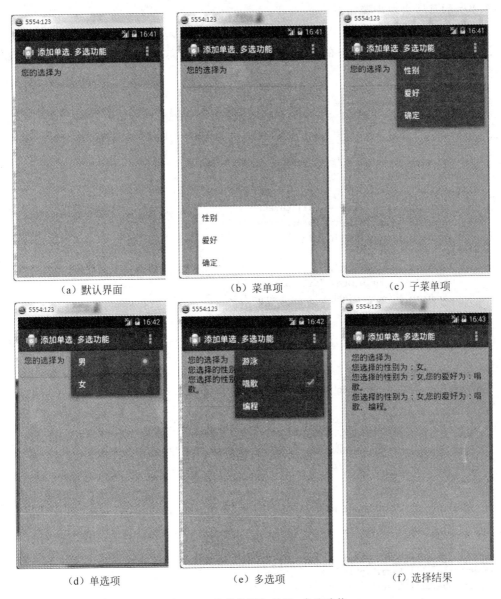

(a)默认界面　　　　　(b)菜单项　　　　　(c)子菜单项

(d)单选项　　　　　(e)多选项　　　　　(f)选择结果

图 3-21　为菜单添加单选、多选功能

3.3.3　添加常用操作实例

上下文与选项菜单不同,选项菜单是为整个 Activity 服务的,而上下文菜单是注册到某个 View 的。

Android 系统中的 ContextMenu(上下文菜单)类似于 PC 中的右键弹出菜单,当一个视图注册到一个上下文菜单时,执行一个在该对象上的"长按"动作,将出现一个提供相关功能的浮动菜单。上下文菜单可以被注册到任何视图对象中,不过,最常见的是用于列表视图 ListView 的 item,在按中列表项时,会转换其背景色而提示将呈现上下文菜单。

本实例通过 ContextMenu 控件,添加常用的上下文快捷菜单操作。通过本实例来展示下 ContextMenu 的基本使用。

本实例的具体实现步骤如下。

(1) 在 Eclipse 中创建一个 Android 应用项目,命名为 ContextMenu_test。

(2) 打开 res\layout 目录下的 main.xml 布局文件,在文件中声明一个 ListView 控件及一个 TextView 控件。代码为:

```xml
<?xml version = "1.0" encoding = "utf - 8"?>
<LinearLayout xmlns:android = "http://schemas.android.com/apk/res/android"
    android:orientation = "vertical"
    android:layout_width = "fill_parent"
    android:layout_height = "fill_parent"
    android:background = "#f0f">
    <TextView
        android:layout_width = "fill_parent"
        android:layout_height = "wrap_content"
        android:text = "添加常用操作"/>
    <ListView
        android:id = "@ + id/lv"
        android:layout_width = "fill_parent"
        android:layout_height = "wrap_content"/>
</LinearLayout>
```

(3) 在 res\menu 目录下新建一个 menun.xml 文件,用于实现添加的菜单选项操作。代码为:

```xml
<?xml version = "1.0" encoding = "utf - 8"?>
<menu
    xmlns:android = "http://schemas.android.com/apk/res/android">
    <item android:id = "@ + id/add" android:title = "增加"/>
    <item android:id = "@ + id/update" android:title = "更新"/>
    <item android:id = "@ + id/delete" android:title = "删除"/>
</menu>
```

(4) 打开 src\fs.contextmenu_test 包下的 MainActivity.java 文件,在文件中实现添加上下文菜单的常用操作。代码为:

```java
package fs.contextmenu_test;
import java.util.ArrayList;
import java.util.List;
import android.app.Activity;
import android.os.Bundle;
import android.view.ContextMenu;
import android.view.ContextMenu.ContextMenuInfo;
import android.view.Menu;
import android.view.MenuInflater;
import android.view.MenuItem;
import android.view.View;
import android.widget.ArrayAdapter;
import android.widget.ListView;
import android.widget.Toast;
public class MainActivity extends Activity {
    ListView lv;
    private ArrayAdapter<String> adapter;
    private List<String> alist = new ArrayList<String>();
    /** 第一次调用 Activity 活动 */
```

```java
    @Override
    public void onCreate(Bundle savedInstanceState) {
        super.onCreate(savedInstanceState);
        setContentView(R.layout.main);
        lv = (ListView)findViewById(R.id.lv);
        alist.add("操作一");
        alist.add("操作二");
        alist.add("操作三");
        adapter = new ArrayAdapter<String>(this,android.R.layout.simple_expandable_list_item_1, alist);
        lv.setAdapter(adapter);
        //注册视图对象,即为 ListView 控件注册上下文菜单
        registerForContextMenu(lv);
    }
    /**
     * 创建上下文菜单选项
     */
    @Override
    public void onCreateContextMenu(ContextMenu menu, View v,
            ContextMenuInfo menuInfo) {
        //1.通过手动添加来配置上下文菜单选项
        //menu.add(0, 1, 0, "修改");
        //menu.add(0, 2, 0, "删除");
        //2.通过 xml 文件来配置上下文菜单选项
        MenuInflater mInflater = getMenuInflater();
        mInflater.inflate(R.menu.menun, menu);
        super.onCreateContextMenu(menu, v, menuInfo);
    }
    /**
     * 当菜单某个选项被单击时调用该方法
     */
    @Override
    public boolean onContextItemSelected(MenuItem item) {
        switch(item.getItemId()){
        case 1:
            Toast.makeText(this, "你选择了手动修改", Toast.LENGTH_SHORT).show();
            break;
        case 2:
            Toast.makeText(this, "你选择了手动删除", Toast.LENGTH_SHORT).show();
            break;
        case R.id.add:
            Toast.makeText(this, "你选择了 XML 增加", Toast.LENGTH_SHORT).show();
            break;
        case R.id.update:
            Toast.makeText(this, "你选择了 XML 更新", Toast.LENGTH_SHORT).show();
            break;
        case R.id.delete:
            Toast.makeText(this, "你选择了 XML 删除", Toast.LENGTH_SHORT).show();
            break;
        }
        return super.onContextItemSelected(item);
    }
    /**
     * 当上下文菜单关闭时调用的方法
     */
```

```
            @Override
            public void onContextMenuClosed(Menu menu) {
                //TODO 自动存根法
                super.onContextMenuClosed(menu);
            }
        }
```

运行程序，默认的界面如图3-22（a）所示，长按ListView控件时，将弹出添加的上下文菜单，效果如图3-22（b）所示。

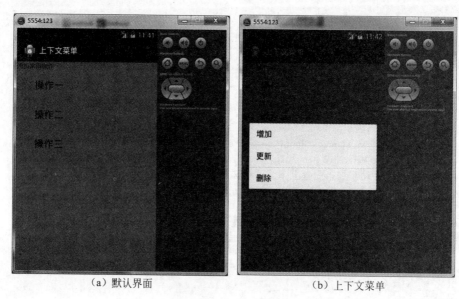

（a）默认界面　　　　　　　　　　（b）上下文菜单

图 3-22　添加常用的菜单操作

从上实例可看出，使用上下文菜单时常用到 Activity 类的成员方法，主要如下。
- registerForContextMenu(View view)：为某个 View 注册菜单。
- onCreateContextMenu(ContextMenu menu, View v, ContextMenuInfo menuInfo)：创建 ContextMenu，会在 menu 第一次显示时调用。
- onContextItemSelected(MenuItem item)：菜单项被选中后处理选中的菜单项。
- onContextMenuClosed(Menu menu)：菜单被关闭的事件。
- openContextMenu(View view)：调用打开菜单。
- closeContextMenu()：调用关闭菜单。

3.4　温馨消息对话框实例

在 Android 开发中，经常会需要在 Android 界面上弹出一些对话框，例如询问用户或者让用户选择。这些功能人们称呼它为 Android Dialog 对话框。

在 Android 中的对话框主要有普通对话框、选项对话框、单选及多选对话框、进度条对话框、日期与时间对话框等，下面分别给予实例介绍。

3.4.1 单击弹出一个对话框实例

对话框实例是通过在界面中添加一个 Button 按钮,并且为该"对话框"按钮添加监听器,当单击"对话框"按钮后弹出对话框。

对话框是 Activity 运行时显示的小窗口,当显示对话框时,当前 Activity 失去焦点而由对话框负责所有的人机交互。

本实例的具体实现步骤如下。

(1) 在 Eclipse 中创建一个 Android 应用项目,命名为 OKDialog_test。

(2) 打开 res\layout 目录下的 main.xml 布局文件,在文件中声明一个 Button 控件。代码为:

```xml
<?xml version = "1.0" encoding = "utf-8"?>
<LinearLayout xmlns:android = "http://schemas.android.com/apk/res/android"
    android:orientation = "vertical"
    android:layout_width = "fill_parent"
    android:layout_height = "fill_parent"
    android:background = "#aabbcc">
<Button
    android:id = "@ + id/Button01"
    android:textSize = "25dip"
    android:layout_width = "wrap_content"
    andro8id:layout_height = "wrap_content"
    android:text = "对话框"/>
</LinearLayout>
```

(3) 打开 src\fs.okdialog_test 包下的 MainActivity.java 文件,在文件中实现当单击界面中的"对话框"按钮时,即弹出一个对话框。代码为:

```java
package fs.okdialog_test;
import android.app.Activity;
import android.app.AlertDialog;
import android.app.Dialog;
import android.app.AlertDialog.Builder;
import android.os.Bundle;
import android.view.View;
import android.view.View.OnClickListener;
import android.widget.Button;
public class MainActivity extends Activity
{
    final int List_DIALOG_MULTIPLE = 0;
    @Override
    public void onCreate(Bundle savedInstanceState)
    {
        super.onCreate(savedInstanceState);
        setContentView(R.layout.main);
        Button bb = (Button)this.findViewById(R.id.Button01);
        bb.setOnClickListener
        (
            //为确定按钮添加监听
            new OnClickListener()
            {
                public void onClick(View v)
```

```
                    {
                        //打开对话框
                        showDialog(List_DIALOG_MULTIPLE);
                    }
                }
            );
        }
        @Override
        public Dialog onCreateDialog(int id)
        {
            Dialog dialog = null;

            switch(id)
            {
                case List_DIALOG_MULTIPLE:              //生成复选列表对话框的代码
                    Builder b = new AlertDialog.Builder(this);
                    b.setIcon(R.drawable.ic_launcher);  //设置图标
                    b.setTitle("对话框");                //设置标题
                    dialog = b.create();
                    break;
            }
            return dialog;
        }
}
```

运行程序,效果如图 3-23(a)所示,当单击界面中的"对话框"按钮时,效果如图 3-23(b)所示。

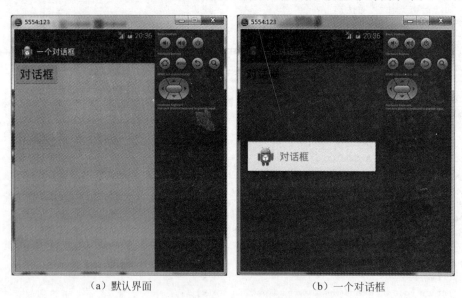

(a) 默认界面 　　　　　　　　　　(b) 一个对话框

图 3-23　单击弹出一个对话框

3.4.2　Android 9 种对话框实例

除了 3.4.1 小节介绍怎样创建一个对话框外,在 Android 中还提供了相应的类用于创建各种类型的对话框。Android 中主要支持以下 9 种对话框,下面通过实例来说明。

本实例通过利用 Android 提供的各类,创建各种类型的对话框,通过本实例主要演示了各种对话框的创建。其具体实现步骤如下。

(1) 在 Eclipse 中创建一个 Android 应用项目,命名为 AllDialog_test。

(2) 打开 res\layout 目录下的 main.xml 布局文件,在文件中定义 9 个 Button 控件及一个 TextView 控件。代码为:

```xml
<?xml version = "1.0" encoding = "utf-8"?>
<LinearLayout xmlns:android = "http://schemas.android.com/apk/res/android"
    android:layout_width = "fill_parent"
    android:layout_height = "fill_parent"
    android:orientation = "vertical"
    android:background = "#77ffff00" >
    <TextView
        android:layout_width = "fill_parent"
        android:layout_height = "wrap_content"
        android:text = "各种 Dialog 合集" />
    <Button
        android:layout_width = "fill_parent"
        android:layout_height = "wrap_content"
        android:text = "普通 Dialog"
        android:id = "@+id/btn_diaNormal"/>
    <Button
        android:layout_width = "fill_parent"
        android:layout_height = "wrap_content"
        android:text = "多按钮 Dialog"
        android:id = "@+id/btn_diaMulti"/>
    <Button
        android:layout_width = "fill_parent"
        android:layout_height = "wrap_content"
        android:text = "列表 Dialog"
        android:id = "@+id/btn_diaList"/>
    <Button
        android:layout_width = "fill_parent"
        android:layout_height = "wrap_content"
        android:text = "单项选择 Dialog"
        android:id = "@+id/btn_diaSigChos"/>
    <Button
        android:layout_width = "fill_parent"
        android:layout_height = "wrap_content"
        android:text = "多项选择 Dialog"
        android:id = "@+id/btn_diaMultiChos"/>
    <Button
        android:layout_width = "fill_parent"
        android:layout_height = "wrap_content"
        android:text = "进度条 Dialog"
        android:id = "@+id/btn_diaReadProcess"/>
    <Button
        android:layout_width = "fill_parent"
        android:layout_height = "wrap_content"
        android:text = "读取中 Dialog"
        android:id = "@+id/btn_diaProcess"/>
    <Button
        android:layout_width = "fill_parent"
        android:layout_height = "wrap_content"
        android:text = "自定义 Dialog"
```

```xml
        android:id = "@ + id/btn_diaCustom"/>
    <Button
        android:layout_width = "fill_parent"
        android:layout_height = "wrap_content"
        android:text = "PopUpWindow实现的dialog"
        android:id = "@ + id/btn_popUpDia"/>
</LinearLayout>
```

(3) 在 res\layout 目录下新建一个 cutom_main.xml 布局文件，在文件中声明一个 TextView 控件。代码为：

```xml
<?xml version = "1.0" encoding = "UTF-8"?>
<LinearLayout xmlns:android = "http://schemas.android.com/apk/res/android"
    android:orientation = "horizontal"
    android:layout_width = "fill_parent"
    android:layout_height = "fill_parent">
<!-- 定义一个 TextView,用于作为列表项的一部分。-->
<TextView android:id = "@ + id/name"
    android:layout_width = "wrap_content"
    android:layout_height = "wrap_content"
    android:textSize = "16dp"
    android:gravity = "center_vertical"
    android:paddingLeft = "10dp"
    android:textColor = "#ff000066"
    android:text = "我是一个自定义对话框"/>
</LinearLayout>
```

(4) 在 res\layout 目录下新建一个 popup.xml 布局文件，在文件声明一个 Button 控件。代码为：

```xml
<?xml version = "1.0" encoding = "utf-8"?>
<LinearLayout xmlns:android = "http://schemas.android.com/apk/res/android"
    android:orientation = "vertical"
    android:layout_width = "fill_parent"
    android:layout_height = "fill_parent"
    android:gravity = "center_horizontal">
<Button
    android:id = "@ + id/close"
    android:layout_width = "wrap_content"
    android:layout_height = "wrap_content"
    android:text = "关闭" />
</LinearLayout>
```

(5) 打开 src\fs.alldialog_test 包下的 MainActivity.java 文件，在文件中实现9种类型的对话框，当单击界面中的任何一个按钮时，即弹出对应类型的对话框，并显示相应的选择提示。代码为：

```java
package fs.alldialog_test;
import java.util.ArrayList;
import android.app.Activity;
import android.app.AlertDialog;
import android.app.ProgressDialog;
import android.content.DialogInterface;
import android.graphics.drawable.BitmapDrawable;
import android.os.Bundle;
```

```java
import android.view.Gravity;
import android.view.LayoutInflater;
import android.view.View;
import android.view.View.OnClickListener;
import android.view.ViewGroup.LayoutParams;
import android.view.WindowManager;
import android.widget.Button;
import android.widget.EditText;
import android.widget.PopupWindow;
import android.widget.Toast;
public class MainActivity extends Activity implements Runnable {
    //定义各种对话框变量
    private Button btn_diaNormal;
    private Button btn_diaMulti;
    private Button btn_diaList;
    private Button btn_diaSinChos;
    private Button btn_diaMultiChos;
    private Button btn_diaProcess;
    private Button btn_diaReadProcess;
    private Button btn_diaCustom;
    private Button btn_popUpDia;
    private PopupWindow window = null;
    private Button cusPopupBtn1;
    private View popupView;
    @Override
    public void onCreate(Bundle savedInstanceState)
    {
        super.onCreate(savedInstanceState);
        setContentView(R.layout.main);
        getView();
        setListener();
    }
    //获取视图
    private void getView()
    {
        btn_diaNormal = (Button)findViewById(R.id.btn_diaNormal);
        btn_diaMulti = (Button)findViewById(R.id.btn_diaMulti);
        btn_diaList = (Button)findViewById(R.id.btn_diaList);
        btn_diaSinChos = (Button)findViewById(R.id.btn_diaSigChos);
        btn_diaMultiChos = (Button)findViewById(R.id.btn_diaMultiChos);
        btn_diaProcess = (Button)findViewById(R.id.btn_diaProcess);
        btn_diaReadProcess = (Button)findViewById(R.id.btn_diaReadProcess);
        btn_diaCustom = (Button)findViewById(R.id.btn_diaCustom);
        btn_popUpDia = (Button)findViewById(R.id.btn_popUpDia);
    }
    //设置监听
    private void setListener()
    {
        btn_diaNormal.setOnClickListener(btnListener);
        btn_diaMulti.setOnClickListener(btnListener);
        btn_diaList.setOnClickListener(btnListener);
        btn_diaSinChos.setOnClickListener(btnListener);
        btn_diaMultiChos.setOnClickListener(btnListener);
        btn_diaProcess.setOnClickListener(btnListener);
        btn_diaReadProcess.setOnClickListener(btnListener);
```

```java
            btn_diaCustom.setOnClickListener(btnListener);
            btn_popUpDia.setOnClickListener(btnListener);
        }
        //设置按钮单击事件
        private Button.OnClickListener btnListener = new Button.OnClickListener()
        {
            public void onClick(View v)
            {
                if(v instanceof Button)
                {
                    int btnId = v.getId();
                    switch(btnId)
                    {
                        case R.id.btn_diaNormal:
                            showNormalDia();
                            break;
                        case R.id.btn_diaMulti:
                            showMultiDia();
                            break;
                        case R.id.btn_diaList:
                            showListDia();
                            break;
                        case R.id.btn_diaSigChos:
                            showSinChosDia();
                            break;
                        case R.id.btn_diaMultiChos:
                            showMultiChosDia();
                            break;
                        case R.id.btn_diaReadProcess:
                            showReadProcess();
                            break;
                        case R.id.btn_diaProcess:
                            showProcessDia();
                            break;
                        case R.id.btn_diaCustom:
                            showCustomDia();
                            break;
                        case R.id.btn_popUpDia:
                            showCusPopUp(v);
                            break;
                        default:
                            break;
                    }
                }
            }
        };
        /*普通的对话框*/
        private void showNormalDia()
        {
//AlertDialog.Builder normalDialog = new AlertDialog.Builder(getApplicationContext());
            AlertDialog.Builder normalDia = new AlertDialog.Builder(MainActivity.this);
            normalDia.setIcon(R.drawable.ic_launcher);
            normalDia.setTitle("普通的对话框");
            normalDia.setMessage("普通对话框的message内容");
            normalDia.setPositiveButton("确定", new DialogInterface.OnClickListener() {
```

```java
            @Override
            public void onClick(DialogInterface dialog, int which) {
                //TODO 自动存根法
                showClickMessage("确定");
            }
        });
        normalDia.setNegativeButton("取消", new DialogInterface.OnClickListener() {
            @Override
            public void onClick(DialogInterface dialog, int which) {
                //TODO 自动存根法
                showClickMessage("取消");
            }
        });
        normalDia.create().show();
    }
    /* 多按钮对话框 */
    private void showMultiDia()
    {
        AlertDialog.Builder multiDia = new AlertDialog.Builder(MainActivity.this);
        multiDia.setTitle("多选项对话框");
        multiDia.setPositiveButton("按钮一", new DialogInterface.OnClickListener() {
            @Override
            public void onClick(DialogInterface dialog, int which) {
                //TODO 自动存根法
                showClickMessage("按钮一");
            }
        });
        multiDia.setNeutralButton("按钮二", new DialogInterface.OnClickListener() {

            @Override
            public void onClick(DialogInterface dialog, int which) {
                //TODO 自动存根法
                showClickMessage("按钮二");
            }
        });
        multiDia.setNegativeButton("按钮三", new DialogInterface.OnClickListener() {
            @Override
            public void onClick(DialogInterface dialog, int which) {
                //TODO 自动存根法
                showClickMessage("按钮三");
            }
        });
        multiDia.create().show();
    }
    /* 列表对话框 */
    private void showListDia()
    {
final String[] mList = {"选项1","选项2","选项3","选项4","选项5","选项6","选项7"};
        AlertDialog.Builder listDia = new AlertDialog.Builder(MainActivity.this);
        listDia.setTitle("列表对话框");
        listDia.setItems(mList, new DialogInterface.OnClickListener() {
            @Override
            public void onClick(DialogInterface dialog, int which) {
                //TODO 自动存根法
                /* 下标是从0开始的 */
```

```java
                showClickMessage(mList[which]);
            }
        });
        listDia.create().show();
    }
    /*单项选择对话框*/
    int yourChose = -1;
    private void showSinChosDia()
    {
        final String[] mList = {"选项1","选项2","选项3","选项4","选项5","选项6","选项7"};
        yourChose = -1;
        AlertDialog.Builder sinChosDia = new AlertDialog.Builder(MainActivity.this);
        sinChosDia.setTitle("单项选择对话框");
        sinChosDia.setSingleChoiceItems(mList, 0, new DialogInterface.OnClickListener() {
            @Override
            public void onClick(DialogInterface dialog, int which) {
                //TODO 自动存根法
                yourChose = which;
            }
        });
        sinChosDia.setPositiveButton("确定", new DialogInterface.OnClickListener() {
            @Override
            public void onClick(DialogInterface dialog, int which) {
                //TODO 自动存根法
                if(yourChose!=-1)
                {
                    showClickMessage(mList[yourChose]);
                }
            }
        });
        sinChosDia.create().show();
    }
    ArrayList<Integer> myChose = new ArrayList<Integer>();
    private void showMultiChosDia()
    {
        final String[] mList = {"选项1","选项2","选项3","选项4","选项5","选项6","选项7"};
        final boolean mChoseSts[] = {false,false,false,false,false,false,false};
        myChose.clear();
        AlertDialog.Builder multiChosDia = new AlertDialog.Builder(MainActivity.this);
        multiChosDia.setTitle("多项选择对话框");
        multiChosDia.setMultiChoiceItems(mList, mChoseSts, new DialogInterface.OnMultiChoiceClickListener() {
            @Override
            public void onClick(DialogInterface dialog, int which, boolean isChecked) {
                //TODO 自动存根法
                if(isChecked)
                {
                    myChose.add(which);
                }
                else
                {
                    myChose.remove(which);
                }
            }
        });
```

```java
        multiChosDia.setPositiveButton("确定", new DialogInterface.OnClickListener() {
            @Override
            public void onClick(DialogInterface dialog, int which) {
                //TODO 自动存根法
                int size = myChose.size();
                String str = "";
                for(int i = 0;i < size;i++)
                {
                    str += mList[myChose.get(i)];
                }
                showClickMessage(str);
            }
        });
        multiChosDia.create().show();
}
//进度读取框需要模拟读取
ProgressDialog mReadProcessDia = null;
public final static int MAX_READPROCESS = 100;
private void showReadProcess()
{
    mReadProcessDia = new ProgressDialog(MainActivity.this);
    mReadProcessDia.setProgress(0);
    mReadProcessDia.setTitle("进度条窗口");
    mReadProcessDia.setProgressStyle(ProgressDialog.STYLE_HORIZONTAL);
    mReadProcessDia.setMax(MAX_READPROCESS);
    mReadProcessDia.show();
    new Thread(this).start();
}
//新开启一个线程,循环的累加,一直到100再停止
@Override
public void run()
{
    int Progress = 0;
    while(Progress < MAX_READPROCESS)
    {
        try {
            Thread.sleep(100);
            Progress++;
            mReadProcessDia.incrementProgressBy(1);
        } catch (InterruptedException e) {
            //TODO 自动存根法
            e.printStackTrace();
        }
    }
    //读取完了以后窗口自动消失
    mReadProcessDia.cancel();
}
/*读取中的对话框*/
private void showProcessDia()
{
    ProgressDialog processDia = new ProgressDialog(MainActivity.this);
    processDia.setTitle("进度条框");
    processDia.setMessage("内容读取中……");
    processDia.setIndeterminate(true);
    processDia.setCancelable(true);
```

```java
            processDia.show();
        }
        /* 自定义对话框 */
        private void showCustomDia()
        {
            AlertDialog.Builder customDia = new AlertDialog.Builder(MainActivity.this);
            final View viewDia = LayoutInflater.from(MainActivity.this).inflate(R.layout.popup,
                null);
            customDia.setTitle("自定义对话框");
            customDia.setView(viewDia);
            customDia.setPositiveButton("确定", new DialogInterface.OnClickListener() {

                @Override
                public void onClick(DialogInterface dialog, int which) {
                    //TODO 自动存根法
                    EditText diaInput = (EditText) viewDia.findViewById(R.id.close);
                    showClickMessage(diaInput.getText().toString());
                }
            });
            customDia.create().show();
        }
        /* popup window 来实现 */
        private void showCusPopUp(View parent)
        {
            if(window == null) {
                popupView = LayoutInflater.from(MainActivity.this).inflate(R.layout.cutom_main, null);
                cusPopupBtn1 = (Button)popupView.findViewById(R.id.name);
                window = new PopupWindow(popupView,LayoutParams.FILL_PARENT,LayoutParams.FILL_
                    PARENT);
            }
            /**
             * 必须调用 setBackgroundDrawable, 因为popupwindow 在初始时, 会检测 background
             * 是否为 null,如果是, onTouch or onKey events 就不会相应,所以必须设置 background */
            window.setFocusable(true);
            window.setBackgroundDrawable(new BitmapDrawable());
            window.update();
            window.showAtLocation(parent, Gravity.CENTER_VERTICAL, 0, 0);
            cusPopupBtn1.setOnClickListener(new OnClickListener() {
                @Override
                public void onClick(View v) {
                    //TODO 自动存根法
                    showClickMessage("popup window 的确定");
                }
            });
        }
        /* 显示单击的内容 */
        private void showClickMessage(String message)
        {
            Toast.makeText(MainActivity.this, "你选择的是: " + message, Toast.LENGTH_SHORT).show();
        }
    }
```

运行程序,默认效果如图 3-24(a)所示,单击界面中的"普通 Dialog"按钮时,效果如图 3-24(b)所示,单击界面中的"列表 Dialog"按钮时,效果如图 3-24(c)所示,单击界面中的"单项选择 Dialog"按钮时,效果如图 3-24(d),单击界面中的"多项选择 Dialog"按钮时,效果如

图 3-24(e)所示,单击界面中的"进度条 Dialog"按钮时,效果如图 3-24(f)所示。

(a) 默认界面　　　　　(b) 普通对话框　　　　　(c) 列表对话框

(d) 单选对话框　　　　(e) 多选对话框　　　　　(f) 进度条对话框

图 3-24　对话框类型

3.4.3　日期选择对话框实例

本实例主要利用 DatePickerDialog 控件创建一个日期选择对话框实例。通过本实例演示了 DatePickerDialog 控件的具体用法。

本实例的具体实现步骤如下。

(1) 在 Eclipse 中创建一个 Android 应用项目,命名为 DatePickerDialog_test。

(2) 打开 res\layout 目录下的 main.xml 布局文件,在布局文件中声明一个 TextView 控件及一个 Button 控件。代码为:

```xml
<?xml version = "1.0" encoding = "utf-8"?>
<LinearLayout xmlns:android = "http://schemas.android.com/apk/res/android"
    android:layout_width = "match_parent"
    android:layout_height = "match_parent"
    android:orientation = "vertical"
    android:background = "#000000">
    <TextView
        android:layout_width = "wrap_content"
        android:layout_height = "wrap_content"
        android:id = "@+id/showtime"
        android:textColor = "#ff000000"
        android:text = ""/>
    <Button
        android:layout_width = "wrap_content"
        android:layout_height = "wrap_content"
        android:id = "@+id/setdate"
        android:text = "设置日期"/>
</LinearLayout>
```

（3）打开 src\fs.datepickerdialog_test 包下的 MainActivity.java 文件，在文件中利用 DatePickerDialog 实现当单击界面中的"设置日期"按钮时，即弹出日期选择对话框，即可在对话框选择对应的日期，即相应的结果显示在对话框上文。代码为：

```java
package fs.datepickerdialog_test;
import java.util.Calendar;
import java.util.Date;
import java.util.Locale;
import android.app.Activity;
import android.os.Bundle;
import android.widget.Button;
import android.widget.DatePicker;
import android.widget.TextView;
import android.view.View;
import android.view.View.OnClickListener;
import android.app.DatePickerDialog;
/**
 * DatePickerDialog 是设置日期对话框，通过 OnDateSetListener 监听并重新设置日期，
 * 当日期被重置后，会执行 OnDateSetListener 类中的方法 onDateSet()
 */
public class MainActivity extends Activity {
    private TextView showdate;
    private Button setdate;
    private int year;
    private int month;
    private int day;
    @Override
    public void onCreate(Bundle savedInstanceState)
    {
        super.onCreate(savedInstanceState);
        setContentView(R.layout.main);
        showdate = (TextView) this.findViewById(R.id.showtime);
        setdate = (Button) this.findViewById(R.id.setdate);
        //初始化 Calendar 日历对象
        Calendar mycalendar = Calendar.getInstance(Locale.CHINA);
        Date mydate = new Date();                              //获取当前日期 Date 对象
```

```java
        mycalendar.setTime(mydate);              //为Calendar对象设置时间为当前日期
        year = mycalendar.get(Calendar.YEAR);    //获取Calendar对象中的年
        month = mycalendar.get(Calendar.MONTH);  //获取Calendar对象中的月
        day = mycalendar.get(Calendar.DAY_OF_MONTH); //获取这个月的第几天
    showdate.setText("当前日期:" + year + " - " + (month + 1) + " - " + day); //显示当前的年月日
    //添加单击事件——设置日期
        setdate.setOnClickListener(new OnClickListener(){
            @Override
            public void onClick(View v)
            {
                /**
                 * 构造函数原型:
                 * public DatePickerDialog (Context context, DatePickerDialog.
                   OnDateSetListener callBack,
                 * int year, int monthOfYear, int dayOfMonth)
                 * content 组件运行 Activity,
                 * DatePickerDialog.OnDateSetListener: 选择日期事件
                 * year: 当前组件上显示的年, monthOfYear: 当前组件上显示的月, dayOfMonth: 当
                   前组件上显示的第几天
                 */
                //创建 DatePickerDialog 对象
                DatePickerDialog dpd = new DatePickerDialog(MainActivity.this, Datelistener,
                year, month, day);
                dpd.show();                      //显示 DatePickerDialog 组件
            }
        });
    }
    private DatePickerDialog.OnDateSetListener Datelistener = new DatePickerDialog.
    OnDateSetListener()
    {
        /** params: view: 该事件关联的组件
         * params: myyear: 当前选择的年
         * params: monthOfYear: 当前选择的月
         * params: dayOfMonth: 当前选择的日
         */
        @Override
    public void onDateSet(DatePicker view, int myyear, int monthOfYear, int dayOfMonth) {
            //修改 year 和 month 和 day 的变量值,以便以后单击按钮时,在 DatePickerDialog 上显
            //示上一次修改后的值
            year = myyear;
            month = monthOfYear;
            day = dayOfMonth;
            //更新日期
            updateDate();
        }
        //当 DatePickerDialog 关闭时,更新日期显示
        private void updateDate()
        {
            //在 TextView 上显示日期
            showdate.setText("当前日期: " + year + " - " + (month + 1) + " - " + day);
        }
    };
}
```

运行程序,默认界面如图 3-25(a)所示,单击界面中的"设置日期"按钮,即弹出日期选择对

话框,如图 3-25(b)所示,选择对应的日期并单击"确定"按钮,即将所选择的结果显示在文本框中,效果如图 3-25(c)所示。

(a)默认界面　　(b)日期选择对话框　　(c)显示选择结果

图 3-25　日期选择对话框

3.4.4　时间日期选择对话框实例

本实例主要结合 DatePickerDialog 与 TimePickerDialog 控件创建一个日期时间选择对话框实例。通过本实例演示了 DatePickerDialog 与 TimePickerDialog 控件的具体用法。

本实例的具体实现步骤如下。

(1) 在 Eclipse 中创建一个 Android 应用项目,命名为 TimePickerDialog_test。

(2) 打开 res\layout 目录下的 main.xml 布局文件,在文件中声明一个 EditView 控件及一个 Button 控件。代码为:

```xml
<?xml version = "1.0" encoding = "utf - 8"?>
<LinearLayout xmlns:android = "http://schemas.android.com/apk/res/android"
    android:orientation = "vertical"
    android:layout_width = "fill_parent"
    android:layout_height = "fill_parent"
    android:gravity = "center_horizontal"
    android:background = "#f00">
<EditText
    android:id = "@ + id/show"
    android:layout_width = "fill_parent"
    android:layout_height = "wrap_content"
    android:editable = "false"/>
<Button
    android:id = "@ + id/dateBn"
    android:layout_width = "wrap_content"
    android:layout_height = "wrap_content"
    android:text = "设置日期与时间"/>
</LinearLayout>
```

(3) 打开 src\fs.timepickerdialog_test 包下的 MainActivity.java 文件,在文件中利用 TimePickerDialog 控件及 DatePickerDialog 控件实现日期时间选择对话框。代码为:

```java
package fs.timepickerdialog_test;
import java.util.Calendar;
```

```java
import android.app.Activity;
import android.app.DatePickerDialog;
import android.app.Dialog;
import android.app.TimePickerDialog;
import android.os.Bundle;
import android.view.View;
import android.view.View.OnClickListener;
import android.widget.Button;
import android.widget.DatePicker;
import android.widget.EditText;
import android.widget.TimePicker;
public class MainActivity extends Activity {
    //用来连接日期和时间,最终用来显示的
    StringBuilder str = new StringBuilder("");
    @Override
    public void onCreate(Bundle savedInstanceState) {
        super.onCreate(savedInstanceState);
        setContentView(R.layout.main);
        Button dateBn = (Button) findViewById(R.id.dateBn);
        //为"设置日期"按钮绑定监听器
        dateBn.setOnClickListener(new OnClickListener() {
        @Override
        public void onClick(View source) {
            Calendar c = Calendar.getInstance();
            //直接创建一个 DatePickerDialog 对话框实例,并将它显示出来
            Dialog dateDialog = new DatePickerDialog(MainActivity.this,
            //绑定监听器
            new DatePickerDialog.OnDateSetListener() {
            @Override
            public void onDateSet(DatePicker dp, int year, int month, int dayOfMonth) {
                str.append(year + "-" + (month + 1) + "-" + dayOfMonth + " ");
                Calendar time = Calendar.getInstance();
                Dialog timeDialog = new TimePickerDialog(MainActivity.this,     //绑定监听器
                    new TimePickerDialog.OnTimeSetListener() {
                    @Override
                    public void onTimeSet(TimePicker tp, int hourOfDay, int minute) {
                        str.append(hourOfDay + ":" + minute);
                        EditText show = (EditText) findViewById(R.id.show);
                        show.setText(str);
                    }
                }
                //设置初始时间
                , time.get(Calendar.HOUR_OF_DAY), time.get(Calendar.MINUTE)
                //true 表示采用 24 小时制
                , true);
                timeDialog.setTitle("请选择时间");
                timeDialog.show();
            }
        }
        //设置初始日期
        , c.get(Calendar.YEAR), c.get(Calendar.MONTH), c.get(Calendar.DAY_OF_MONTH));
        dateDialog.setTitle("请选择日期");
        dateDialog.show();
        }
    });
```

            }
        }

运行程序，默认效果如图 3-26(a)所示，当单击界面中的"设置日期与时间"按钮时，即弹出日期选择对话框，效果如图 3-26(b)所示，设置好日期后单击对话框的"确定"按钮，即接着弹出时间选择对话框，效果如图 3-26(c)所示，设置好时间后单击对话框的"确定"按钮，即在界面中把设置好的日期与时间显示在编辑框中，效果如图 3-26(d)所示。

(a) 默认界面　　　　　　　　　　(b) 日期选择对话框

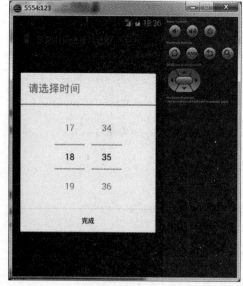 

(c) 时间选择对话框　　　　　　　　(d) 显示结果

图 3-26　日期时间选择对话框

第 4 章

Android 动态效果实例

Android 系统提供了 ImageView 显示普通静态图片，也提供了 AnimationDrawable 来开发逐帧动画，还可通过 Animation 对普通图片使用补间动画。图形和图像处理不仅对 Android 系统的应用界面非常重要，而且也是 Android 系统上益智类游戏和 2D 游戏大量需要的。所谓游戏，本质就是提供更逼真和能模拟某种环境的用户界面，并根据某种规则来响应用户操作。为了提供更逼真的用户界面，需要借助于图形处理。

而 Android 中的动画主要包括两大类：一类为帧动画；另一类为补间动画。

(1) 帧动画：通过若干帧图片的轮流显示来实现的。

(2) 补间动画：主要包括对位置、角度和尺寸等属性的变换。

4.1 基本二维图形实例

本实例主要通过 Canvas 及 Paint 类，在 Android 中绘制基本的二维图形。通过本实例主要细致和全面介绍了 Canvas 及 Paint 类的具体用法。

Paint 类主要用于设置绘制风格，包括画笔颜色、画笔笔触粗细和填充风格等。

本实例的具体实现步骤如下。

(1) 在 Eclipse 中创建一个 Android 应用项目，命名为 Paint_test。

(2) 打开 src\fs.paint_test 包下的 MainActivity.java 文件，在文件中实现利用 Paint 及 Canvas 类绘制不同风格的基本二维图形。代码为：

```
package fs.paint_test;
import android.app.Activity;
import android.os.Bundle;
import android.view.View;
import android.content.Context;
import android.graphics.Canvas;
import android.graphics.Color;
import android.graphics.Paint;
import android.graphics.RectF;
import android.graphics.Path;
import android.graphics.Shader;
import android.graphics.LinearGradient;
/**
 * 主程序中继承自 Android.view.View 的 MyView 类，重写 MyView 的 onDraw()方法，
```

```java
 * 一开始就会运行绘制的工作,在 onDraw()中以 Paint 将几何图形绘制在 Canvas 上,
 * 以 paint.setColor()改变图形颜色,以 paint.setStyle()的设置来控制画出的图形是空心
 * 还是实心.程序的最后一段,就是直接在 Canvas 写上文字了,随着 Paint 对象里的
 * 属性设置,也会有不同的外观模式.
 */
public class MainActivity extends Activity {
    /** 第一次调用 Activity 活动 */
    @Override
    public void onCreate(Bundle savedInstanceState) {
        super.onCreate(savedInstanceState);
        /* 设置 ContentView 为自定义的 MyView */
        MyView myView = new MyView(this);
        setContentView(myView);
    }
    /* 自定义继承 View 的 MyView */
    private class MyView extends View {
        public MyView(Context context){
            super(context) ;
        }
        /* 重写 onDraw() */
        @Override
        protected void onDraw(Canvas canvas)
        {
            super.onDraw(canvas);
            /* 设置背景为白色 */
            canvas.drawColor(Color.WHITE);
            Paint paint = new Paint();
            /* 去锯齿 */
            paint.setAntiAlias(true);
            /* 设置 paint 的颜色 */
            paint.setColor(Color.RED);
            /* 设置 paint 的 style 为 STROKE:空心 */
            paint.setStyle(Paint.Style.STROKE);
            /* 设置 paint 的外框宽度 */
            paint.setStrokeWidth(3);
            /* 画一个空心圆形 */
            canvas.drawCircle(40, 40, 30, paint);
            /* 画一个空心正方形 */
            canvas.drawRect(10, 90, 70, 150, paint);
            /* 画一个空心长方形 */
            canvas.drawRect(10, 170, 70,200, paint);
            /* 画一个空心椭圆形 */
            canvas.drawOval(new RectF(10,220,70,250), paint);
            /* 画一个空心三角形 */
            Path path = new Path();
            path.moveTo(10, 330);
            path.lineTo(70,330);
            path.lineTo(40,270);
            path.close();
            canvas.drawPath(path, paint);
            /* 画一个空心梯形 */
            Path path1 = new Path();
            path1.moveTo(10, 410);
            path1.lineTo(70,410);
            path1.lineTo(55,350);
```

```java
path1.lineTo(25, 350);
path1.close();
canvas.drawPath(path1, paint);
/*设置paint的颜色*/
paint.setColor(Color.BLUE);
/*设置paint的style为FILL:实心*/
paint.setStyle(Paint.Style.FILL);
/*画一个实心圆*/
canvas.drawCircle(120,40,30, paint);
/*画一个实心正方形*/
canvas.drawRect(90, 90, 150, 150, paint);
/*画一个实心长方形*/
canvas.drawRect(90, 170, 150,200, paint);
/*画一个实心椭圆*/
RectF re2 = new RectF(90,220,150,250);
canvas.drawOval(re2, paint);
/*画一个实心三角形*/
Path path2 = new Path();
path2.moveTo(90, 330);
path2.lineTo(150,330);
path2.lineTo(120,270);
path2.close();
canvas.drawPath(path2, paint);
/*画一个实心梯形*/
Path path3 = new Path();
path3.moveTo(90, 410);
path3.lineTo(150,410);
path3.lineTo(135,350);
path3.lineTo(105, 350);
path3.close();
canvas.drawPath(path3, paint);
/*设置渐变色*/
Shader mShader = new LinearGradient(0,0,100,100,
new int[]{Color.RED,Color.GREEN,Color.BLUE,Color.YELLOW},
        null,Shader.TileMode.REPEAT);
//Shader.TileMode 3种模式
//REPEAT:沿着渐变方向循环重复
//CLAMP:如果在预先定义的范围外画的话,就重复边界的颜色
//MIRROR:与REPEAT一样都是循环重复,但这个会对称重复
paint.setShader(mShader);            //用Shader中定义的颜色来画
/*画一个渐变色圆*/
canvas.drawCircle(200,40,30, paint);
/*画一个渐变色正方形*/
canvas.drawRect(170, 90, 230, 150, paint);
/*画一个渐变色长方形*/
canvas.drawRect(170, 170, 230,200, paint);
/*画一个渐变色椭圆*/
RectF re3 = new RectF(170,220,230,250);
canvas.drawOval(re3, paint);
/*画一个渐变色三角形*/
Path path4 = new Path();
path4.moveTo(170,330);
path4.lineTo(230,330);
path4.lineTo(200,270);
path4.close();
```

```
                canvas.drawPath(path4, paint);
                /*画一个渐变色梯形*/
                Path path5 = new Path();
                path5.moveTo(170, 410);
                path5.lineTo(230, 410);
                path5.lineTo(215, 350);
                path5.lineTo(185, 350);
                path5.close();
                canvas.drawPath(path5, paint);
                /*写字*/
                paint.setTextSize(24);
                canvas.drawText("圆形", 240, 50, paint);
                canvas.drawText("正方形", 240, 120, paint);
                canvas.drawText("长方形", 240, 190, paint);
                canvas.drawText("椭圆形", 240, 250, paint);
                canvas.drawText("三角形", 240, 320, paint);
                canvas.drawText("梯形", 240, 390, paint);
            }
        }
    }
```

运行程序，效果如图 4-1 所示。

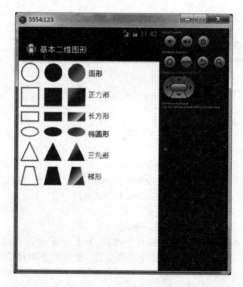

图 4-1　不同风格的二维图形

由上实例可看出，Paint 有其自身的特有的方法，主要内容如下。

- setARGB(int a, int r, int g, int b)/setColor(int color)：设置颜色。
- setAlpha(int a)：设置透明度。
- setAntiAlias(boolean aa)：设置是否抗锯齿。
- setPathEffect(PathEffect effect)：设置绘制路径时的路径效果。
- setShader(Shader shader)：设置画笔的填充效果。
- setShadowLayer(float radius, float dx, float dy, int color)：设置阴影。
- setStrokeWidth(float width)：设置画笔的笔触宽度。
- setStrokeJoin(Paint.Join join)：设置画笔转弯处的连接风格。

- setStyle(Paint.Style style)：设置 Paint 的填充风格。
- setTextAlign(Paint.Align align)：设置绘制文本时的文字的对齐方式。
- setTextSize(float textSize)：设置绘制文本时的文字大小。

4.2 绘制路径实例

Android 提供了一个非常有用的类，即为 Path，它可以预先在 View 上将 N 个点连成一条"路径"，然后调用 Canvas 的 drawPath(path,paint) 即可沿着路径绘制图形。实际上 Android 还为路径绘制提供了 PathEffect 来定义绘制效果，PathEffect 包含了如下子类（每个子类代表一种绘制效果）：

- ComposePathEffect
- CornerPathEffect
- DashPathEffect
- DiscretePathEffect
- PathDashPathEffect
- SumPathEffect

本实例通过 Path 中的几个子类，绘制几条不同效果的路径动态图。通过本实例演示了 Path 类的具体用法。

本实例的具体实现步骤如下。

(1) 在 Eclipse 中创建一个 Android 应用项目，命名为 Path_test。

(2) 打开 src\fs.path_test 包下的 MainActivity.java 文件，在文件中利用 Path 类绘制 7 种不同类型的动态效果路径。代码为：

```
package fs.path_test;
import android.app.Activity;
import android.content.Context;
import android.graphics.Canvas;
import android.graphics.Color;
import android.graphics.ComposePathEffect;
import android.graphics.CornerPathEffect;
import android.graphics.DashPathEffect;
import android.graphics.DiscretePathEffect;
import android.graphics.Paint;
import android.graphics.Path;
import android.graphics.PathDashPathEffect;
import android.graphics.SumPathEffect;
import android.graphics.PathEffect;
import android.os.Bundle;
import android.view.View;
public class MainActivity extends Activity
{
    @Override
    protected void onCreate(Bundle savedInstanceState)
    {
        super.onCreate(savedInstanceState);
        setContentView(new MyView(this));
    }
```

```java
class MyView extends View
{
    float phase;
    PathEffect[] effects = new PathEffect[7];
    int[] colors;
    private Paint paint;
    Path path;
    public MyView(Context context)
    {
        super(context);
        paint = new Paint();
        paint.setStyle(Paint.Style.STROKE);
        paint.setStrokeWidth(4);
        //创建并初始化 Path
        path = new Path();
        path.moveTo(0, 0);
        for (int i = 1; i <= 15; i++)
        {
            //生成15个点,随机生成它们的Y坐标.并将它们连成一条 Path
            path.lineTo(i * 20, (float) Math.random() * 60);
        }
        //初始化7个颜色
        colors = new int[] {Color.BLACK, Color.BLUE, Color.CYAN
            , Color.GREEN, Color.MAGENTA, Color.RED , Color.YELLOW};
    }
    @Override
    protected void onDraw(Canvas canvas)
    {
        //将背景填充成白色
        canvas.drawColor(Color.WHITE);
        //初始化7种路径效果
        //不使用路径效果
        effects[0] = null;
        //使用 CornerPathEffect 路径效果
        effects[1] = new CornerPathEffect(10);
        //初始化 DiscretePathEffect
        effects[2] = new DiscretePathEffect(3.0f , 5.0f);
        //初始化 DashPathEffect
        effects[3] = new DashPathEffect(new float[]
            { 20, 10, 5, 10 }, phase);
        //初始化 PathDashPathEffect
        Path p = new Path();
        p.addRect(0 , 0, 8, 8, Path.Direction.CCW);
        effects[4] = new PathDashPathEffect(p, 12, phase,
            PathDashPathEffect.Style.ROTATE);
        //初始化 PathDashPathEffect
        effects[5] = new ComposePathEffect(effects[2], effects[4]);
        effects[6] = new SumPathEffect(effects[4], effects[3]);
        //将画布移动到(8,8)处开始绘制
        canvas.translate(8, 8);
        //使用7种不同路径效果和7种不同的颜色来绘制路径
        for (int i = 0; i < effects.length; i++)
        {
            paint.setPathEffect(effects[i]);
            paint.setColor(colors[i]);
```

```
                    canvas.drawPath(path, paint);
                    canvas.translate(0, 60);
                }
        //改变 phase 值,形成动画效果
                phase += 1;
                invalidate();
            }
        }
    }
```

运行程序,效果如图 4-2 所示。

图 4-2 7 种不同风格的路径图

4.3 绘制路径文本实例

    Android 的 Canvas 还提供了一个 drawTextOnPath(String text, Path path, float hOffset, float vOffset, Paint paint)方法,用于沿着 Path 绘制文本。其中,hOffset 参数指定水平偏移,vOffset 参数指定垂直偏移。

    本实例用于使用 drawTextOnPath 绘制沿路径绘制文本。通过本实例演示 drawTextOnPath 的具体用法。

本实例的具体实现步骤如下。

(1) 在 Eclipse 中创建一个 Android 应用项目,命名为 drawtext_test。

(2) 打开 src\fs.drawtext_test 包下的 MainActivity.java 文件,在文件中实现文本绘制。代码为:

```
package fs.drawtext_test;
import android.app.Activity;
import android.content.Context;
import android.graphics.Canvas;
import android.graphics.Color;
import android.graphics.Paint;
```

```java
import android.graphics.Path;
import android.graphics.RectF;
import android.os.Bundle;
import android.view.View;
public class MainActivity extends Activity
{
    @Override
    public void onCreate(Bundle savedInstanceState)
    {
        super.onCreate(savedInstanceState);
        setContentView(new TextView(this));
    }
    class TextView extends View
    {
        final String DRAW_STR = "Android精要介绍";
        Path[] paths = new Path[3];
        Paint paint;
        public TextView(Context context)
        {
            super(context);
            paths[0] = new Path();
            paths[0].moveTo(0, 0);
            for (int i = 1; i <= 7; i++)
            {
                //生成7个点,随机生成它们的Y坐标。并将它们连成一条Path
                paths[0].lineTo(i * 30, (float) Math.random() * 30);
            }
            paths[1] = new Path();
            RectF rectF = new RectF(0 , 0 , 200 , 120);
            paths[1].addOval(rectF, Path.Direction.CCW);
            paths[2] = new Path();
            paths[2].addArc(rectF , 60, 180);
            //初始化画笔
            paint = new Paint();
            paint.setAntiAlias(true);
            paint.setColor(Color.RED);
            paint.setStrokeWidth(1);
        }
        @Override
        protected void onDraw(Canvas canvas)
        {
            canvas.drawColor(Color.WHITE);
            canvas.translate(40, 40);
            //设置从右边开始绘制(右对齐)
            paint.setTextAlign(Paint.Align.RIGHT);
            paint.setTextSize(20);
            //绘制路径
            paint.setStyle(Paint.Style.STROKE);
            canvas.drawPath(paths[0], paint);
            //沿着路径绘制一段文本
            paint.setStyle(Paint.Style.FILL);
            canvas.drawTextOnPath(DRAW_STR, paths[0], -8 , 20 , paint);
            //画布下移120
            canvas.translate(0, 60);
            //绘制路径
```

```
            paint.setStyle(Paint.Style.STROKE);
            canvas.drawPath(paths[1], paint);
            //沿着路径绘制一段文本
            paint.setStyle(Paint.Style.FILL);
            canvas.drawTextOnPath(DRAW_STR, paths[1], -20 , 20 , paint);
            //画布下移 120
            canvas.translate(0, 120);
            //绘制路径
            paint.setStyle(Paint.Style.STROKE);
            canvas.drawPath(paths[2], paint);
            //沿着路径绘制一段文本
            paint.setStyle(Paint.Style.FILL);
            canvas.drawTextOnPath(DRAW_STR, paths[2]
                , -10 , 20 , paint);
        }
    }
}
```

运行程序,效果如图 4-3 所示。

图 4-3 绘制文本

4.4 电影式播放实例

本实例主要利用 Android 提供的 AnimationDrawable 类实现利用电影式播放动画效果的帧动画。通过本实例具体演示了怎样在 Android 中实现帧动画。

帧动画是比较传统的动画方式,帧动画将一系列的图片文件像放电影般依次进行播放,帧动画主要用到的类是 AnimationDrawable,每帧动画都是一个 AnimationDrawable 对象。

本实例的具体实现步骤如下。

(1) 在 Eclipse 中创建一个 Android 应用项目,命名为 FrameAnimation_test。

(2) 打开 res\layout 目录下的 main.xml 文件,在文件中声明一个 ImageView 控件及 3 个 Button 控件。代码为:

```xml
<?xml version="1.0" encoding="utf-8"?>
<LinearLayout xmlns:android="http://schemas.android.com/apk/res/android"
    android:layout_width="fill_parent"
    android:layout_height="fill_parent"
    android:orientation="vertical"
    android:background="#aabbcc">
    <ImageView
            android:id="@+id/animationIV"
            android:layout_width="wrap_content"
            android:layout_height="wrap_content"
            android:padding="5px"
            android:src="@drawable/animation1"/>
    <Button
            android:id="@+id/buttonA"
            android:layout_width="wrap_content"
            android:layout_height="wrap_content"
            android:padding="5px"
            android:text="顺序显示" />
    <Button
            android:id="@+id/buttonB"
            android:layout_width="wrap_content"
            android:layout_height="wrap_content"
            android:padding="5px"
            android:text="停止" />
    <Button
            android:id="@+id/buttonC"
            android:layout_width="wrap_content"
            android:layout_height="wrap_content"
            android:padding="5px"
            android:text="倒序显示" />
</LinearLayout>
```

（3）在 res\drawable-mdpi 目录下创建两个文件，分别为 animation1.xml 及 animation2.xml。animation1.xml 文件用于顺序显示动画文件。代码为：

```xml
<?xml version="1.0" encoding="utf-8"?>
<!-- 根标签为 animation-list,其中 oneshot 代表着是否只展示一遍,设置为 false 会不停地循环播放动画。根标签下,通过 item 标签对动画中的每一个图片进行声明 android:duration 表示展示所用的该图片的时间长度 -->
<animation-list
    xmlns:android="http://schemas.android.com/apk/res/android"
    android:oneshot="true">
    <item
        android:drawable="@drawable/ab"
        android:duration="150"></item>
    <item
        android:drawable="@drawable/ac"
        android:duration="150"></item>
    <item
        android:drawable="@drawable/ad"
        android:duration="150"></item>
    <item
        android:drawable="@drawable/ae"
        android:duration="150"></item>
    <item
```

```
            android:drawable = "@drawable/af"
            android:duration = "150"></item>
        <item
            android:drawable = "@drawable/ag"
            android:duration = "150"></item>
</animation-list>
```

animation2.xml 文件用于倒序显示动画文件。代码为：

```
<?xml version = "1.0" encoding = "utf-8"?>
<!-- 根标签为 animation-list,其中 oneshot 代表着是否只展示一遍,设置为 false 即会不停地循环
播放动画根标签,通过 item 标签对动画中的每一个图片进行声明 android:duration 表示展示所用的该
图片的时间长度 -->
<animation-list
    xmlns:android = "http://schemas.android.com/apk/res/android"
    android:oneshot = "true">
        <item
            android:drawable = "@drawable/ag"
            android:duration = "150"></item>
        <item
            android:drawable = "@drawable/af"
            android:duration = "150"></item>
        <item
            android:drawable = "@drawable/ae"
            android:duration = "150"></item>
        <item
            android:drawable = "@drawable/ad"
            android:duration = "150"></item>
        <item
            android:drawable = "@drawable/ac"
            android:duration = "150"></item>
        <item
            android:drawable = "@drawable/ab"
            android:duration = "150"></item>
</animation-list>
```

（4）打开 src\fs.frameanimation_test 包下的 MainActivity.java 文件,在文件中实现帧动画。代码为：

```
package fs.frameanimation_test;
import android.app.Activity;
import android.graphics.drawable.AnimationDrawable;
import android.os.Bundle;
import android.view.View;
import android.view.View.OnClickListener;
import android.view.Window;
import android.widget.Button;
import android.widget.ImageView;
public class MainActivity extends Activity
{
    private ImageView animationIV;
    private Button buttonA, buttonB, buttonC;
    private AnimationDrawable animationDrawable;
    @Override
    public void onCreate(Bundle savedInstanceState) {
        super.onCreate(savedInstanceState);
        requestWindowFeature(Window.FEATURE_NO_TITLE);
```

```java
        setContentView(R.layout.main);
        animationIV = (ImageView) findViewById(R.id.animationIV);
        buttonA = (Button) findViewById(R.id.buttonA);
        buttonB = (Button) findViewById(R.id.buttonB);
        buttonC = (Button) findViewById(R.id.buttonC);
        buttonA.setOnClickListener(new OnClickListener()
        {
            @Override
            public void onClick(View v) {
                //TODO 自动存根法
                animationIV.setImageResource(R.drawable.animation1);
                animationDrawable = (AnimationDrawable) animationIV.getDrawable();
                animationDrawable.start();
            }
        });
        buttonB.setOnClickListener(new OnClickListener()
        {
            @Override
            public void onClick(View v) {
                //TODO 自动存根法
                animationDrawable = (AnimationDrawable) animationIV.getDrawable();
                animationDrawable.stop();
            }
        });
        buttonC.setOnClickListener(new OnClickListener()
        {
            @Override
            public void onClick(View v) {
                //TODO 自动存根法
                animationIV.setImageResource(R.drawable.animation2);
                animationDrawable = (AnimationDrawable) animationIV.getDrawable();
                animationDrawable.start();
            }
        });
    }
}
```

运行程序,效果如图 4-4 所示。

图 4-4　以电影方式播放动画

由以上实例可知,在帧动画的 XML 文件中主要用到的标记及其属性值主要如表 4-1 所示。

表 4-1 帧动画中标记及其属性说明

标记名称	属性值	说 明
\<animation—list\>	android:oneshot:如果设置为 true,则该动画只播放一次,然后停止在最后一帧	Frame Animation 的根标记,包含若干\<item\>标记
\<item\>	android:drawable:图片帧的引用; android:duration:图片帧的停留时间; android:visible:图片帧是否可见	每个\<item\>标记定义一个图片帧,其中包含图片资源的引用等属性

4.5 平面贴图实例

本实例实现手机中的平面贴图效果,在界面中包含未处理的贴图、旋转一定角度的贴图和半透明效果的贴图。

平面贴图可以通过继承并扩展 SurfaceView 类来实现。在进行平面贴图时需要声明 Paint 对象,使用 Bitmap 创建位图,绘制时需要指定画笔和位图即可。图片旋转效果的实现首先创建 Matrix 对象,然后调用 Matrix 的 setRoate()方法,半透明效果的绘制是通过设置图片的 Alpha 值确定。

本实例的具体实现步骤如下:

(1) 在 Eclipse 中创建一个 Android 应用项目,命名为 Surface_rotate。

(2) res\layout 目录下的 main.xml 文件保持默认代码。

(3) 打开 src\fs.surface_rotate 包下的 MainActivity.java 文件,在该文件中实现调用 Activity 活动。代码为:

```java
package fs.surface_rotate;
import android.app.Activity;
import android.os.Bundle;
public class MainActivity extends Activity {
    @Override
    public void onCreate(Bundle savedInstanceState) {
        super.onCreate(savedInstanceState);
        Surface_rotate mySurfaceView = new Surface_rotate(this);
        this.setContentView(mySurfaceView);
    }
}
```

(4) 在 src\fs.surface_rotate 包下新建一个 Surface_rotate.java 文件,在文件中实现继承并扩展 SurfaceView 类,实现声明周期回调接口,并实现图片的旋转及半透明度。代码为:

```java
package fs.surface_rotate;
import android.annotation.SuppressLint;
import android.graphics.Bitmap;
import android.graphics.BitmapFactory;
import android.graphics.Canvas;
import android.graphics.Matrix;
```

```java
import android.graphics.Paint;
import android.view.SurfaceHolder;
import android.view.SurfaceView;
@SuppressLint("WrongCall")
public class Surface_rotate extends SurfaceView
implements SurfaceHolder.Callback                  //实现生命周期回调接口
{
    MainActivity activity;
    Paint paint;                                    //画笔
    public Surface_rotate(MainActivity activity) {
        super(activity);
        this.activity = activity;
        this.getHolder().addCallback(this);         //设置生命周期回调接口的实现者
        paint = new Paint();                        //创建画笔
        paint.setAntiAlias(true);                   //打开抗锯齿
    }
    public void onDraw(Canvas canvas){
        //进行平面贴图
        //加载图片
        Bitmap bitmapTmp = BitmapFactory.decodeResource(activity.getResources(), R.drawable.g3);
        //在屏幕的 20,180 位置贴图
        canvas.drawBitmap(bitmapTmp, 20, 130, paint);
        //将图片旋转 45 度并移动到 200,100 位置贴图
        Matrix m1 = new Matrix();
        m1.setTranslate(360,80);
        Matrix m2 = new Matrix();
        m2.setRotate(45);
        Matrix mz = new Matrix();
        mz.setConcat(m1, m2);
        canvas.drawBitmap(bitmapTmp, mz, paint);
        //改变图片透明度并在 250,10 位置贴图
        paint.setAlpha(128);
        canvas.drawBitmap(bitmapTmp, 290, 10, paint);
    }
    public void surfaceChanged(SurfaceHolder arg0, int arg1, int arg2, int arg3) {
    }
    public void surfaceCreated(SurfaceHolder holder) {    //创建时被调用
        Canvas canvas = holder.lockCanvas();              //获取画布
        try{
            synchronized(holder){
                onDraw(canvas);                           //绘制
            }
        }
        finally{
            if(canvas != null){
                holder.unlockCanvasAndPost(canvas);
            }
        }
    }
    public void surfaceDestroyed(SurfaceHolder arg0) {    //销毁时被调用
    }
}
```

运行程序,效果如图 4-5 所示。

图 4-5　平面贴图

4.6　图像淡入淡出实例

本实例主要实现简单的图片淡入淡出效果，本实现首先通过 BitmapFactory.decodeResource()方法加载位图，在 surfaceCreated()方法中设定线程动态改变图片的 Alpha 值，改变图片的透明度，从而实现图片淡淡消失的效果。

游戏的开始界面经常会出现淡入淡出效果的界面，这样的效果大大增加了游戏的吸引力。本实例的具体实现步骤如下。

（1）在 Eclipse 中创建一个 Android 应用项目，命名为 Alpha_test。

（2）打开 src\fs.alpha_test 包下的 MainActivity.java 文件，在该文件中实现横屏与全屏的设置，并获取手机屏幕的分辨率。代码为：

```
package fs.alpha_test;
import android.app.Activity;
import android.content.pm.ActivityInfo;
import android.os.Bundle;
import android.util.DisplayMetrics;
import android.view.Window;
import android.view.WindowManager;
public class MainActivity extends Activity {
    /** 第一次调用 Activity 活动 */
    static float screenHeight;          //屏幕高度
    static float screenWidth;           //屏幕宽度
    StartView startView;
    @Override
    public void onCreate(Bundle savedInstanceState)
    {
        super.onCreate(savedInstanceState);
//设置为横屏
this.setRequestedOrientation(ActivityInfo.SCREEN_ORIENTATION_LANDSCAPE);
```

```java
        requestWindowFeature(Window.FEATURE_NO_TITLE);        //设置全屏
        getWindow().setFlags(WindowManager.LayoutParams.FLAG_FULLSCREEN, WindowManager.
LayoutParams.FLAG_FULLSCREEN);
        DisplayMetrics dm = new DisplayMetrics();              //获取手机分辨率
        getWindowManager().getDefaultDisplay().getMetrics(dm);
        screenHeight = dm.heightPixels;
        screenWidth = dm.widthPixels;
        startView = new StartView(this);                       //开始界面
        this.setContentView(startView);
    }
}
```

(3) 在 src\fs.alpha_test 包下新建一个 StartView.java 文件,实现图像的淡出淡入效果。代码为:

```java
package fs.alpha_test;
import android.annotation.SuppressLint;
import android.graphics.Bitmap;
import android.graphics.BitmapFactory;
import android.graphics.Canvas;
import android.graphics.Color;
import android.graphics.Paint;
import android.view.SurfaceHolder;
import android.view.SurfaceView;
public class StartView extends SurfaceView
        implements SurfaceHolder.Callback{              //实现生命周期回调接口
    MainActivity activity;
    Paint paint;
    int currentAloha = 0;                               //当前的透明度
    int screenWidth = (int)MainActivity.screenWidth;
    int screenHeight = (int)MainActivity.screenHeight;
    int sleepSpan = 50;
    Bitmap[] logos = new Bitmap[2];
    int pic1;
    int pic2;
    Bitmap currentLogo;                                 //当前 Logo 图片引用
    int currentX;
    int currentY;
    public StartView(MainActivity activity) {
        super(activity);
        this.activity = activity;
        this.getHolder().addCallback(this);             //设置生命周期回调接口的实现者
        paint = new Paint();                            //创建画笔
        paint.setAntiAlias(true);                       //打开抗锯齿
        pic1 = R.drawable.a04;
        pic2 = R.drawable.a02;
        logos[0] = BitmapFactory.decodeResource(activity.getResources(), pic1);
        logos[1] = BitmapFactory.decodeResource(activity.getResources(), pic2);
    }
    @Override
    public void onDraw(Canvas canvas)
    {
        try
        {
            //绘制黑色填充矩形背景
```

```java
                paint.setColor(Color.BLACK);
                paint.setAlpha(255);
                canvas.drawRect(0, 0,screenWidth, screenHeight ,paint);
                //进行平面贴图
                if(currentLogo == null)return;
                paint.setAlpha(currentAloha);
                canvas.drawBitmap(currentLogo, 0, 0,paint);
            }
            catch(Exception e)
            {
                e.printStackTrace();
            }
    }
    public void surfaceChanged(SurfaceHolder arg0, int arg1, int arg2, int arg3) {
    }
    public void surfaceCreated(SurfaceHolder arg0) {          //创建时被调用
        new Thread()
        {
            @SuppressLint("WrongCall")
            public void run()
            {
                for(Bitmap bm:logos)
                {
                    currentLogo = bm;
                    //图片的位置
                    currentX = screenWidth/2 - bm.getWidth()/2;   //X坐标位置
                    currentY = screenHeight/2 - bm.getHeight()/2; //Y坐标位置
                    for(int i = 255;i > - 10;i = i - 10) //动态更改图片的透明度值并不断重绘
                    {
                        currentAloha = i;
                        if(currentAloha < 0)
                        {
                            currentAloha = 0;
                        }
                        SurfaceHolder myholder = StartView.this.getHolder();
                        Canvas canvas = myholder.lockCanvas(); //获取画布
                        try
                        {
                            synchronized(myholder)
                            {
                                onDraw(canvas);                      //绘制
                            }
                        }
                        finally
                        {
                            if(canvas != null)
                            {
                                myholder.unlockCanvasAndPost(canvas);}
                        }
                        try
                        {
                            if(i == 255)
                            {
                                Thread.sleep(1000);
                            }
```

```
                            Thread.sleep(sleepSpan);
                        }catch(Exception e)
                        {
                            e.printStackTrace();
                        }
                    }
                }
            }
        }.start();
    }
    public void surfaceDestroyed(SurfaceHolder arg0) {              //销毁时被调用
    }
}
```

运行程序,得到的横屏淡入效果如图4-6(a)所示,图4-6(b)为淡入另一幅图片。

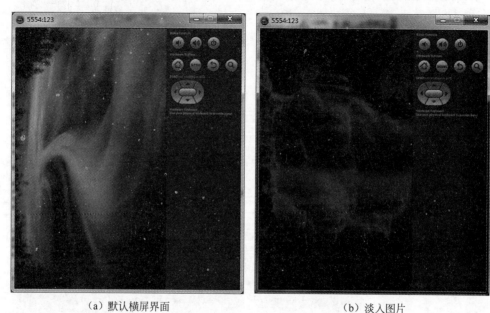

(a)默认横屏界面　　　　　　　　　　(b)淡入图片

图4-6　图片的淡出淡入效果

4.7　图像变大变小实例

Android手机中,经常有缩小或放大图像的动作,这样使动画效果更逼真。本实例通过ScaleAnimation类实现图像的尺寸变大变小,通过本实例演示了ScaleAnimation类的具体用法。

本实例的具体实现步骤如下。

(1) 在Eclipse中创建一个Android应用项目,命名为Scale_test。

(2) 打开res\layout目录下的main.xml文件,总的布局方式为帧布局,摆放方向为横向,然后设置LinearLayout布局,并在该布局中设置两个Button按钮的位置,最后设置另一个LinearLayout布局,并在该布局中设置ImageView的具体属性。代码为:

```
<RelativeLayout xmlns:android = "http://schemas.android.com/apk/res/android"
```

```
        xmlns:tools = "http://schemas.android.com/tools"
        android:layout_width = "match_parent"
        android:layout_height = "match_parent"
        android:paddingBottom = "@dimen/activity_vertical_margin"
        android:paddingLeft = "@dimen/activity_horizontal_margin"
        android:paddingRight = "@dimen/activity_horizontal_margin"
        android:paddingTop = "@dimen/activity_vertical_margin"
        tools:context = ".MainActivity"
        android:background = "#0f0">
        <Button
            android:id = "@+id/button1"
            android:layout_width = "wrap_content"
            android:layout_height = "wrap_content"
            android:layout_alignParentBottom = "true"
            android:layout_alignParentLeft = "true"
            android:layout_alignParentRight = "true"
            android:layout_marginBottom = "73dp"
            android:textSize = "25dip"
            android:text = "变小" />
        <Button
            android:id = "@+id/Button01"
            android:layout_width = "wrap_content"
            android:layout_height = "wrap_content"
            android:layout_alignLeft = "@+id/button1"
            android:layout_alignParentBottom = "true"
            android:layout_alignParentRight = "true"
            android:textSize = "25dip"
            android:text = "变大" />
        <ImageView
            android:id = "@+id/ImageView1"
            android:layout_width = "wrap_content"
            android:layout_height = "wrap_content"/>
</RelativeLayout>
```

(3) 打开 src\fs.scale_test 目录下的 MainActivity.java 文件,实现当单击屏幕中的"放大"或"缩小"按钮时,实现图片的放大和缩小功能。代码为:

```
package fs.scale_test;
import android.app.Activity;
import android.graphics.Bitmap;
import android.graphics.BitmapFactory;
import android.graphics.Matrix;
import android.os.Bundle;
import android.util.DisplayMetrics;
import android.view.View;
import android.view.View.OnClickListener;
import android.widget.Button;
import android.widget.ImageView;
public class MainActivity extends Activity
{
    ImageView iv;                    //声明 ImageView 的引用
    Bitmap bmp;                      //声明 Bitmap 的引用
    int screenWidth;                 //屏幕的宽度
    int secrenHeight;                //屏幕的高度
    Button b2;
```

```java
@Override
public void onCreate(Bundle savedInstanceState)
{
    super.onCreate(savedInstanceState);
    setContentView(R.layout.main);                              //切屏到主界面
    //创建 Bitmap 对象
    bmp = BitmapFactory.decodeResource(getResources(), R.drawable.kt1);
    DisplayMetrics dm = new DisplayMetrics();                   //创建矩阵
    getWindowManager().getDefaultDisplay().getMetrics(dm);
    screenWidth = dm.widthPixels;                               //得到屏幕的宽度
    screenWidth = dm.heightPixels - 80;                         //得到屏幕的高度
    iv = (ImageView)findViewById(R.id.ImageView1);              //ImageView 控件
    Button b1 = (Button)findViewById(R.id.button1);             //缩小按钮
    b2 = (Button)findViewById(R.id.Button01);                   //放大按钮
    iv.setImageBitmap(bmp);                                     //为 ImageView 设置图片
    b1.setOnClickListener                                       //设置监听
    (
        new OnClickListener()
        {
            public void onClick(View v)
            {
                //单击缩小按钮,即图像缩小到原来的 60 %
                iv.setImageBitmap(scaleToFit(bmp,0.8f));
                bmp = scaleToFit(bmp,0.8f); }
        }
    );
    b2.setOnClickListener//设置监听
    (
        new OnClickListener()
        {
            public void onClick(View v)
            {
                //单击放大按钮,即图像放大到原来的 150 %
                iv.setImageBitmap(scaleToFit(bmp,1.5f));
                bmp = scaleToFit(bmp,1.5f);
            }
        }
    );
}
public static Bitmap scaleToFit(Bitmap bm,float scale)          //缩放图片的方法
{
    Bitmap bmResult = null;
    if((bm.getWidth()<280)&&(bm.getWidth()>0)&&(bm.getHeight()>0))
    {
        int width = bm.getWidth();                              //图片宽度
        int height = bm.getHeight();                            //图片高度
        Matrix matrix = new Matrix();                           //创建矩阵
        matrix.postScale(scale, scale);                         //图片等比例缩小
        //声明位图
        bmResult = Bitmap.createBitmap(bm, 0, 0, width, height, matrix, true);
    }
    else
    {
        System.out.println();
    }
```

```
            return bmResult;
        }
}
```

运行程序,默认效果如图 4-7(a)所示,单击界面中的"变小"按钮时,即图像缩小,效果如图 4-7(b)所示,当单击"变大"按钮时,即图像放大,效果如图 4-7(c)所示。

（a）默认界面　　　　　　　（b）图像变小　　　　　　　（c）图像变大

图 4-7　缩放图像

4.8　图像移动实例

在 Android 中提供了 Translate 对象用于实现图像的平移。平移变化的动画,创建该动画时只要指定动画开始时的位置(以 X、Y 坐标来表示)和结束的位置(以 X、Y 坐标来表示),并指定动画持续时间即可。

本实例主要利用 Translate 对象实现 Android 界面图像的平移。通过本实例演示了 Translate 对象的具体用法。本实例的具体实现步骤如下。

（1）在 Eclipse 中创建一个 Android 应用项目,命名为 Translate_test。

（2）打开 res\layout 目录下的 main.xml 布局文件,在文件中声明两个 Button 控件和一个 ImageView 控件。代码为：

```
<LinearLayout xmlns:android = "http://schemas.android.com/apk/res/android"
    android:orientation = "vertical"
    android:layout_width = "fill_parent"
    android:layout_height = "fill_parent"
    android:background = "#0f0">
    <Button
        android:id = "@ + id/bt1"
        android:layout_width = "match_parent"
        android:layout_height = "wrap_content"
        android:layout_marginTop = "26dp"
        android:text = "开始动画" />
```

```xml
<Button
    android:id = "@+id/bt2"
    android:layout_width = "match_parent"
    android:layout_height = "wrap_content"
    android:layout_below = "@+id/bt1"
    android:text = "取消动画" />
<ImageView
    android:id = "@+id/imgView"
    android:layout_width = "wrap_content"
    android:layout_height = "wrap_content"
    android:layout_alignLeft = "@+id/bt1"
    android:layout_below = "@+id/bt2"
    android:layout_marginTop = "67dp"
    android:src = "@drawable/kt1" />
</LinearLayout>
```

(3) 打开 src\fs.translate_test 包下的 MainActivity.java 文件，在文件中实现当单击界面中的"开始动画"按钮时，即实现图像的平移，当单击"取消动画"按钮时，即取消图像平移动画效果，图像并返回初始位置。代码为：

```java
package fs.translate_test;
import android.app.Activity;
import android.graphics.Bitmap;
import android.graphics.BitmapFactory;
import android.graphics.Matrix;
import android.os.Bundle;
import android.util.DisplayMetrics;
import android.view.View;
import android.view.View.OnClickListener;
import android.view.animation.Animation;
import android.view.animation.TranslateAnimation;
import android.widget.Button;
import android.widget.ImageView;
public class MainActivity extends Activity {
    ImageView image;
    Button start;
    Button cancel;
    @Override
    public void onCreate(Bundle savedInstanceState) {
        super.onCreate(savedInstanceState);
        setContentView(R.layout.main);
        image = (ImageView) findViewById(R.id.imgView);
        start = (Button) findViewById(R.id.bt1);
        cancel = (Button) findViewById(R.id.bt2);
        /** 设置位移动画 向右位移 150 */
        final TranslateAnimation animation = new TranslateAnimation(0, 150, 0, 0);
        animation.setDuration(2000);                          //设置动画持续时间
        animation.setRepeatCount(2);                          //设置重复次数
        animation.setRepeatMode(Animation.REVERSE);           //设置反方向执行
        start.setOnClickListener(new OnClickListener() {
            public void onClick(View arg0) {
                image.setAnimation(animation);
                /** 开始动画 */
                animation.startNow();
            }
```

```
            });
            cancel.setOnClickListener(new OnClickListener() {
                public void onClick(View v) {
                    /** 结束动画 */
                    animation.cancel();
                }
            });
        }
    }
```

运行程序，默认效果如图 4-8(a)所示，当单击界面中的"开始动画"按钮时，效果如图 4-8(b)所示。

（a）默认界面

（b）图像移动

图 4-8　图像移动效果

由以上实例可知，Translate 中定义可实现平移的关键属性如下。
- float fromXDelta：动画开始的点离当前 View X 坐标上的差值。
- float toXDelta：动画结束的点离当前 View X 坐标上的差值。
- float fromYDelta：动画开始的点离当前 View Y 坐标上的差值。
- float toYDelta：动画结束的点离当前 View Y 坐标上的差值。

4.9　动画综合实例

前面的实例都是将一种动画效果应用于 Button 控件上，本实例将把 4 种动画效果应用界面上。

本实例主要用于在 Android 中实现奔跑的野猪的动画。

其实现的具体操作如下。

(1) 在新建项目的 res 目录中，创建一个名称为 anim 的目录，并在该目录中创建实现野猪做向右奔跑动作和向左奔跑动作的逐帧动画资源文件。

① 在 anim 目录下创建名称为 right.xml 的 XML 资源文件，在该文件中定义一个野猪向

右跑动作的动画,该动画由两帧组成,也即是由两个预先定义好的图片组成的。操作为:

```xml
<?xml version = "1.0" encoding = "utf-8"?>
<animation-list xmlns:android = "http://schemas.android.com/apk/res/android" >
    <item android:drawable = "@drawable/pig1" android:duration = "30" />
    <item android:drawable = "@drawable/pig2" android:duration = "30" />
</animation-list>
```

② 在 anim 目录下创建名称为 left.xml 的 XML 资源文件,在该文件中定义一个野猪向左跑动作的动画,该动画由两帧组成,也即是由两个预先定义好的图片组成。操作为:

```xml
<?xml version = "1.0" encoding = "utf-8"?>
<animation-list xmlns:android = "http://schemas.android.com/apk/res/android" >
    <item android:drawable = "@drawable/pig3" android:duration = "30" />
    <item android:drawable = "@drawable/pig4" android:duration = "30" />
</animation-list>
```

(2) 在 anim 目录中,创建实现野猪向右奔跑和左奔跑的补间动画资源文件。

① 在 anim 目录下创建名称为 rotright.xml 的 XML 资源文件,在文件中定义一个实现野猪向右奔跑的补间动画,该动画为水平方向上向右平移 660 像素,持续时间为 3 秒。代码为:

```xml
<?xml version = "1.0" encoding = "utf-8"?>
<set xmlns:android = "http://schemas.android.com/apk/res/android">
    <translate
        android:fromXDelta = "0"
        android:toXDelta = "660"
        android:fromYDelta = "0"
        android:toYDelta = "0"
        android:duration = "3000">
    </translate>
</set>
```

② 在 anim 目录下创建名称为 rotleft.xml 的 XML 资源文件,在文件中定义一个实现野猪向左奔跑的补间动画,该动画为水平方向上向左平移 660 像素,持续时间为 3 秒。代码为:

```xml
<?xml version = "1.0" encoding = "utf-8"?>
<set xmlns:android = "http://schemas.android.com/apk/res/android" >
    <translate
        android:fromXDelta = "660"
        android:toXDelta = "0"
        android:fromYDelta = "0"
        android:toYDelta = "0"
        android:duration = "3000">
    </translate>
</set>
```

(3) 修改新建项目的 res\layout 目录下的布局文件 main.xml,将默认添加的 TextView 组件删除,接着在默认添加的线性布局管理器中添加一个 ImageView 组件,并设置该组件的背景为逐帧动画资源 right.xml,最后设置 ImageView 组件的顶外边距和左外边距。代码为:

```xml
<?xml version = "1.0" encoding = "utf-8"?>
<LinearLayout xmlns:android = "http://schemas.android.com/apk/res/android"
    android:id = "@ + id/linearLayout1"
    android:background = "@drawable/bg"
```

```xml
android:layout_width = "fill_parent"
android:layout_height = "fill_parent"
android:orientation = "vertical" >
< ImageView
    android:id = "@ + id/imageView1"
    android:layout_width = "wrap_content"
    android:layout_height = "wrap_content"
    android:background = "@anim/right"
    android:layout_marginTop = "280px"
    android:layout_marginLeft = "30px"/>
</LinearLayout >
```

(4) 打开默认创建的 MainActivity, 在 onCreate() 方法中, 先获取要应用动画效果的 ImageView, 并获取向右奔跑和向左奔跑的补间动画资源, 然后获取 ImagView 应用的逐帧动画及线性布局管理器, 并显示一个消息提示框, 再为线性布局管理器添加触摸监听器, 在重写 onTouch() 方法中, 开始播放逐帧动画并播放向右奔跑的补间动画, 最后为向右奔跑和向左奔跑的动画添加动画监听器, 并在重写的 onAnimationEnd() 方法中改变要使用的逐帧动画、补间动画和播放动画, 实现野猪来回奔跑的动画效果。代码为:

```java
public class MainActivity extends Activity
{
    private AnimationDrawable anim;
    @Override
    public void onCreate(Bundle savedInstanceState)
    {
        super.onCreate(savedInstanceState);
        setContentView(R.layout.main);
        //获取要应用动画效果的 ImageView
        final ImageView iv = (ImageView)findViewById(R.id.imageView1);
        //获取"向右奔跑"的动画资源
        final Animation tright = AnimationUtils.loadAnimation(this, R.anim.rotright);
        //获取"向左奔跑"的动画资源
        final Animation tleft = AnimationUtils.loadAnimation(this, R.anim.rotleft);
        anim = (AnimationDrawable)iv.getBackground();          //获取应用的帧动画
        //获取线性布局管理器
        LinearLayout ll = (LinearLayout)findViewById(R.id.linearLayout1);
        //显示一个消息提示框
        Toast.makeText(this,"触摸屏幕开始播放……", Toast.LENGTH_SHORT).show();
        ll.setOnTouchListener(new OnTouchListener() {
            @Override
            public boolean onTouch(View v, MotionEvent event) {
                anim.start();                            //开始播放帧动画
                iv.startAnimation(tright);               //播放"向右奔跑"的动画
                return false;
            }
        });
        tright.setAnimationListener(new AnimationListener() {
            @Override
            public void onAnimationStart(Animation animation) {}
            @Override
            public void onAnimationRepeat(Animation animation) {}
            @Override
            public void onAnimationEnd(Animation animation) {
                //重新设置 ImageView 应用的帧动画
```

```
                    iv.setBackgroundResource(R.anim.left);
                    iv.startAnimation(tleft);                        //播放"向左奔跑"的动画
                    anim = (AnimationDrawable)iv.getBackground();    //获取应用的帧动画
                    anim.start();                                    //开始播放帧动画
                }
            });
            tleft.setAnimationListener(new AnimationListener() {
                @Override
                public void onAnimationStart(Animation animation) {}
                @Override
                public void onAnimationRepeat(Animation animation) {}
                @Override
                public void onAnimationEnd(Animation animation) {
                    //重新设置 ImageView 应用的帧动画
                    iv.setBackgroundResource(R.anim.right);
                    iv.startAnimation(tright);                       //播放"向右奔跑"的动画
                    anim = (AnimationDrawable)iv.getBackground();    //获取应用的帧动画
                    anim.start();                                    //开始播放帧动画
                }
            });
    }
}
```

(5) 修改动画的标题,打开 res\value 目录下的 string.xml 文件。代码为:

```xml
<?xml version = "1.0" encoding = "utf-8"?>
<resources>
    <string name = "app_name">奔跑的野猪</string>
    <string name = "action_settings">Settings</string>
    <string name = "hello_world">Hello world!</string>
</resources>
```

(6) 运行程序,单击屏幕后,屏幕中的野猪将从左侧奔跑到右侧,效果如图 4-9 所示,撞到右侧后,转身向左侧奔跑,直到撞上左侧的栅栏,再转身向右侧奔跑,反复此动画。

图 4-9 奔跑的野猪

4.10 图像特效实例

在编程中有时候需要对图片做特殊的处理,例如将图片做出黑白的,或者老照片的效果,有时候还要对图片进行变换、拉伸和扭曲等。这些效果在 Android 中有很好的支持。

本实例通过利用 Android 提供的 Matrix 工具类,实现对图像进行特效处理。通过本实例演示了 Matrix 工具类的具体用法。

本实例的具体实现步骤如下。

(1) 在 Eclipse 中创建一个 Android 应用项目,命名为 Matrix_test。

(2) 打开 res\layout 目录下的 main.xml 布局文件,文件代码为:

```xml
<?xml version = "1.0" encoding = "utf-8"?>
<LinearLayout xmlns:android = "http://schemas.android.com/apk/res/android"
    android:orientation = "vertical"
    android:layout_width = "fill_parent"
    android:layout_height = "fill_parent"
    android:background = "#f0f">
<fs.matrix_test.ViewMatrix
    android:layout_width = "fill_parent"
    android:layout_height = "fill_parent" />
</LinearLayout>
```

(3) 打开 src\fs.matrix_test 包下的 MainActivity.java 文件,在文件中实现主界面的调用。代码为:

```java
package fs.matrix_test;
import android.os.Bundle;
import android.app.Activity;
import android.view.Menu;
public class MainActivity extends Activity
{
    @Override
    public void onCreate(Bundle savedInstanceState)
    {
        super.onCreate(savedInstanceState);
        setContentView(R.layout.main);
    }
}
```

(4) 在 src\fs.matrix_test 包下新建一个 ViewMatrix.java 文件,在文件中自定义一个 View,在该自定义的 View 中可以检测到用户的键盘事件,当用户单击手机的方向键时,该自定义 View 会用 Matrix 对绘制的图形进行旋转和倾斜等变换。代码为:

```java
package fs.matrix_test;
import android.content.Context;
import android.graphics.Bitmap;
import android.graphics.Canvas;
import android.graphics.Matrix;
import android.graphics.drawable.BitmapDrawable;
import android.util.AttributeSet;
import android.view.KeyEvent;
```

```java
import android.view.View;
public class ViewMatrix extends View
{
    //初始的图片资源
    private Bitmap bitmap;
    //Matrix 实例
    private Matrix matrix = new Matrix();
    //设置倾斜度
    private float sx = 0.0f;
    //位图宽和高
    private int width, height;
    //缩放比例
    private float scale = 1.0f;
    //判断缩放还是旋转
    private boolean isScale = false;
    public ViewMatrix(Context context , AttributeSet set)
    {
        super(context , set);
        //获得位图
        bitmap = ((BitmapDrawable) context.getResources().getDrawable(
            R.drawable.g2)).getBitmap();
        //获得位图宽
        width = bitmap.getWidth();
        //获得位图高
        height = bitmap.getHeight();
        //使当前视图获得焦点
        this.setFocusable(true);
    }
    @Override
    protected void onDraw(Canvas canvas)
    {
        super.onDraw(canvas);
        //重置 Matrix
        matrix.reset();
        if (!isScale)
        {
            //旋转 Matrix
            matrix.setSkew(sx, 0);
        }
        else
        {
            //缩放 Matrix
            matrix.setScale(scale, scale);
        }
        //根据原始位图和 Matrix 创建新图片
        Bitmap bitmap2 = Bitmap.createBitmap(bitmap, 0, 0, width, height,
            matrix, true);
        //绘制新位图
        canvas.drawBitmap(bitmap2, matrix, null);
    }
    @Override
    public boolean onKeyDown(int keyCode, KeyEvent event)
    {
        switch(keyCode)
        {
```

```
            //向左倾斜
            case KeyEvent.KEYCODE_DPAD_LEFT:
                isScale = false;
                sx += 0.1;
                postInvalidate();
                break;
            //向右倾斜
            case KeyEvent.KEYCODE_DPAD_RIGHT:
                isScale = false;
                sx -= 0.1;
                postInvalidate();
                break;
            //放大
            case KeyEvent.KEYCODE_DPAD_UP:
                isScale = true;
                if (scale < 2.0)
                    scale += 0.1;
                postInvalidate();
                break;
            //缩小
            case KeyEvent.KEYCODE_DPAD_DOWN:
                isScale = true;
                if (scale > 0.5)
                    scale -= 0.1;
                postInvalidate();
                break;
        }
        return super.onKeyDown(keyCode, event);
    }
}
```

由以上实例可知，Matrix 工具类中提供了如下方法来控制平移、旋转和缩放。

- setTranslate(float dx,float dy)：控制 Matrix 进行平移。
- setSkew(float kx,float ky,float px,float py)：控制 Matrix 以 px 和 py 为轴进行倾斜。kx 和 ky 为 X 和 Y 方向上的倾斜距离。
- setRotate(float degrees,float px,float py)：设置以 px 和 py 为轴心进行旋转，degrees 控制旋转的角度。
- setScale(float sx,float sy,float px,float py)：设置 Matrix 以 px 和 py 为轴心进行缩放，sx 和 sy 控制 X 和 Y 方向上的缩放比例。

4.11 图像扭曲实例

Canvas 提供了 drawBitmapMesh 方法，总体来说这个方法就是在重新操作像素，每个像素按照自己的想法来处理，达到拉伸、变形和扭曲等效果，貌似水纹波浪的效果这个方法完全可以完成。

drawBitmapMesh 方法的调用格式为：

drawBitmapMesh(Bitmap bitmap, int meshWidth, int meshHeight, float[] verts, int vertOffset, int[] colors, int colorffset, Paint paint)

其中,参数 bitmap 为需要扭曲的源位图;meshWidth 为控制在横向上把该源位图划成成多少格;meshHeight 为控制在纵向上把该源位图划成成多少格;verts 为长度为(meshWidth + 1) * (meshHeight + 1) * 2 的数组,它记录了扭曲后的位图各顶点位置;vertOffset 为控制 verts 数组中从第几个数组元素开始才对 bitmap 进行扭曲。

本实例通过利用 drawBitmapMesh 方法实现图像的扭曲处理,通过该实例演示了 drawBitmapMesh 方法的具体用法。

本实例的具体实现步骤如下。

(1) 在 Eclipse 中创建一个 Android 应用项目,命名为 drawBitmapMesh_test。

(2) 打开 src\fs.drawbitmapmesh_test 包下的 MainActivity.java 文件,在文件中实现当单击图像或图像边缘时,即实现图像的扭曲效果。代码为:

```java
package fs.drawbitmapmesh_test;
import android.app.Activity;
import android.content.Context;
import android.graphics.Bitmap;
import android.graphics.BitmapFactory;
import android.graphics.Canvas;
import android.graphics.Color;
import android.os.Bundle;
import android.util.AttributeSet;
import android.view.MotionEvent;
import android.view.View;
public class MainActivity extends Activity {
    /** 第一次调用 Activity 活动 */
    private Bitmap bitmap;
    @Override
    public void onCreate(Bundle savedInstanceState) {
        super.onCreate(savedInstanceState);
        setContentView(new MyView(this,R.drawable.gud4));
    }
    private class MyView extends View
    {
        //定义两个常量,这两个常量指定该图片横向和纵向上都被划分为 20 格
        private final int WIDTH = 20;
        private final int HEIGHT = 20;
        //记录该图片上包含 441 个顶点
        private final int COUNT = (WIDTH + 1) * (HEIGHT + 1);
        //定义一个数组,记录 Bitmap 上的 21 * 21 个点的坐标
        private final float[] verts = new float[COUNT * 2];
        //定义一个数组,记录 Bitmap 上的 21 * 21 个点经过扭曲后的坐标
        //对图片扭曲的关键就是修改该数组里元素的值
        private final float[] orig = new float[COUNT * 2];
        public MyView(Context context, int drawableId) {
            super(context);
            setFocusable(true);
            //根据指定资源加载图片
            bitmap = BitmapFactory.decodeResource(getResources(), drawableId);
            //获取图片宽度和高度
            float bitmapWidth = bitmap.getWidth();
            float bitmapHeight = bitmap.getHeight();
            int index = 0;
            for(int y = 0; y <= HEIGHT; y++)
```

```java
    {
        float fy = bitmapHeight * y / HEIGHT;
        for(int x = 0 ; x <= WIDTH; x++)
        {
            float fx = bitmapWidth * x / WIDTH;
            //初始化 orig 和 verts 数组
            //初始化,orig 和 verts 两个数组均匀地保存了 21 * 21 个点的(x,y)坐标
            orig[index * 2 + 0] = verts[index * 2 + 0] = fx;
            orig[index * 2 + 1] = verts[index * 2 + 1] = fy;
            index += 1;
        }
    }
    //设置背景色
    setBackgroundColor(Color.WHITE);
}
protected void onDraw(Canvas canvas)
{
    //对 bitmap 按 verts 数组进行扭曲
    //从第 1 个点(由第 5 个参数 0 控制)开始扭曲
    canvas.drawBitmapMesh(bitmap, WIDTH, HEIGHT, verts, 0, null, 0, null);
}
//工具方法,用于根据触摸事件的位置计算 verts 数组里各元素的值
private void warp(float cx, float cy)
{
    for(int i = 0; i < COUNT * 2; i += 2)
    {
        float dx = cx - orig[i + 0];
        float dy = cy - orig[i + 1];
        float dd = dx * dx + dy * dy;
        //计算每个坐标点与当前点(cx,cy)之间的距离
        float d = (float)Math.sqrt(dd);
        //计算扭曲度,距离当前点(cx,cy)越远,扭曲度越小
        float pull = 80000 / ((float)(dd * d));
        //对 verts 数组(保存 bitmap 上 21 * 21 个点经过扭曲后的坐标)重新赋值
        if(pull >= 1)
        {
            verts[i + 0] = cx;
            verts[i + 1] = cy;
        }
        else
        {
            //控制各顶点向触摸事件发生点偏移
            verts[i + 0] = orig[i + 0] + dx * pull;
            verts[i + 1] = orig[i + 1] + dx * pull;
        }
    }
    //通知 View 组件重绘
    invalidate();
}
public boolean onTouchEvent(MotionEvent event)
{
    //调用 warp 方法根据触摸屏事件的坐标点来扭曲 verts 数组
    warp(event.getX() , event.getY());
    return true;
}
```

}
}

运行程序，默认效果如图 4-10(a)所示，当单击图像边缘时，即实现图像的扭曲处理，效果如图 4-10(b)和 4-10(c)所示。

　　　　(a) 默认图像　　　　　　(b) 扭曲效果1　　　　　　(c) 扭曲效果2

图 4-10　图像的扭曲处理

4.12　图像渲染实例

本实例通过利用 Android 提供的 Shader 类，实现图像几种不同类型的渲染效果，通过本实例演示了 Shader 类具体实现用法。

本实例的具体实现步骤如下。

(1) 在 Eclipse 中创建一个 Android 应用项目，命名为 Shader_test。

(2) 打开 src\fs.shader_test 包下的 MainActivity.java 文件，在文件中实现调用自定义的 GameView 类。代码为：

```
package fs.shader_test;
import android.app.Activity;
import android.os.Bundle;
import android.view.KeyEvent;
public class MainActivity extends Activity {
    private GameView mGameView = null;
    @Override
    public void onCreate(Bundle savedInstanceState) {
        super.onCreate(savedInstanceState);
        mGameView = new GameView(this);
        setContentView(mGameView);
    }
    public boolean onKeyUp(int keyCode, KeyEvent event) {
        super.onKeyUp(keyCode, event);
        return true;
```

```
            }
            public boolean onKeyDown(int keyCode, KeyEvent event) {
                if (mGameView == null) {
                    return false;
                }
                if (keyCode == KeyEvent.KEYCODE_BACK) {
                    this.finish();
                    return true;
                }
                return mGameView.onKeyDown(keyCode, event);
            }
        }
```

(3) 在 src\fs.shader_test 包下创建一个 GameView 类，实现图像几种渲染效果。代码为：

```
        package fs.shader_test;
        import android.content.Context;
        import android.graphics.Bitmap;
        import android.graphics.BitmapShader;
        import android.graphics.Canvas;
        import android.graphics.Color;
        import android.graphics.ComposeShader;
        import android.graphics.LinearGradient;
        import android.graphics.Paint;
        import android.graphics.PorterDuff;
        import android.graphics.RadialGradient;
        import android.graphics.Shader;
        import android.graphics.SweepGradient;
        import android.graphics.drawable.BitmapDrawable;
        import android.graphics.drawable.ShapeDrawable;
        import android.graphics.drawable.shapes.OvalShape;
        import android.view.KeyEvent;
        import android.view.MotionEvent;
        import android.view.View;
        public class GameView extends View implements Runnable
        {
            /* 声明 Bitmap 对象 */
            BitmapmBitQQ = null;
            intBitQQwidth = 0;
            intBitQQheight = 0;
            Paint mPaint = null;
            /* Bitmap 渲染 */
            Shader mBitmapShader = null;
            /* 线性渐变渲染 */
            Shader mLinearGradient = null;
            /* 混合渲染 */
            Shader mComposeShader = null;
            /* 唤醒渐变渲染 */
            Shader mRadialGradient = null;
            /* 梯度渲染 */
            Shader mSweepGradient = null;
            ShapeDrawable mShapeDrawableQQ = null;
            public GameView(Context context)
            {
```

```
        super(context);
        /* 装载资源 */
        mBitQQ = ((BitmapDrawable) getResources().getDrawable(R.drawable.gud2)).getBitmap();
        /* 得到图片的宽度和高度 */
        BitQQwidth = mBitQQ.getWidth();
        BitQQheight = mBitQQ.getHeight();
        /* 创建 BitmapShader 对象 */
        mBitmapShader = new BitmapShader(mBitQQ,Shader.TileMode.REPEAT,Shader.TileMode.
MIRROR);
        /* 创建 LinearGradient 并设置渐变的颜色数组,说明参数如下
         * 第1个:起始的x坐标
         * 第2个:起始的y坐标
         * 第3个:结束的x坐标
         * 第4个:结束的y坐标
         * 第5个:颜色数组
         * 第6个:这个也是用一个数组来指定颜色数组的相对位置的,如果为null就沿坡度线均
           匀分布
         * 第7个:渲染模式
         * */
        mLinearGradient = new LinearGradient(0,0,100,100,
                new int[]{Color.RED,Color.GREEN,Color.BLUE,Color.WHITE},
                                            null,Shader.TileMode.REPEAT);
        /* 这里理解为混合渲染 */
        mComposeShader = new ComposeShader(mBitmapShader, mLinearGradient, PorterDuff.Mode.
DARKEN);
        /* 构建 RadialGradient 对象,设置半径的属性 */
        //这里使用了 BitmapShader 和 LinearGradient 进行混合
        //当然也可以使用其他的组合
        //混合渲染的模式很多,可以根据自己需要来选择
        mRadialGradient = new RadialGradient(50,200,50,
                new int[]{Color.GREEN,Color.RED,Color.BLUE,Color.WHITE},
                                            null,Shader.TileMode.REPEAT);
        /* 构建 SweepGradient 对象 */
        mSweepGradient = new SweepGradient(30,30,new int[]{Color.GREEN,Color.RED,Color.BLUE,
Color.WHITE},null);
        mPaint = new Paint();
        /* 开启线程 */
        new Thread(this).start();
    }
    public void onDraw(Canvas canvas)
    {
        super.onDraw(canvas);
        //将图片裁剪为椭圆形
        /* 构建 ShapeDrawable 对象并定义形状为椭圆 */
        mShapeDrawableQQ = new ShapeDrawable(new OvalShape());
        /* 设置要绘制的椭圆形的为 ShapeDrawable 图片 */
        mShapeDrawableQQ.getPaint().setShader(mBitmapShader);
        /* 设置显示区域 */
        mShapeDrawableQQ.setBounds(0,0, BitQQwidth, BitQQheight);
        /* 绘制 ShapeDrawableQQ */
        mShapeDrawableQQ.draw(canvas);
        //绘制渐变的矩形
        mPaint.setShader(mLinearGradient);
        canvas.drawRect(BitQQwidth, 0, 320, 156, mPaint);
        //显示混合渲染效果
```

```java
            mPaint.setShader(mComposeShader);
            canvas.drawRect(0, 300, BitQQwidth, 300 + BitQQheight, mPaint);
            //绘制环形渐变
            mPaint.setShader(mRadialGradient);
            canvas.drawCircle(50, 200, 50, mPaint);
            //绘制梯度渐变
            mPaint.setShader(mSweepGradient);
            canvas.drawRect(150, 160, 300, 300, mPaint);
    }
    //触笔事件
    public boolean onTouchEvent(MotionEvent event)
    {
            return true;
    }
    //按键按下事件
    public boolean onKeyDown(int keyCode, KeyEvent event)
    {
            return true;
    }
    //按键弹起事件
    public boolean onKeyUp(int keyCode, KeyEvent event)
    {
            return false;
    }
    public boolean onKeyMultiple(int keyCode, int repeatCount, KeyEvent event)
    {
            return true;
    }
    /**
     * 线程处理
     */
    public void run()
    {
        while (!Thread.currentThread().isInterrupted())
        {
            try
            {
                Thread.sleep(100);
            }
            catch (InterruptedException e)
            {
                Thread.currentThread().interrupt();
            }

            //使用postInvalidate可以直接在线程中更新界面
            postInvalidate();
        }
    }
}
```

运行程序,图像的几种渲染效果如图4-11所示。

由以上实例可知,Shader 类包括了 5 个直接子类,分别为 BitmapShader(图像渲染)、LinearGradient(线性渲染)、ComposeShader(混合渲染)、RadialGradient(环形渲染)以及

图 4-11 图像的渲染效果

SweepGradient(梯度渲染)。

使用 Shader 类进行图像渲染时,首先需要构建 Shader 对象,然后通过 Paint 的 setShader() 方法来设置渲染对象,最后将这个 Paint 对象绘制到屏幕上即可。

注意:使用不同的方式渲染图像时需要构建不同的对象。

1) BitmapShader(图像渲染)

BitmapShader 的作用是使用一张位图作为纹理来对某一区域进行填充。可以想象成在一块区域内铺瓷砖,只是这里的瓷砖是一张张位图而已。

BitmapShader 函数原型为:

public BitmapShader (Bitmap bitmap, Shader.TileMode tileX, Shader.TileMode tileY);

其中,参数 bitmap 表示用来作为纹理填充的位图;参数 tileX 表示在位图 X 方向上位图衔接形式;参数 tileY 表示在位图 Y 方向上位图衔接形式。

2) LinearGradient(线性渲染)

LinearGradient 的作用是实现某一区域内颜色的线性渐变效果。

LinearGradient 的函数原型为:

public LinearGradient (float x0, float y0, float x1, float y1, int[] colors, float[] positions, Shader.TileMode tile);

其中,参数 $x0$ 表示渐变的起始点 x 坐标;参数 $y0$ 表示渐变的起始点 y 坐标;参数 $x1$ 表示渐变的终点 x 坐标;参数 $y1$ 表示渐变的终点 y 坐标;参数 colors 表示渐变的颜色数组;参数 positions 用来指定颜色数组的相对位置;参数 tile 表示平铺方式。

3) ComposeShader(混合渲染)

ComposeShader 的作用是实现渲染效果的叠加,如 BitmapShader 与 LinearGradient 的混合渲染效果等。

ComposeShader 的函数原型为:

public ComposeShader (Shader shaderA, Shader shaderB, PorterDuff.Mode mode);

其中,参数 shaderA 表示某一种渲染效果;参数 shaderB 也表示某一种渲染效果;参数 mode 表示两种渲染效果的叠加模式。

4) RadialGradient(环形渲染)

RadialGradient 的作用是在某一区域内实现环形的渐变效果。

RadialGradient 的函数原型为:

```
public RadialGradient (float x, float y, float radius, int[] colors, float[] positions, Shader.
TileMode tile);
```

其中,参数 x 表示环形的圆心 x 坐标;参数 y 表示环形的圆心 y 坐标;参数 radius 表示环形的半径;参数 colors 表示环形渐变的颜色数组;参数 positions 用来指定颜色数组的相对位置;参数 tile 表示平铺的方式。

5) SweepGradient(梯度渲染)

SweepGradient 也称为扫描渲染,是指在某一中心以 x 轴正方向逆时针旋转一周而形成的扫描效果的渲染形式。

SweepGradient 的函数原型为:

```
public SweepGradient (float cx, float cy, int[] colors, float[] positions);
```

其中,参数 cx 表示扫描的中心 x 坐标;参数 cy 表示扫描的中心 y 坐标;参数 colors 表示梯度渐变的颜色数组;参数 positions 用来指定颜色数组的相对位置。

4.13 示波器实例

SurfaceView 一般会与 SurfaceHolder 结合使用,SurfaceHolder 用于在与之关联的 SurfaceView 上绘图,调用 SurfaceView 的 getHolder()方法即可获取 SurfacView 关联的 SurfaceHolder。

本实例利用 SurfaceView 来实现一个绘制示波器程序,通过本实例演示了 SurfaceView 的具体用法。

本实例的具体实现步骤如下。

(1) 在 Eclipse 中创建一个 Android 应用项目,命名为 SurfaceView_test。

(2) 打开 res\layout 目录下的 main.xml 布局文件,在文件中声明两个 Button 控件及一个 SurfaceView 控件。代码为:

```xml
<?xml version = "1.0" encoding = "utf-8"?>
<LinearLayout xmlns:android = "http://schemas.android.com/apk/res/android"
        android:layout_width = "fill_parent"
        android:layout_height = "fill_parent"
        android:orientation = "vertical">
    <LinearLayout
        android:id = "@ + id/LinearLayout01"
        android:layout_width = "wrap_content"
        android:layout_height = "wrap_content">
        <Button
            android:id = "@ + id/Button01"
            android:layout_width = "wrap_content"
```

```xml
            android:layout_height = "wrap_content"
            android:text = "简单绘画"/>
        <Button
            android:id = "@ + id/Button02"
            android:layout_width = "wrap_content"
            android:layout_height = "wrap_content"
            android:text = "定时器绘画"/>
    </LinearLayout>
    <SurfaceView
        android:id = "@ + id/SurfaceView01"
        android:layout_width = "fill_parent"
        android:layout_height = "fill_parent"/>
</LinearLayout>
```

（3）打开 src\fs.surfaceview_test 包下的 MainActivity.java 文件，在文件中实现当单击界面中的"简单绘画"按钮时，即实现快速绘制正弦曲线，单击界面中的"定时器绘画"按钮时，即实现慢慢绘制余弦曲线。代码为：

```java
package fs.surfaceview_test;
import java.util.Timer;
import java.util.TimerTask;
import android.app.Activity;
import android.graphics.Canvas;
import android.graphics.Color;
import android.graphics.Paint;
import android.graphics.Rect;
import android.os.Bundle;
import android.util.Log;
import android.view.SurfaceHolder;
import android.view.SurfaceView;
import android.view.View;
import android.widget.Button;
public class MainActivity extends Activity {
    /** 第一次调用 Activity 活动 */
    Button btnSimpleDraw, btnTimerDraw;
    SurfaceView sfv;
    SurfaceHolder sfh;
    private Timer mTimer;
    private MyTimerTask mTimerTask;
    int Y_axis[],                         //保存正弦波的 Y 轴上的点
    centerY,                              //中心线
    oldX, oldY,                           //上一个 X 和 Y 点
    currentX;                             //当前绘制到的 X 轴上的点
    @Override
    public void onCreate(Bundle savedInstanceState) {
        super.onCreate(savedInstanceState);
        setContentView(R.layout.main);
        btnSimpleDraw = (Button) this.findViewById(R.id.Button01);
        btnTimerDraw = (Button) this.findViewById(R.id.Button02);
        btnSimpleDraw.setOnClickListener(new ClickEvent());
        btnTimerDraw.setOnClickListener(new ClickEvent());
        sfv = (SurfaceView) this.findViewById(R.id.SurfaceView01);
        sfh = sfv.getHolder();
        //动态绘制正弦波的定时器
        mTimer = new Timer();
```

```java
        mTimerTask = new MyTimerTask();
        //初始化Y轴数据
        centerY = (getWindowManager().getDefaultDisplay().getHeight() - sfv
                .getTop()) / 2;
        Y_axis = new int[getWindowManager().getDefaultDisplay().getWidth()];
        for (int i = 1; i < Y_axis.length; i++) {                //计算正弦波
            Y_axis[i - 1] = centerY
                    - (int) (100 * Math.sin(i * 2 * Math.PI / 180));
        }
    }
    class ClickEvent implements View.OnClickListener {
        @Override
        public void onClick(View v) {
            if (v == btnSimpleDraw) {
                SimpleDraw(Y_axis.length - 1);                   //直接绘制正弦波

            } else if (v == btnTimerDraw) {
                oldY = centerY;
                mTimer.schedule(mTimerTask, 0, 5);               //动态绘制正弦波
            }
        }
    }
    class MyTimerTask extends TimerTask {
        @Override
        public void run() {
            SimpleDraw(currentX);
            currentX++;                                          //前进
            if (currentX == Y_axis.length - 1) {                 //如果到了终点,则清屏重来
                ClearDraw();
                currentX = 0;
                oldY = centerY;
            }
        }
    }
    /**
     * 绘制指定区域
     */
    void SimpleDraw(int length) {
        if (length == 0)
            oldX = 0;
        Canvas canvas = sfh.lockCanvas(new Rect(oldX, 0, oldX + length,
                getWindowManager().getDefaultDisplay().getHeight()));    //关键:获取画布
        Log.i("Canvas:",
                String.valueOf(oldX) + "," + String.valueOf(oldX + length));
        Paint mPaint = new Paint();
        mPaint.setColor(Color.GREEN);                            //画笔为绿色
        mPaint.setStrokeWidth(2);                                //设置画笔粗细
        int y;
        for (int i = oldX + 1; i < length; i++) {                //绘画正弦波
            y = Y_axis[i - 1];
            canvas.drawLine(oldX, oldY, i, y, mPaint);
            oldX = i;
            oldY = y;
        }
        sfh.unlockCanvasAndPost(canvas);                         //解锁画布,提交画好的图像
```

```
    }
    void ClearDraw() {
        Canvas canvas = sfh.lockCanvas(null);
        canvas.drawColor(Color.BLACK);                              //清除画布
        sfh.unlockCanvasAndPost(canvas);
    }
}
```

运行程序,默认效果如图 4-12(a)所示,当单击界面中的"简单绘画"按钮时,效果如图 4-12(b)所示,当单击"定时器绘画"按钮时,效果如图 4-12(c)所示。

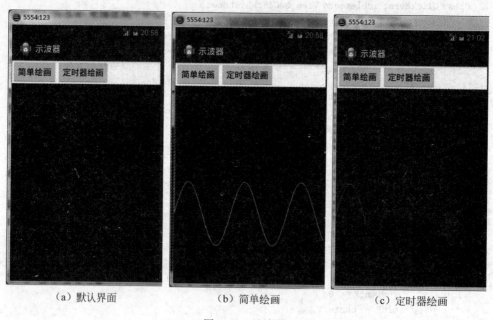

(a) 默认界面　　　　　　(b) 简单绘画　　　　　　(c) 定时器绘画

图 4-12　示波器

第 5 章 Android 通信服务实例

本章主要介绍 Android 通信服务及其控制,Android 的通信服务及控制功能是非常强大的,如何巧妙地运用这些功能呢? 本章将进行介绍。

5.1 还原桌面实例

手机的桌面背景可以进行设置,同样可以将其还原,那么怎样在 Android 中实现背景的还原呢? 本实例进行演示。

本实例通过使用 clearWallpaper()方法,实现自动将更改的桌面背景还原为默认背景。通过本实例演示了 clearWallpaper()方法的具体用法。

本实例的具体实现步骤如下。

(1) 在 Eclipse 中新建一个 Android 应用项目,命名为 clearWallpaper_test。

(2) 打开 res\layout 目录下的 main.xml 布局文件,在文件中定义一个 Button 控件。代码为:

```xml
<?xml version = "1.0" encoding = "utf - 8"?>
<LinearLayout xmlns:android = "http://schemas.android.com/apk/res/android"
    android:orientation = "vertical"
    android:layout_width = "fill_parent"
    android:layout_height = "fill_parent"
    android:background = "#f0f">
<Button
android:text = "还原默认桌面"
android:id = "@ + id/Button01"
android:layout_width = "fill_parent"
android:layout_height = "wrap_content"/>
</LinearLayout>
```

(3) 打开 src\fs.clearwallpaper_test 包下的 MainActivity.java 文件,在文件中实现还原桌面背景,并实现 Toast 提示。代码为:

```java
package fs.clearwallpaper_test;
import android.app.Activity;
import android.os.Bundle;
import android.view.View;
import android.view.View.OnClickListener;
```

```java
import android.widget.Button;
import android.widget.Toast;
public class MainActivity extends Activity {
    Button button;
    @Override
    public void onCreate(Bundle savedInstanceState) {
        super.onCreate(savedInstanceState);
        setContentView(R.layout.main);
        button = (Button)this.findViewById(R.id.Button01);          //获取 Button 对象
        button.setOnClickListener
        (
                new OnClickListener()
                {
                    public void onClick(View v) {
                        try
                        {
                            MainActivity.this.clearWallpaper();
                            Toast.makeText(
                                    MainActivity.this,
                                    "桌面背景已还原为默认设置!!",
                                    Toast.LENGTH_SHORT).show();
                        }catch(Exception e)
                        {
                            e.printStackTrace();
                        }
                    }

                }
        );
    }
}
```

(4) 打开 AndroidManifest.xml 文件，为文件添加权限。代码为：

```xml
<?xml version = "1.0" encoding = "utf-8"?>
<manifest xmlns:android = "http://schemas.android.com/apk/res/android"
    package = "fs.clearwallpaper_test"
    android:versionCode = "1"
    android:versionName = "1.0" >
    <uses-sdk
        android:minSdkVersion = "8"
        android:targetSdkVersion = "18" />
    <application
        android:allowBackup = "true"
        android:icon = "@drawable/ic_launcher"
        android:label = "@string/app_name"
        android:theme = "@style/AppTheme" >
        <activity
            android:name = "fs.clearwallpaper_test.MainActivity"
            android:label = "@string/app_name" >
            <intent-filter >
                <action android:name = "android.intent.action.MAIN" />
                <category android:name = "android.intent.category.LAUNCHER" />
            </intent-filter>
        </activity>
    </application>
```

```
<!-- 设置权限 -->
<uses-permission android:name = "android.permission.SET_WALLPAPER"></uses-permission>
</manifest>
```

运行程序，默认效果如图 5-1(a)所示，单击界面中的"还原默认桌面"按钮，即显示 Toast 提示，效果如图 5-1(b)所示。

（a）默认界面　　　　　　　　　　（b）Toast 提示

图 5-1　还原桌面背景

5.2　数据交换实例

在 Android 中，提供了 Activity 实现活动，它是 Android 中最基本的模块，提供了和用户交互的可视化界面。一个 Android 应用程序可以只有一个 Activity，也可以包含多个，每个 Activity 的作用及其数目，取决于应用程序及其设计。

当在一个 Activity 中启动另一个 Activity 时，经常需要传递一些数据过去。这时就可以通过 Intent 来实现，因为 Intent 通常被称为是两个 Activity 之间的信使，通过将要传递的数据保存在 Intent 中，就可以将其传递到另一个 Activity 中了。

在 Android 中，可以将要保存的数据存放在 Bundle 对象中，然后通过 Intent 提供的 putExtras 方法将要携带的数据保存到 Intent 中。

本实例使用 Bundle 在 Activity 之间交换数据，实现根据输入的性别和身高计算标准体重。通过本实例演示了 Android 数据交换的典型应用。

本实例的具体实现步骤如下。

（1）在 Eclipse 中创建一个 Android 应用项目，命名为 Activity_test。

（2）打开 res\layout 目录下的 main.xml 布局文件，在文件中声明一个 Button 控件、一个 EditText 控件及一组 RadioButton 控件。代码为：

```
<?xml version = "1.0" encoding = "utf-8"?>
<LinearLayout xmlns:android = "http://schemas.android.com/apk/res/android"
```

```xml
    android:layout_width = "fill_parent"
    android:layout_height = "fill_parent"
    android:orientation = "vertical"
    android:background = "#f0f">
<TextView
    android:layout_width = "fill_parent"
    android:layout_height = "wrap_content"
    android:layout_gravity = "center_horizontal"
    android:padding = "20px"
    android:text = "计算标准体重" />
<LinearLayout
    android:id = "@+id/linearLayout1"
    android:layout_width = "match_parent"
    android:layout_height = "wrap_content"
    android:gravity = "center_vertical" >
    <TextView
        android:id = "@+id/textView1"
        android:layout_width = "wrap_content"
        android:layout_height = "wrap_content"
        android:text = "性别: " />
    <RadioGroup
        android:id = "@+id/sex"
        android:layout_width = "wrap_content"
        android:layout_height = "wrap_content"
        android:orientation = "horizontal" >
        <RadioButton
            android:id = "@+id/radio0"
            android:layout_width = "wrap_content"
            android:layout_height = "wrap_content"
            android:checked = "true"
            android:text = "男" />
        <RadioButton
            android:id = "@+id/radio1"
            android:layout_width = "wrap_content"
            android:layout_height = "wrap_content"
            android:text = "女" />
    </RadioGroup>
</LinearLayout>
<LinearLayout
    android:id = "@+id/linearLayout1"
    android:layout_width = "match_parent"
    android:layout_height = "wrap_content"
    android:gravity = "center_vertical" >
    <TextView
        android:id = "@+id/textView1"
        android:layout_width = "wrap_content"
        android:layout_height = "wrap_content"
        android:text = "身高: " />
    <EditText
        android:id = "@+id/stature"
        android:layout_width = "wrap_content"
        android:layout_height = "wrap_content"
        android:minWidth = "100px"/>
    <TextView
        android:id = "@+id/textView2"
```

```
            android:layout_width = "wrap_content"
            android:layout_height = "wrap_content"
            android:text = "cm" />
    </LinearLayout>
    <Button
        android:id = "@ + id/button1"
        android:layout_width = "wrap_content"
        android:layout_height = "wrap_content"
        android:text = "计算" />
</LinearLayout>
```

(3) 在 res\layout 目录下新建一个 result.xml 布局文件，在文件中声明 3 个 TextView 组件，分别用于显示性别、身高和计算后的标准体重。代码为：

```
<?xml version = "1.0" encoding = "utf - 8"?>
<LinearLayout xmlns:android = "http://schemas.android.com/apk/res/android"
    android:layout_width = "match_parent"
    android:layout_height = "match_parent"
    android:orientation = "vertical"
    android:background = "#0f0" >
    <TextView
        android:id = "@ + id/sex"
        android:layout_width = "wrap_content"
        android:layout_height = "wrap_content"
        android:padding = "10px"
        android:text = "性别" />
    <TextView
        android:id = "@ + id/stature"
        android:layout_width = "wrap_content"
        android:layout_height = "wrap_content"
        android:padding = "10px"
        android:text = "身高" />
    <TextView
        android:id = "@ + id/weight"
        android:padding = "10px"
        android:layout_width = "wrap_content"
        android:layout_height = "wrap_content"
        android:text = "标准体重" />
</LinearLayout>
```

(4) 打开 src\fs.activity_test 包下的 MainActivity.java 文件，在文件中实例化一个保存性别和身高的可序列化对象 info，并判断输入的身高是否为空。然后实例化一个 Bundle 对象，并将输入的身高和性别保存到 Bundle 对象中。接着再创建一个启动显示结果 Activity 的 intent 对象，并将 Bundle 对象保存到该 intent 对象中。最后启动 intent 对应的 Activity。代码为：

```
package fs.activity_test;
import android.app.Activity;
import android.content.Intent;
import android.os.Bundle;
import android.view.View;
import android.view.View.OnClickListener;
import android.widget.Button;
import android.widget.EditText;
```

```java
import android.widget.RadioButton;
import android.widget.RadioGroup;
import android.widget.Toast;
public class MainActivity extends Activity {
    /** 第一次调用 Activity 活动 */
    @Override
    public void onCreate(Bundle savedInstanceState) {
        super.onCreate(savedInstanceState);
        setContentView(R.layout.main);
        Button button = (Button)findViewById(R.id.button1);
        button.setOnClickListener(new OnClickListener() {
            @Override
            public void onClick(View v) {
                Info info = new Info();                    //实例化一个保存输入基本信息的对象
                if("".equals(((EditText)findViewById(R.id.stature)).getText().toString())){
                    Toast.makeText(MainActivity.this, "请输入身高,否则不能计算!", Toast.
                    LENGTH_SHORT).show();
                    return;
                }
                int stature = Integer.parseInt(((EditText)findViewById(R.id.stature)).
                getText().toString());
                RadioGroup sex = (RadioGroup)findViewById(R.id.sex);//获取设置性别的单选按钮组
                //获取单选按钮组的值
                for(int i = 0;i < sex.getChildCount();i++){
                    RadioButton r = (RadioButton)sex.getChildAt(i); //根据索引值获取单选按钮
                    if(r.isChecked()){                               //判断单选按钮是否被选中
                        info.setSex(r.getText().toString());         //获取被选中的单选按钮的值
                        break;                                       //跳出 for 循环
                    }
                }
                info.setStature(stature);                  //设置身高
                Bundle bundle = new Bundle();              //实例化一个 Bundle 对象
                bundle.putSerializable("info", info);      //将输入的基本信息保存到 Bundle 对象中
                Intent intent = new Intent(MainActivity.this,ResultActivity.class);
                intent.putExtras(bundle);                  //将 bundle 保存到 Intent 对象中
                startActivity(intent);                     //启动 intent 对应的 Activity
            }
        });
    }
}
```

(5) 在 src\fs.activity_test 包下新建一个 info.java 文件,在文件中创建两个变量,一个用于保存性别,一个用于保存身高,并为这两个属性添加对应的 setter 和 getter 方法。代码为:

```java
package fs.activity_test;
import java.io.Serializable;
public class Info implements Serializable {
    private static final long serialVersionUID = 1L;
    private String sex = "";                       //性别
    private int stature = 0;                       //身高
    public String getSex() {
        return sex;
    }
    public void setSex(String sex) {
        this.sex = sex;
```

```
    }
    public int getStature() {
        return stature;
    }
    public void setStature(int stature) {
        this.stature = stature;
    }
}
```

(6) 在 src\fs.activity_test 包下新建一个 ResultActivity.java 文件,用于根据身高和性别计算标准体重。代码为:

```
package fs.activity_test;
import java.text.DecimalFormat;
import java.text.NumberFormat;
import android.app.Activity;
import android.content.Intent;
import android.os.Bundle;
import android.widget.TextView;
public class ResultActivity extends Activity {
    @Override
    protected void onCreate(Bundle savedInstanceState) {
        super.onCreate(savedInstanceState);
        setContentView(R.layout.result);                           //设置该 Activity 使用的布局
        TextView sex = (TextView)findViewById(R.id.sex);           //获取显示性别的文本框
        TextView stature = (TextView)findViewById(R.id.stature);   //获取显示身高的文本框
        TextView weight = (TextView)findViewById(R.id.weight);     //获取显示标准体重的文本框
        Intent intent = getIntent();                               //获取 Intent 对象
        Bundle bundle = intent.getExtras();                        //获取传递的数据包
        Info info = (Info)bundle.getSerializable("info");          //获取一个可序列化的 info 对象
        sex.setText("您是一位" + info.getSex() + "士");              //获取性别并显示到相应文本框中
        stature.setText("身高是" + info.getStature() + "厘米");     //获取身高并显示到相应文本框中
        //显示计算后的标准体重
        weight.setText("标准体重是" + getWeight(info.getSex(),info.getStature()) + "公斤");
    }
    /**
     * 功能:计算标准体重
     */
    private String getWeight(String sex,float stature){
        String weight = "";                                        //保存体重
        NumberFormat format = new DecimalFormat();

        if(sex.equals("男")){                                      //计算男士标准体重
            weight = format.format((stature - 80) * 0.7);
        }else{                                                     //计算女士标准体重
            weight = format.format((stature - 70) * 0.6);
        }
        return weight;
    }
}
```

(7) 打开 AndroidManifest.xml 文件,在文件中配置 ResultActivity,配置的主要属性有 Activity 使用标准、图标和实现类。代码为:

```
<?xml version = "1.0" encoding = "utf - 8"?>
```

```xml
<manifest xmlns:android = "http://schemas.android.com/apk/res/android"
    package = "fs.activity_test"
    android:versionCode = "1"
    android:versionName = "1.0" >
    <uses-sdk
        android:minSdkVersion = "8"
        android:targetSdkVersion = "18" />
    <application
        android:allowBackup = "true"
        android:icon = "@drawable/ic_launcher"
        android:label = "@string/app_name"
        android:theme = "@style/AppTheme" >
        <activity
            android:name = "fs.activity_test.MainActivity"
            android:label = "@string/app_name" >
            <intent-filter>
                <action android:name = "android.intent.action.MAIN" />
                <category android:name = "android.intent.category.LAUNCHER" />
            </intent-filter>
        </activity>
        <activity
            android:label = "显示结果"
            android:icon = "@drawable/ic_launcher"
            android:name = ".ResultActivity">
        </activity>
    </application>
</manifest>
```

运行程序，默认界面如图5-2(a)所示，选择相应的性别及输入对应的身高，效果如图5-2(b)所示，最后单击界面中的"计算"按钮，即显示对应的标准体重，效果如图5-2(c)所示。

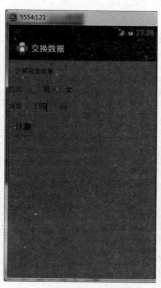

(a) 默认界面　　　　　(b) 选择性别及输入身高　　　　　(c) 显示标准体重

图5-2　计算标准体重

5.3 查询星座实例

在占星学上,黄道十二星座是宇宙方位的代名词,十二星座代表了 12 种基本性格原型,一个人出生时各星体落入黄道上的位置,正是说明着一个人的先天性格及天赋。因此,现在很多人都希望知道自己的星座。

本实例将实现根据文本的输入的阳历生日判断所属星座。通过本实例演示了 Activity 活动的具体用法。

本实例的具体实现步骤如下。

(1) 在 Eclipse 中创建一个 Android 应用项目,命名为 Constellation_test。

(2) 打开 res\layout 目录下的 main.xml 布局文件,在文件中声明 3 个 TextView 控件、一个 EditText 控件及一个 Button 控件。代码为:

```xml
<?xml version = "1.0" encoding = "utf - 8"?>
<LinearLayout xmlns:android = "http://schemas.android.com/apk/res/android"
    android:layout_width = "fill_parent"
    android:layout_height = "fill_parent"
    android:orientation = "vertical"
    android:background = "#f0f">
    <TextView
        android:layout_width = "fill_parent"
        android:layout_height = "wrap_content"
        android:layout_gravity = "center_horizontal"
        android:padding = "20dp"
        android:text = "计算星座" />
    <LinearLayout
        android:id = "@ + id/linearLayout1"
        android:gravity = "center_vertical"
        android:layout_width = "match_parent"
        android:layout_height = "wrap_content" >
        <TextView
            android:id = "@ + id/textView1"
            android:layout_width = "wrap_content"
            android:layout_height = "wrap_content"
            android:text = "阳历生日: " />
        <EditText
            android:id = "@ + id/birthday"
            android:minWidth = "100dp"
            android:layout_width = "wrap_content"
            android:layout_height = "wrap_content"/>
        <TextView
            android:id = "@ + id/textView2"
            android:layout_width = "wrap_content"
            android:layout_height = "wrap_content"
            android:text = "格式: YYYY - MM - DD 例如: 2012 - 01 - 01" />
    </LinearLayout>
    <Button
        android:id = "@ + id/button1"
        android:layout_width = "wrap_content"
        android:layout_height = "wrap_content"
        android:text = "查询" />
```

```
    </LinearLayout>
```

(3) 在 res\layout 目录下新建一个 result.xml 布局文件,在文件中定义两个 TextView 控件,分别用于显示生日和计算结果。代码为:

```
<?xml version = "1.0" encoding = "utf - 8"?>
<LinearLayout xmlns:android = "http://schemas.android.com/apk/res/android"
    android:layout_width = "match_parent"
    android:layout_height = "match_parent"
    android:orientation = "vertical"
    android:background = "#0f0" >
    <TextView
        android:id = "@ + id/birthday"
        android:layout_width = "wrap_content"
        android:layout_height = "wrap_content"
        android:padding = "10px"
        android:text = "阳历生日" />
    <TextView
        android:id = "@ + id/result"
        android:padding = "10px"
        android:layout_width = "wrap_content"
        android:layout_height = "wrap_content"
        android:text = "星座" />
</LinearLayout>
```

(4) 打开 src\fs.constellation_test 包下的 MainActivity.java 文件,在文件中实现单击事件监听器,并实例化一个保存生日的可序列化对象 info,并判断是否输入生日,如果没有输入,则给出消息提示,并返回,否则,首先获取生日并保存到 info 中,然后实例化一个 Bundle 对象,并将输入的生日保存到 Bundle 对象中,接着再创建一个启动显示结果 Activity 的 intent 对象,并将 Bundle 对象保存到该 intent 对象中,最后启动 intent 对应的 Activity。代码为:

```
package fs.constellation_test;
import android.app.Activity;
import android.content.Intent;
import android.os.Bundle;
import android.view.View;
import android.view.View.OnClickListener;
import android.widget.Button;
import android.widget.EditText;
import android.widget.Toast;
public class MainActivity extends Activity {
    @Override
    public void onCreate(Bundle savedInstanceState) {
        super.onCreate(savedInstanceState);
        setContentView(R.layout.main);
        Button button = (Button)findViewById(R.id.button1);
        button.setOnClickListener(new OnClickListener() {
            @Override
            public void onClick(View v) {
                Info info = new Info();           //实例化一个保存输入基本信息的对象
                if("".equals(((EditText)findViewById(R.id.birthday)).getText().toString())){
                    Toast.makeText(MainActivity.this, "请输入您的阳历生日,否则不能计算!",
                    Toast.LENGTH_SHORT).show();
                    return;
```

```
                    }
                    String birthday = ((EditText)findViewById(R.id.birthday)).getText().toString();
                      info.setBirthday(birthday);          //设置生日
                      Bundle bundle = new Bundle();         //实例化一个 Bundle 对象
                      bundle.putSerializable("info", info);
                                                           //将输入的基本信息保存到 Bundle 对象中
                      Intent intent = new Intent(MainActivity.this,ResultActivity.class);
                      intent.putExtras(bundle);            //将 bundle 保存到 Intent 对象中
                      startActivity(intent);               //启动 intent 对应的 Activity
                }
            });
       }
    }
```

(5) 在 src\fs.constellation_test 包下新建一个 Info.java 文件,用于保存生日的属性。代码为:

```
package fs.constellation_test;
import java.io.Serializable;
public class Info implements Serializable {
    private static final long serialVersionUID = 1L;
    private String birthday = "";              //生日
    public String getBirthday() {
        return birthday;
    }
    public void setBirthday(String birthday) {
        this.birthday = birthday;
    }
}
```

(6) 在 src\fs.constellation_test 包下新建一个 ResultActivity.java 文件,在文件中首先设置该 Activity 使用的布局文件 result.xml 中定义的布局,然后获取生日和显示结果文本框,再获取 intent 对象以及传递的数据包,最后将传递过来的生日和判断结果显示到对应的文本框中。代码为:

```
package fs.constellation_test;
import android.app.Activity;
import android.content.Intent;
import android.os.Bundle;
import android.widget.TextView;
public class ResultActivity extends Activity {
    @Override
    protected void onCreate(Bundle savedInstanceState) {
        super.onCreate(savedInstanceState);
        setContentView(R.layout.result);                        //设置该 Activity 使用的布局
TextView birthday = (TextView) findViewById(R.id.birthday);     //获取显示生日的文本框
    TextView result = (TextView) findViewById(R.id.result);     //获取显示星座的文本框
        Intent intent = getIntent();                            //获取 Intent 对象
        Bundle bundle = intent.getExtras();                     //获取传递的数据包
        Info info = (Info) bundle.getSerializable("info");      //获取一个可序列化的 info 对象
        birthday.setText("您的阳历生日是" + info.getBirthday());
                                                                //获取性别并显示到相应文本框中
        result.setText( query(info.getBirthday()));             //显示计算后的星座
    }
```

```java
/**
 * 根据生日查询星座
 */
public String query(String birthday) {
    int month = 0;                                                      //月 ·
    int day = 0;                                                        //日
    try{                                                                //捕获异常
        month = Integer.parseInt(birthday.substring(5, 7));             //获取输入的月份
        day = Integer.parseInt(birthday.substring(8, 10));              //获取输入的日
    }catch(Exception e){
        e.printStackTrace();
    }
    String name = "";                                                   //提示信息
    if (month > 0 && month < 13 && day > 0 && day < 32) { //如果输入的月和日有效
        if ((month == 3 && day > 20) || (month == 4 && day < 21)) {
            name = "您是白羊座!";
        } else if ((month == 4 && day > 20) || (month == 5 && day < 21)) {
            name = "您是金牛座!";
        } else if ((month == 5 && day > 20) || (month == 6 && day < 22)) {
            name = "您是双子座!";
        } else if ((month == 6 && day > 21) || (month == 7 && day < 23)) {
            name = "您是巨蟹座!";
        } else if ((month == 7 && day > 22) || (month == 8 && day < 23)) {
            name = "您是狮子座!";
        } else if ((month == 8 && day > 22) || (month == 9 && day < 23)) {
            name = "您是处女座!";
        } else if ((month == 9 && day > 22) || (month == 10 && day < 23)) {
            name = "您是天秤座!";
        } else if ((month == 10 && day > 22) || (month == 11 && day < 22)) {
            name = "您是天蝎座!";
        } else if ((month == 11 && day > 21) || (month == 12 && day < 22)) {
            name = "您是射手座!";
        } else if ((month == 12 && day > 21) || (month == 1 && day < 20)) {
            name = "您是摩羯座!";
        } else if ((month == 1 && day > 19) || (month == 2 && day < 19)) {
            name = "您是水牛座!";
        } else if ((month == 2 && day > 18) || (month == 3 && day < 21)) {
            name = "您是双鱼座!";
        }
        name = month + "月" + day + "日" + name;
    } else {//如果输入的月和日无效
        name = "您输入的生日格式不正确或者不是真实生日!";
    }
    return name;                                                        //返回星座或提示信息
}
```

(7) 打开 AndroidManifest.xml 文件,在文件中配置的主属性有 Activity 使用的标签、图标和实现类。代码为:

```xml
<?xml version = "1.0" encoding = "utf-8"?>
<manifest xmlns:android = "http://schemas.android.com/apk/res/android"
    package = "fs.constellation_test"
    android:versionCode = "1"
    android:versionName = "1.0" >
    <uses-sdk
```

```
            android:minSdkVersion = "8"
            android:targetSdkVersion = "18" />
    <application
        android:allowBackup = "true"
        android:icon = "@drawable/ic_launcher"
        android:label = "@string/app_name"
        android:theme = "@style/AppTheme" >
        <activity
            android:name = "fs.constellation_test.MainActivity"
            android:label = "@string/app_name" >
            <intent-filter>
                <action android:name = "android.intent.action.MAIN" />
                <category android:name = "android.intent.category.LAUNCHER" />
            </intent-filter>
        </activity>
        <activity
            android:label = "显示结果"
            android:icon = "@drawable/ic_launcher"
            android:name = ".ResultActivity">
        </activity>
    </application>
</manifest>
```

运行程序,默认界面如图 5-3(a)所示,在界面的文本框中输入对应的阳历,效果如图 5-3(b)所示,并单击界面中的"查询"按钮,即显示对应的查询结果,如图 5-3(c)所示。

(a)默认界面　　　　(b)输入对应生日　　　　(c)查询结果

图 5-3　星座查询

5.4　Intent 发短信实例

Intent 是一种运行时绑定(run-time binding)机制,它能在程序运行过程中连接两个不同的组件。通过 Intent,程序可以向 Android 表达某种请求或者意愿,Android 会根据意愿的内容选择适当的组件来完成请求。

Android 的 3 个基本组件——Activity、Service 及 BroadcastReceiver，都是通过 Intent 机制激活的，不同类型的组件有不同的传递 Intent 方式。

- Activity：通过将一个 Intent 对象传递给 Context.startActivity() 或 Activity.startActivityForResult()，启动一个活动或者使一个已存在的活动去做新的事情。
- Service：通过将一个 Intent 对象传递给 Context.startService()，初始化一个 Service 或者传递一个新的指令给正在运行的 Service；类似地，通过将一个 Intent 对象传递给 Context.bindService()，可以建立调用组件和目标服务之间的连接。
- BroadcastReceiver：通过将一个 Intent 对象传递给任何广播方法（如 Context.sendBroadcast()、Context.sendOrderedBroadcast() 和 Context.sendStickyBroadcast() 等），都可以传递到所有感兴趣的广播接收者。

下面实例使用 Intent 机制实现发送短信。通过本实例演示了 Intent 机制的具体用法。本实例的具体实现步骤如下。

（1）在 Eclipse 中创建一个 Android 应用项目，命名为 Intent_send。

（2）打开 res\layout 目录下的 main.xml 布局文件，在文件中增加文本框和按钮等控件。代码为：

```xml
<?xml version = "1.0" encoding = "utf-8"?>
<LinearLayout xmlns:android = "http://schemas.android.com/apk/res/android"
    android:layout_width = "fill_parent"
    android:layout_height = "fill_parent"
    android:background = "#f0f0"
    android:orientation = "vertical">
    <LinearLayout
        android:layout_width = "match_parent"
        android:layout_height = "wrap_content" >
        <TextView
            android:layout_width = "wrap_content"
            android:layout_height = "wrap_content"
            android:text = "号码"
            android:textColor = "@android:color/white"
            android:textSize = "25dp" />
        <EditText
            android:id = "@+id/number"
            android:layout_width = "0dip"
            android:layout_height = "wrap_content"
            android:layout_weight = "1"
            android:inputType = "number"
            android:textColor = "@android:color/white"
            android:textSize = "25dp" >
            <requestFocus />
        </EditText>
    </LinearLayout>
    <LinearLayout
        android:layout_width = "match_parent"
        android:layout_height = "wrap_content" >
        <EditText
            android:id = "@+id/message"
            android:layout_width = "0dip"
            android:layout_height = "wrap_content"
```

```
            android:layout_weight = "1"
            android:hint = "短信内容!"
            android:inputType = "textMultiLine"
            android:textColor = "@android:color/white"
            android:textSize = "25dp" />
    </LinearLayout>
    <Button
        android:id = "@ + id/send"
        android:layout_width = "wrap_content"
        android:layout_height = "wrap_content"
        android:text = "发送"
        android:textColor = "@android:color/white"
        android:textSize = "25dp" />
</LinearLayout>
```

（3）打开 src\fs.intent_send 包下的 MainActivity.java 文件，在文件中编写 SMSSenderActivity，通过为按钮增加单击事件监听器来完成发送短信功能。代码为：

```
package fs.intent_send;
import android.app.Activity;
import android.content.Intent;
import android.net.Uri;
import android.os.Bundle;
import android.view.View;
import android.widget.Button;
import android.widget.EditText;
public class MainActivity extends Activity {
    /** 第一次调用 Activity 活动 */
    @Override
    public void onCreate(Bundle savedInstanceState) {
        super.onCreate(savedInstanceState);
        setContentView(R.layout.main);                          //设置页面布局
        //通过 id 值获得文本框对象
        final EditText numberET = (EditText) findViewById(R.id.number);
        //通过 id 值获得文本框对象
        final EditText messageET = (EditText) findViewById(R.id.message);
        Button call = (Button) findViewById(R.id.send);         //通过 id 值获得按钮对象
        call.setOnClickListener(new View.OnClickListener() {
            public void onClick(View v) {
                String number = numberET.getText().toString();  //获得用户输入的号码
                String message = messageET.getText().toString(); //获得用户输入的短信
                Intent intent = new Intent();                    //创建 Intent 对象
                intent.setData(Uri.parse("smsto:" + number));    //设置要发送的号码
                intent.putExtra("sms_body", message);            //设置要发送的信息内容
                startActivity(intent);                           //将 Intent 传递给 Activity
            }
        });
    }
}
```

（4）打开 AndroidManifest.xml 文件，为发送短信添加权限。代码为：

```
<?xml version = "1.0" encoding = "utf-8"?>
<manifest xmlns:android = "http://schemas.android.com/apk/res/android"
    package = "fs.intent_send"
    android:versionCode = "1"
```

```xml
        android:versionName = "1.0" >
    < uses – sdk
        android:minSdkVersion = "8"
        android:targetSdkVersion = "18" />
<! -- 设置权限 -->
 < uses – permission android:name = "android.permission.SEND_SMS" />
 < application
     android:allowBackup = "true"
     android:icon = "@drawable/ic_launcher"
     android:label = "@string/app_name"
     android:theme = "@style/AppTheme" >
     < activity
         android:name = "fs.intent_send.MainActivity"
         android:label = "@string/app_name" >
         < intent – filter >
             < action android:name = "android.intent.action.MAIN" />
             < category android:name = "android.intent.category.LAUNCHER" />
         </ intent – filter >
     </ activity >
 </ application >
</ manifest >
```

运行程序，默认界面如图 5-4 所示，在界面中输入对应的号码及短信内容，单击界面中的"发送"按钮，即实现短信的发送。

图 5-4 Intent 发送短信

5.5 发送短信实例

发送短信是手机的一个重要功能，在 Android 平台下，也能够自己设计一个具有个性的短信发送软件。

本实例介绍怎样自定义一个短信编辑系统，然后调用 sendTextMessage() 方法发送已编辑的短信。本实例的具体实现步骤如下。

(1) 在 Eclipse 中创建一个 Android 应用项目,命名为 send_test。

(2) 打开 res\layout 目录下的 main.xml 布局文件,在文件中声明两个 TextView 控件、两个 EditText 控件及一个 Button 控件。代码为:

```xml
<?xml version = "1.0" encoding = "utf-8"?>
<LinearLayout xmlns:android = "http://schemas.android.com/apk/res/android"
    android:orientation = "vertical"
    android:layout_width = "fill_parent"
    android:layout_height = "fill_parent"
    android:background = "#f0f">
    <TextView
        android:layout_width = "fill_parent"
        android:layout_height = "wrap_content"
        android:text = "@string/str_input_phone_number"/>
    <EditText
        android:layout_width = "fill_parent"
        android:layout_height = "wrap_content"
        android:id = "@+id/phone_number_editText"/>
    <TextView
        android:layout_width = "fill_parent"
        android:layout_height = "wrap_content"
        android:text = "@string/str_input_sms_content"/>
    <EditText
        android:layout_width = "fill_parent"
        android:layout_height = "wrap_content"
        android:id = "@+id/sms_content_editText"/>
    <Button
        android:layout_width = "wrap_content"
        android:layout_height = "wrap_content"
        android:text = "@string/str_send_sms"
        android:id = "@+id/send_sms_button"/>
</LinearLayout>
```

(3) 打开 src\fs.send_test 包下的 MainActivity.java 文件,在文件中实现当输入号码及短信内容时,单击界面中的"发送短信"按钮,即实现短信发送。代码为:

```java
package fs.send_test;
import java.util.List;
import android.app.Activity;
import android.os.Bundle;
import android.telephony.SmsManager;
import android.view.View;
import android.view.View.OnClickListener;
import android.widget.Button;
import android.widget.EditText;
import android.widget.Toast;
public class MainActivity extends Activity {
    /** 第一次调用 Activity 活动 */
    @Override
    public void onCreate(Bundle savedInstanceState) {
        super.onCreate(savedInstanceState);
        setContentView(R.layout.main);
        phone_number_editText = (EditText) findViewById(R.id.phone_number_editText);
        sms_content_editText = (EditText) findViewById(R.id.sms_content_editText);
        send_sms_button = (Button) findViewById(R.id.send_sms_button);
```

```java
send_sms_button.setOnClickListener(new OnClickListener() {
    @Override
    public void onClick(View arg0) {
        String phone_number = phone_number_editText.getText().toString().trim();
        String sms_content = sms_content_editText.getText().toString().trim();
        if(phone_number.equals("")) {
            Toast.makeText(MainActivity.this, R.string.str_remind_input_phone_number, Toast.LENGTH_LONG).show();
        } else {
            SmsManager smsManager = SmsManager.getDefault();
            if(sms_content.length() > 70) {
                List<String> contents = smsManager.divideMessage(sms_content);
                for(String sms : contents) {
                    smsManager.sendTextMessage(phone_number, null, sms, null, null);
                }
            } else {
                smsManager.sendTextMessage(phone_number, null, sms_content, null, null);
            }
            Toast.makeText(MainActivity.this, R.string.str_remind_sms_send_finish, Toast.LENGTH_SHORT).show();
        }
    }
});
}
private EditText phone_number_editText;
private EditText sms_content_editText;
private Button send_sms_button;
}
```

（4）打开 AndroidManifest.xml 文件，为发送短信添加权限。代码为：

```xml
<?xml version="1.0" encoding="utf-8"?>
<manifest xmlns:android="http://schemas.android.com/apk/res/android"
    package="fs.send_test"
    android:versionCode="1"
    android:versionName="1.0" >
    <uses-sdk
        android:minSdkVersion="8"
        android:targetSdkVersion="18" />
    <application
        android:allowBackup="true"
        android:icon="@drawable/ic_launcher"
        android:label="@string/app_name"
        android:theme="@style/AppTheme" >
        <activity
            android:name="fs.send_test.MainActivity"
            android:label="@string/app_name" >
            <intent-filter>
                <action android:name="android.intent.action.MAIN" />
                <category android:name="android.intent.category.LAUNCHER" />
            </intent-filter>
        </activity>
    </application>
    <!-- 设置此应用程序具有发短信权限 -->
    <uses-permission android:name="android.permission.SEND_SMS" />
</manifest>
```

运行程序,默认界面如图 5-5(a)所示,输入对应的手机号及短信内容,效果如图 5-5(b)所示,单击界面中的"发送短信"按钮,即完成相应短信发送,效果如图 5-5(c)所示。

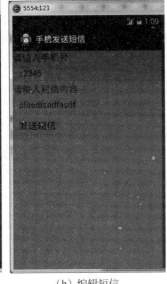

　　(a)默认界面　　　　　　(b)编辑短信　　　　　　(c)发送短信

图 5-5　手机发送短信

5.6　隐式 Intent 实例

Intent 可以分为以下两类。

(1) 显式 Intent 通过组件名称来指定目标组件。由于其他应用程序的组件名称对于开发人员通常是未知的,显式 Intent 通常用于应用程序内部消息,例如 Activity 启动子 Service 或其他 Activity。

(2) 隐式 Intent 不指定组件名称。隐式 Intent 通常用于激活其他应用程序中的组件。

当在 Android 程序中使用显式 Intent 时,Intent 对象中只用组件名字内容就可以决定哪个组件应该获得这个 Intent,而不用其他内容。而使用隐式 Intent 时,由于默认指定目标,Android 程序必须查找一个最适合的组件(一些组件)去处理 Intent,一个活动或服务去执行请求动作,或一组广播接收者去响应广播声明,该过程是通过比较 Intent 对象的内容和 Intent 过滤器(Intent Filters)来完成的。Intent 过滤器关联到潜在的接收 Intent 组件,过滤器声明组件的能力和界定它能处理的 Intent,它们打开组件接收声明的 Intent 类型的隐式 Intent。如果一个组件没有任何 Intent 过滤器,它仅能接收显式的 Intent,而声明了 Intent 过滤器的组件可以接收显式和隐式的 Intent。

本实例实现在 Activity 中使用包含预定义动作的隐式 Intent 启动另一个 Activity。通过本实例演示隐式 Intent 的具体用法。

本实例的具体实现步骤如下。

(1) 在 Eclipse 中创建一个 Android 应用项目,命名为 Intent_Predefined。

(2) 打开 res\layout 目录下的 main.xml,它是第一个 Activity 页面布局文件,在文件中声明一个 Button 控件。代码为：

```xml
<?xml version = "1.0" encoding = "utf-8"?>
<LinearLayout xmlns:android = "http://schemas.android.com/apk/res/android"
    android:layout_width = "fill_parent"
    android:layout_height = "fill_parent"
    android:orientation = "vertical"
    android:background = "#e0e" >
    <Button
        android:id = "@+id/button"
        android:layout_width = "wrap_content"
        android:layout_height = "wrap_content"
        android:text = "进入下一个 Activity 页面"
        android:textColor = "@android:color/white" />
</LinearLayout>
```

（3）在 res\layout 目录下新建一个 second_main.xml，它是第二个 Activity 页面布局文件，在文件中声明一个 TextView 控件。代码为：

```xml
<?xml version = "1.0" encoding = "utf-8"?>
<LinearLayout xmlns:android = "http://schemas.android.com/apk/res/android"
    android:layout_width = "fill_parent"
    android:layout_height = "fill_parent"
    android:background = "#00f"
    android:orientation = "vertical" >
    <TextView
        android:id = "@+id/textView"
        android:layout_width = "wrap_content"
        android:layout_height = "wrap_content"
        android:text = "第二个 Activity 页面"
        android:textColor = "@android:color/white"
        android:textSize = "25dp" />
</LinearLayout>
```

（4）打开 src\fs.intent_predefined 包下的 MainActivity 文件，在文件中实现第一个 Activity 界面，并为按钮控件添加单击监听事件，实现单击"进入一个 Activity 页面"按钮时，即跳转到下一个 Activity 页面。代码为：

```java
package fs.intent_predefined;
import android.app.Activity;
import android.content.Intent;
import android.os.Bundle;
import android.view.View;
import android.widget.Button;
public class MainActivity extends Activity {
    @Override
    protected void onCreate(Bundle savedInstanceState) {
        super.onCreate(savedInstanceState);
        setContentView(R.layout.main);                              //设置页面布局
        Button button = (Button) findViewById(R.id.button);         //通过 id 值获得按钮对象
        button.setOnClickListener(new View.OnClickListener() {
            //为按钮增加单击事件监听器
            public void onClick(View v) {
                Intent intent = new Intent();                       //创建 Intent 对象
                intent.setAction(Intent.ACTION_VIEW);                //为 Intent 设置动作
```

```
            startActivity(intent);                              //将 Intent 传递给 Activity
        }
    });
    }
}
```

(5) 在 src\fs.intent_predefined 包下新建一个 SecondMainActivity.java 文件，在文件中实现第二个 Activity 页面。代码为：

```
package fs.intent_predefined;
import android.app.Activity;
import android.os.Bundle;
public class SecondMainActivity extends Activity {
    @Override
    protected void onCreate(Bundle savedInstanceState) {
        super.onCreate(savedInstanceState);
        setContentView(R.layout.second_main);                   //设置页面布局
    }
}
```

(6) 打开 AndroidManifest.xml 文件，为两个 Activity 设置不同的 Intent 过滤器。代码为：

```xml
<?xml version = "1.0" encoding = "utf - 8"?>
<manifest xmlns:android = "http://schemas.android.com/apk/res/android"
    package = "fs.intent_predefined"
    android:versionCode = "1"
    android:versionName = "1.0" >
    <uses - sdk
        android:minSdkVersion = "8"
        android:targetSdkVersion = "18" />
    <application
        android:allowBackup = "true"
        android:icon = "@drawable/ic_launcher"
        android:label = "@string/app_name"
        android:theme = "@style/AppTheme" >
        <activity
            android:name = "fs.intent_predefined.MainActivity"
            android:label = "@string/app_name" >
            <intent - filter >
                <action android:name = "android.intent.action.MAIN" />
                <category android:name = "android.intent.category.LAUNCHER" />
            </intent - filter >
        </activity>
        <activity android:name = ".SecondMainActivity" >
            <intent - filter >
                <action android:name = "android.intent.action.VIEW" />
                <category android:name = "android.intent.category.DEFAULT" />
            </intent - filter >
        </activity>
    </application>
</manifest>
```

运行程序，默认界面如图 5-6(a)所示，当单击界面中的"进入下一个 Activity 页面"按钮时，弹出如图 5-6(b)所示的页面，在页面中选择"预定义隐式 Intent"项，并单击 Just once 按钮,即进入第二个 Activity 页面,效果如图 5-6(c)所示。

(a)第一个Activity页面

(b)应用选择页面

(c)第二个Activity页面

图 5-6　预定义隐式 Intent

上实例实现了预定义动作的隐式 Intent,本实例将实现在 Activity 中使用包含自定义动作的隐式 Intent 启动另外一个 Activity。通过本实例进一步演示了隐式 Intent 的用法。

本实例的具体实现步骤如下。

(1) 在 Eclipse 中创建一个 Android 应用项目,命名为 Intent_Custom。

(2) 打开 res\layout 目录下的 main.xml,它是第一个 Activity 页面布局文件,在文件中声明一个 Button 控件。代码为:

```
<?xml version = "1.0" encoding = "utf - 8"?>
<LinearLayout xmlns:android = "http://schemas.android.com/apk/res/android"
    android:layout_width = "fill_parent"
    android:layout_height = "fill_parent"
    android:background = "#ccc"
    android:orientation = "vertical" >
    <Button
        android:id = "@ + id/button"
        android:layout_width = "wrap_content"
        android:layout_height = "wrap_content"
        android:text = "进入下一个 Activity 页面"
        android:textColor = "@android:color/white" />
</LinearLayout>
```

(3) 在 res\layout 目录下新建一个 second_main.xml,它是第二个 Activity 页面布局文件,在文件中声明一个 TextView 控件。代码为:

```
<?xml version = "1.0" encoding = "utf - 8"?>
<LinearLayout xmlns:android = "http://schemas.android.com/apk/res/android"
    android:layout_width = "fill_parent"
    android:layout_height = "fill_parent"
    android:background = "#00e"
```

```xml
            android:orientation = "vertical" >
    < TextView
            android:id = "@ + id/textView"
            android:layout_width = "wrap_content"
            android:layout_height = "wrap_content"
            android:text = "第二个 Activity 页面"
            android:textColor = "@android:color/white"
            android:textSize = "25dp" />
</LinearLayout>
```

(4) 打开 src\fs.intent_custom 包下的 MainActivity.java 文件，在文件中实现第一个 Activity 页面，并为按钮控件添加单击监听事件，实现单击"进入一个 Activity 页面"按钮时，即跳转到下一个 Activity 页面。代码为：

```java
package fs.intent_custom;
import android.app.Activity;
import android.content.Intent;
import android.os.Bundle;
import android.view.View;
import android.widget.Button;
public class MainActivity extends Activity {
    @Override
    protected void onCreate(Bundle savedInstanceState) {
        super.onCreate(savedInstanceState);
        setContentView(R.layout.main);                              //设置页面布局
        Button button = (Button) findViewById(R.id.button);         //通过 id 值获得按钮对象
        button.setOnClickListener(new View.OnClickListener() {
        //为按钮增加单击事件监听器
            public void onClick(View v) {
                Intent intent = new Intent();                       //创建 Intent 对象
                intent.setAction("test_action");                    //为 Intent 设置动作
                startActivity(intent);                              //将 Intent 传递给 Activity
            }
        });
    }
}
```

(5) 在 src\fs.intent_predefined 包下新建一个 SecondMainActivity.java 文件，在文件中实现第二个 Activity 页面。代码为：

```java
package fs.intent_custom;
import android.app.Activity;
import android.os.Bundle;
public class SecondMainActivity extends Activity {
    @Override
    protected void onCreate(Bundle savedInstanceState) {
        super.onCreate(savedInstanceState);
        setContentView(R.layout.second_main);                       //设置页面布局
    }
}
```

(6) 打开 AndroidManifest.xml 文件，为两个 Activity 设置不同的 Intent 过滤器。代码为：

```xml
<?xml version = "1.0" encoding = "utf - 8"?>
```

```xml
<manifest xmlns:android="http://schemas.android.com/apk/res/android"
    package="fs.intent_custom"
    android:versionCode="1"
    android:versionName="1.0" >
    <uses-sdk
        android:minSdkVersion="8"
        android:targetSdkVersion="18" />
    <application
        android:allowBackup="true"
        android:icon="@drawable/ic_launcher"
        android:label="@string/app_name"
        android:theme="@style/AppTheme" >
        <activity
            android:name="fs.intent_custom.MainActivity"
            android:label="@string/app_name" >
            <intent-filter>
                <action android:name="android.intent.action.MAIN" />
                <category android:name="android.intent.category.LAUNCHER" />
            </intent-filter>
        </activity>
        <activity android:name=".SecondMainActivity" >
            <intent-filter>
                <action android:name="test_action" />
                <category android:name="android.intent.category.DEFAULT" />
            </intent-filter>
        </activity>
    </application>
</manifest>
```

运行程序,默认界面如图 5-7(a)所示,单击界面中的"进入下一个 Activity 页面"按钮时,效果如图 5-7(b)所示。

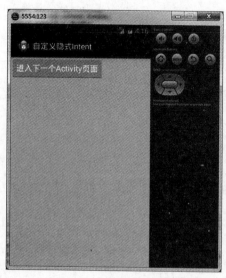

(a) 第一个Activity页面　　　　　　　　(b) 第二个Activity页面

图 5-7　自定义隐式 Intent

5.7 电话拨号实例

拨打电话手机最基本的功能之一，那么电话的拨打在 Android 中是如何实现的呢？本实例将介绍 Intent 在电话拨号中的应用。

本实例演示了如何利用 Intent 在 Android 中实现拨打电话功能。本实例的具体实现步骤如下。

(1) 在 Eclipse 中创建一个 Android 应用项目，命名为 telephone_Intent。

(2) 打开 res\layout 目录下的 main.xml 布局文件，在文件中声明一个 TextView 控件、一个 EditText 控件及一个 Button 控件。代码为：

```xml
<?xml version = "1.0" encoding = "utf-8"?>
<LinearLayout xmlns:android = "http://schemas.android.com/apk/res/android"
    android:orientation = "vertical"
    android:layout_width = "fill_parent"
    android:layout_height = "fill_parent"
    android:background = "#0cc">
    <TextView
      android:layout_width = "fill_parent"
      android:layout_height = "wrap_content"
      android:text = "输入电话号码："/>
    <EditText
      android:id = "@ + id/EditText01"
      android:layout_width = "200dip"
      android:layout_height = "wrap_content"/>
    <Button
      android:text = "拨打"
      android:id = "@ + id/Button01"
      android:layout_width = "wrap_content"
      android:layout_height = "wrap_content">
    </Button>
</LinearLayout>
```

(3) 打开 src\fs.telephone_intent 包下的 MainActivity.java 文件，在文件中实现在文本框中输入电话号码时，单击界面中的"拨打"按钮，当号码格式正确时，即实现电话拨打功能，当号码格式错误时，即显示相应的 Toast 提示。代码为：

```java
package fs.telephone_intent;
import java.util.regex.Matcher;
import java.util.regex.Pattern;
import android.app.Activity;
import android.content.Intent;
import android.net.Uri;
import android.os.Bundle;
import android.view.View;
import android.view.View.OnClickListener;
import android.widget.Button;
import android.widget.EditText;
import android.widget.Toast;
public class MainActivity extends Activity {
    private EditText et;                                    //声明 EditText 引用
```

```java
        private Button bOk;                                                //声明 Button 引用
        @Override
        public void onCreate(Bundle savedInstanceState) {
            super.onCreate(savedInstanceState);
            setContentView(R.layout.main);
            et = (EditText)this.findViewById(R.id.EditText01);             //创建对象
            bOk = (Button)this.findViewById(R.id.Button01);                //创建对象
            bOk.setOnClickListener
            (
                    new OnClickListener()
                    {
                            public void onClick(View v) {
                            String number = et.getText().toString().trim();  //获取输入的手机号码
                                boolean flag = phoneNumber(number);
                                if(flag)
                                {
Intent intent = //构建一个新的 intent,运行 action.CALL 的常数,通过 Uri 将字符创传入
    new Intent("android.intent.action.CALL",Uri.parse("tel:" + number));
                                    startActivity(intent);
                                    et.setText("");
                                }else
                                {
                                    Toast.makeText(MainActivity.this, "您输入的电话号码格式不正
                                    确",Toast.LENGTH_SHORT ).show();
                                    et.setText("");                        //将 EditText 字符设为空
                                }
                            }
                    }
            );
        }
        public boolean phoneNumber(String number)
        {
            boolean flag = false;
            String pare = "\\d{11}";                                       //11 个整数的手机号码正则式
            String pare2 = "\\d{12}";                                      //12 个整数的座机号码正则式
            CharSequence num = number;                                     //获取电话号码
            Pattern pattern = Pattern.compile(pare);                       //判断是否为手机号码
            Matcher matcher = pattern.matcher(num);
            Pattern pattern2 = Pattern.compile(pare2);                     //判断是否为座机号码
            Matcher matcher2 = pattern2.matcher(num);
            if(matcher.matches()||matcher2.matches())                      //如果符合格式
            {
                flag = true;                                               //标志位设为 true
            }
            return flag;
        }
    }
```

(4) 打开 AndroidManifest.xml 文件,设置拨打电话权限。代码为:

```xml
<?xml version = "1.0" encoding = "utf - 8"?>
< manifest xmlns:android = "http://schemas.android.com/apk/res/android"
    package = "fs.telephone_intent"
    android:versionCode = "1"
    android:versionName = "1.0" >
    < uses - sdk
```

```xml
        android:minSdkVersion = "8"
        android:targetSdkVersion = "18" />
    <application
        android:allowBackup = "true"
        android:icon = "@drawable/ic_launcher"
        android:label = "@string/app_name"
        android:theme = "@style/AppTheme" >
        <activity
            android:name = "fs.telephone_intent.MainActivity"
            android:label = "@string/app_name" >
            <intent-filter>
                <action android:name = "android.intent.action.MAIN" />
                <category android:name = "android.intent.category.LAUNCHER" />
            </intent-filter>
        </activity>
    </application>
    <!-- 添加拨打电话的权限 -->
    <uses-permission android:name = "android.permission.CALL_PHONE"/>
</manifest>
```

运行程序,效果如图 5-8 所示。

图 5-8　Intent 拨打电话

5.8　自定义拨打电话实例

Android 自带一套拨号系统,在 Android 中是允许对其进行更改的。对于没有键盘的手机,每次拨打电话总会使用系统自带的拨号按钮,如果觉得手机的拨号按钮不好看,那么也可以自己制作一个个性的拨号系统。

本实例用于演示一个自制的电话拨号系统,通过本实例演示如何替换系统的拨号系统,并使用自定义的个性拨号系统。

本实例的具体实现步骤如下。

(1) 在 Eclipse 中创建一个 Android 应用项目，命名为 Custom_Phone。

(2) 打开 res\layout 目录下的 main.xml 文件，在文件中实现 6 个线性布局，声明一个 EditText 控件及 10 个 Button 控件。代码为：

```xml
<?xml version = "1.0" encoding = "utf-8"?>
<LinearLayout xmlns:android = "http://schemas.android.com/apk/res/android"
    android:orientation = "vertical"
    android:layout_width = "fill_parent"
    android:layout_height = "fill_parent"
    android:background = "@drawable/kp">
  <LinearLayout
    android:id = "@ + id/LinearLayout6"
    android:orientation = "horizontal"
    android:layout_width = "fill_parent"
    android:layout_height = "wrap_content">
     <EditText
        android:text = "@string/default_number"
        android:id = "@ + id/EditText1"
        android:layout_width = "260dip"
        android:textSize = "24dip"
        android:editable = "false"
        android:enabled = "false"
        android:singleLine = "true"
        android:background = "#FFFFFF"
        android:textColor = "#000000"
        android:layout_marginRight = "6dip"
        android:layout_marginLeft = "10dip"
        android:layout_height = "wrap_content"/>
     <Button
        android:text = " "
        android:id = "@ + id/Button_del"
        android:textSize = "24dip"
        android:layout_width = "wrap_content"
        android:layout_height = "wrap_content"
        android:background = "@drawable/deldown"/>
  </LinearLayout>
  <LinearLayout
    android:id = "@ + id/LinearLayout1"
    android:orientation = "vertical"
    android:layout_width = "fill_parent"
    android:layout_height = "wrap_content">
   <LinearLayout
      android:id = "@ + id/LinearLayout2"
      android:orientation = "horizontal"
      android:gravity = "center_horizontal"
      android:layout_width = "fill_parent"
      android:layout_height = "wrap_content">
       <Button
          android:text = "1"
          android:id = "@ + id/Button1"
          android:textSize = "54dip"
          android:textStyle = "bold"
          android:typeface = "serif"
          android:layout_width = "wrap_content"
          android:layout_height = "wrap_content"
```

```xml
            android:background = "@drawable/c1"/>
        <Button
            android:text = "2"
            android:id = "@+id/Button2"
            android:textSize = "54dip"
            android:textStyle = "bold"
            android:typeface = "serif"
            android:layout_width = "wrap_content"
            android:layout_height = "wrap_content"
            android:layout_marginLeft = "20dip"
            android:layout_marginRight = "20dip"
            android:background = "@drawable/c1"/>
        <Button
            android:text = "3"
            android:id = "@+id/Button3"
            android:textSize = "54dip"
            android:textStyle = "bold"
            android:typeface = "serif"
            android:layout_width = "wrap_content"
            android:layout_height = "wrap_content"
            android:background = "@drawable/c1"/>
</LinearLayout>
<LinearLayout
    android:id = "@+id/LinearLayout3"
    android:orientation = "horizontal"
    android:gravity = "center_horizontal"
    android:layout_width = "fill_parent"
    android:layout_height = "wrap_content"
    android:layout_marginTop = "20dip">
        <Button
            android:text = "4"
            android:id = "@+id/Button4"
            android:textSize = "54dip"
            android:textStyle = "bold"
            android:typeface = "serif"
            android:layout_width = "wrap_content"
            android:layout_height = "wrap_content"
            android:background = "@drawable/c1"/>
        <Button
            android:text = "5"
            android:id = "@+id/Button5"
            android:textSize = "54dip"
            android:textStyle = "bold"
            android:typeface = "serif"
            android:layout_width = "wrap_content"
            android:layout_height = "wrap_content"
            android:layout_marginLeft = "20dip"
            android:layout_marginRight = "20dip"
            android:background = "@drawable/c1"/>
        <Button
            android:text = "6"
            android:id = "@+id/Button6"
            android:textSize = "54dip"
            android:textStyle = "bold"
            android:typeface = "serif"
```

```xml
            android:layout_width = "wrap_content"
            android:layout_height = "wrap_content"
            android:background = "@drawable/c1"/>
</LinearLayout>
<LinearLayout
    android:id = "@ + id/LinearLayout4"
    android:orientation = "horizontal"
    android:gravity = "center_horizontal"
    android:layout_width = "fill_parent"
    android:layout_height = "wrap_content"
    android:layout_marginTop = "20dip">
      <Button
        android:text = "7"
        android:id = "@ + id/Button7"
        android:textSize = "54dip"
        android:textStyle = "bold"
        android:typeface = "serif"
        android:layout_width = "wrap_content"
        android:layout_height = "wrap_content"
        android:background = "@drawable/c1"/>
      <Button
        android:text = "8"
        android:id = "@ + id/Button8"
        android:textSize = "54dip"
        android:textStyle = "bold"
        android:typeface = "serif"
        android:layout_width = "wrap_content"
        android:layout_height = "wrap_content"
        android:layout_marginLeft = "20dip"
        android:layout_marginRight = "20dip"
        android:background = "@drawable/c1"/>
      <Button
        android:text = "9"
        android:id = "@ + id/Button9"
        android:textSize = "54dip"
        android:textStyle = "bold"
        android:typeface = "serif"
        android:layout_width = "wrap_content"
        android:layout_height = "wrap_content"
        android:background = "@drawable/c1"/>
</LinearLayout>
<LinearLayout
    android:id = "@ + id/LinearLayout5"
    android:orientation = "horizontal"
    android:gravity = "center_horizontal"
    android:layout_width = "fill_parent"
    android:layout_height = "wrap_content"
    android:layout_marginTop = "20dip">
      <Button
        android:text = " "
        android:id = "@ + id/Button_dial"
        android:textSize = "54dip"
        android:textStyle = "bold"
        android:typeface = "serif"
        android:layout_width = "wrap_content"
```

```xml
            android:layout_height = "wrap_content"
            android:background = "@drawable/dial"/>
        <Button
            android:text = "0"
            android:id = "@ + id/Button0"
            android:textSize = "54dip"
            android:textStyle = "bold"
            android:typeface = "serif"
            android:layout_width = "wrap_content"
            android:layout_height = "wrap_content"
            android:layout_marginLeft = "20dip"
            android:layout_marginRight = "20dip"
            android:background = "@drawable/ic_launcher"/>
        <Button
            android:text = " "
            android:id = "@ + id/Button_cancel"
            android:textSize = "54dip"
            android:textStyle = "bold"
            android:typeface = "serif"
            android:layout_width = "wrap_content"
            android:layout_height = "wrap_content"
            android:background = "@drawable/dialcancel"/>
    </LinearLayout>
  </LinearLayout>
</LinearLayout>
```

(3) 打开 src\fs.custom_phone 包下的 MainActivity.java 文件,实现自定义拨号功能。代码为:

```java
package fs.custom_phone;
import android.app.Activity;
import android.content.Intent;
import android.net.Uri;
import android.os.Bundle;
import android.view.View;
import android.widget.Button;
import android.widget.EditText;
public class MainActivity extends Activity {
    //数字按钮的 id 数组
    int[] numButtonIds =
    {
        R.id.Button0,R.id.Button1,R.id.Button2,
        R.id.Button3,R.id.Button4,R.id.Button5,
        R.id.Button6,R.id.Button7,R.id.Button8,
        R.id.Button9
    };
    @Override
    public void onCreate(Bundle savedInstanceState) {
        super.onCreate(savedInstanceState);
        setContentView(R.layout.main);
        //为删除按钮添加监听器
        Button bDel = (Button)this.findViewById(R.id.Button_del);
        bDel.setOnClickListener(
            //OnClickListener 为 View 的内部接口,其实现者负责监听单击事件
            new View.OnClickListener()
```

```java
            {
                public void onClick(View v)
                {
                    EditText et = (EditText)findViewById(R.id.EditText1);
                    String num = et.getText().toString();
                    num = (num.length()>1)?num.substring(0,num.length()-1):"";
                    et.setText(num);
                }
        });
        //为拨号按钮添加监听器
        Button bDial = (Button)this.findViewById(R.id.Button_dial);
        bDial.setOnClickListener(
         //OnClickListener 为 View 的内部接口,其实现者负责监听单击事件
                new View.OnClickListener()
                {
                  public void onClick(View v)
                  {
                        //获取输入的电话号码
                        EditText et = (EditText)findViewById(R.id.EditText1);
                        String num = et.getText().toString();
                        //根据获取的电话号码创建 Intent 拨号
                        Intent dial = new Intent();
                        dial.setAction("android.intent.action.CALL");
                        dial.setData(Uri.parse("tel://" + num));
                        startActivity(dial);
                  }
        });
        //为退出按钮添加监听器
        Button bCancel = (Button)this.findViewById(R.id.Button_cancel);
        bCancel.setOnClickListener(
            //OnClickListener 为 View 的内部接口,其实现者负责监听单击事件
                new View.OnClickListener()
                {
                  public void onClick(View v)
                  {
                        MainActivity.this.finish();
                  }
        });
        //为 0~9 数字按钮创建监听器
        View.OnClickListener numListener = new View.OnClickListener()
        {
            public void onClick(View v)
            {
                Button tempb = (Button)v;
                EditText et = (EditText)findViewById(R.id.EditText1);
                et.append(tempb.getText());
            }
        };
        //为所有数字按钮添加监听器
        for(int id:numButtonIds)
        {
            Button tempb = (Button)this.findViewById(id);
            tempb.setOnClickListener(numListener);
        }
```

            }
       }

运行程序,效果如图 5-9 所示。

图 5-9 自定义拨号功能

5.9 邮箱实例

电子邮箱是通过网络电子邮局为网络客户提供的网络交流电子信息空间。现在电子邮箱被越来越多的人使用了,Android 手机都提供了相应的电子邮箱系统,用于发送 E-mail。

本实例通过自定义 Intent 对象,使用 Android.content.Intent.ACTION_SEND 的参数来实现通过手机发送 E-mail 的服务。

本实例的具体实现步骤如下。

(1) 在 Eclipse 中创建一个 Android 应用项目,命名为 Email_test。

(2) 打开 res\layout 目录下的 main.xml 布局文件,在文件中声明收件人的地址、发件人地址、主题、邮件的内容及发送按钮。代码为:

```
<?xml version = "1.0" encoding = "utf - 8"?>
< LinearLayout xmlns:android = "http://schemas.android.com/apk/res/android"
      android:orientation = "vertical"
      android:layout_width = "fill_parent"
      android:layout_height = "fill_parent"
      android:background = "#0aa">
      < LinearLayout
         android:layout_width = "fill_parent"
         android:layout_height = "wrap_content"
         android:orientation = "horizontal">
           < TextView
             android:text = "收件人地址: "
             android:id = "@ + id/TextView01"
             android:textColor = "#222222"
```

```xml
            android:layout_width = "wrap_content"
            android:layout_height = "wrap_content"/>
        <EditText
            android:text = "fsyaohua168@126.com"
            android:id = "@+id/EditText01"
            android:textColor = "#222222"
            android:layout_width = "fill_parent"
            android:layout_height = "wrap_content"/>
    </LinearLayout>
    <LinearLayout
        android:layout_width = "fill_parent"
        android:layout_height = "wrap_content"
        android:orientation = "horizontal">
        <TextView
            android:text = "发件人地址："
            android:id = "@+id/TextView04"
            android:textColor = "#222222"
            android:layout_width = "wrap_content"
            android:layout_height = "wrap_content"/>
        <EditText
            android:text = "fs3344168@126.com"
            android:id = "@+id/EditText04"
            android:textColor = "#222222"
            android:layout_width = "fill_parent"
            android:layout_height = "wrap_content"/>
    </LinearLayout>
    <LinearLayout
        android:layout_width = "fill_parent"
        android:layout_height = "wrap_content"
        android:orientation = "horizontal">
        <TextView
            android:text = "邮件主题："
            android:id = "@+id/TextView02"
            android:textColor = "#222222"
            android:layout_width = "wrap_content"
            android:layout_height = "wrap_content"/>
        <EditText
            android:id = "@+id/EditText02"
            android:textColor = "#222222"
            android:layout_width = "fill_parent"
            android:layout_height = "wrap_content"/>
    </LinearLayout>
    <TextView
        android:text = "邮件内容："
        android:textColor = "#222222"
        android:id = "@+id/TextView03"
        android:layout_width = "wrap_content"
        android:layout_height = "wrap_content"/>
    <EditText
        android:id = "@+id/EditText03"
        android:textColor = "#222222"
        android:layout_width = "fill_parent"
        android:layout_height = "100dip"
        android:gravity = "top|left"/>
    <Button
```

```
            android:text = "发送"
            android:textColor = "#222222"
            android:id = "@ + id/Button01"
            android:layout_width = "wrap_content"
            android:layout_height = "wrap_content"/>
</LinearLayout>
```

(3) 打开 src\fs.email_test 包下的 MainActivity.java 文件，在文件中实现当单击界面中的"发送"按钮时，即系统会自动检测收件地址和发件人地址格式填写是否正确，如果不正确则使用 Toast 提示用户填写错误，如果填写正确，则正常发送邮件。代码为：

```java
package fs.email_test;
import android.app.Activity;
import android.content.Intent;
import android.os.Bundle;
import android.view.View;
import android.view.View.OnClickListener;
import android.widget.Button;
import android.widget.EditText;
import android.widget.Toast;
public class MainActivity extends Activity {
    EditText etReceiver;                                            //收件人
    EditText etSender;                                              //发件人
    EditText etTheme;                                               //主题
    EditText etMessage;                                             //内容
    Button bSend;                                                   //发送按钮
    String strReceiver;                                             //收件人信息
    String strSender;                                               //发件人信息
    String strTheme;                                                //主题信息
    String strMessage;                                              //内容信息
    @Override
    public void onCreate(Bundle savedInstanceState) {
        super.onCreate(savedInstanceState);
        setContentView(R.layout.main);
        etReceiver = (EditText)this.findViewById(R.id.EditText01);   //获取对象
        etSender = (EditText)this.findViewById(R.id.EditText04);     //获取对象
        etTheme = (EditText)this.findViewById(R.id.EditText02);      //获取对象
        etMessage = (EditText)this.findViewById(R.id.EditText03);    //获取对象
        bSend = (Button)this.findViewById(R.id.Button01);            //发送按钮
        bSend.setOnClickListener
        (
            new OnClickListener()
            {
                public void onClick(View v) {
                    strReceiver = etReceiver.getText().toString().trim();   //获取收件人
                    strSender = etSender.getText().toString().trim();       //获取发件人
                    strTheme = etTheme.getText().toString().trim();         //获取主题
                    strMessage = etMessage.getText().toString().trim();     //获取内容
                    String parent = "^[a-zA-Z][\\w\\.-]*[a-zA-Z0-9]@[a-zA-Z0-9]
                        [\\w\\.-]*[a-zA-Z0-9]\\.[a-zA-Z][a-zA-Z\\.]*[a-zA-Z]$";
                    if(!strReceiver.matches(parent))//查看收件人地址是否符合格式
                    {
```

```
                                Toast.makeText(MainActivity.this, "收件人地址格式错误", Toast.
                                LENGTH_SHORT).show();
                        }else if(!strSender.matches(parent))//查看发件人地址是否符合格式
                        {
            Toast.makeText(MainActivity.this, "发件人地址格式错误", Toast.LENGTH_SHORT).show();
                        }else//若都符合格式,则发送邮件
                        {
            Intent intent = new Intent(android.content.Intent.ACTION_SEND);        //发送邮件功能
                                intent.setType("plain/text");
                    intent.putExtra(android.content.Intent.EXTRA_EMAIL, strReceiver);
            intent.putExtra(android.content.Intent.EXTRA_CC, strSender);
                                intent.putExtra(android.content.Intent.EXTRA_SUBJECT,
                                strTheme);
            intent.putExtra(android.content.Intent.EXTRA_TEXT, strMessage);
    startActivity(Intent.createChooser(intent, getResources().getString(R.string.start)));
                        }
                    }
                }
            );
        }
    }
```

(4) 打开 res\values 目录下的 strings.xml 文件,为变量声明赋值。代码为:

```
<?xml version = "1.0" encoding = "utf-8"?>
<resources>
    <string name = "app_name">邮箱</string>
    <string name = "action_settings">Settings</string>
    <string name = "hello_world">Hello world!</string>
    <string name = "start">邮件发送中……</string>
</resources>
```

运行程序,效果如图 5-10 所示。

图 5-10 邮箱

5.10 保护视力实例

Service 可以在很多场合的应用中使用,例如播放多媒体的时候用户启动了其他 Activity 这个时候程序要在后台继续播放,例如检测 SD 卡上文件的变化,再或者在后台记录地理信息位置的改变等,总之服务总是藏在后台的。

服务从本质上可分为两类。

- Started(启动):当应用程序组件(如 Activity)通过调用 startService()方法启动服务时,服务处于 started 状态。一旦启动,服务能在后台无限期运行,即使启动它的组件已经被销毁。通常,启动服务执行单个操作并且不会向调用者返回结果。例如,它可能通过网络下载或上传文件。如果操作完成,服务需要停止自身。
- Bound(绑定):当应用程序组件通过调用 bindService()方法绑定服务时,服务处于 bound 状态。绑定服务提供客户端/服务器接口,以允许组件与服务交互、发送请求和获得结果,甚至使得进程间通信(IPC)跨进程完成这些操作。仅当其他应用程序组件之间绑定时,绑定服务才运行。多个组件可以一次绑定到一个服务上,当它们都解绑定时,服务被销毁。

本实例利用 Service 组件实现一个视力保护程序,通过本实例演示 Service 组件的具体用法。本实例的具体实现步骤如下。

(1) 在 Eclipse 中创建一个 Android 应用项目,命名为 Eye_protection。

(2) 打开 res\layout 目录下的 main.xml 布局文件,在文件中声明一个 TextView 控件。代码为:

```
<?xml version = "1.0" encoding = "utf - 8"?>
<LinearLayout xmlns:android = "http://schemas.android.com/apk/res/android"
    android:layout_width = "fill_parent"
    android:layout_height = "fill_parent"
    android:background = "#0aa"
    android:orientation = "vertical" >
    <TextView
        android:id = "@ + id/textView"
        android:layout_width = "fill_parent"
        android:layout_height = "wrap_content"
        android:gravity = "center"
        android:text = "视力保护程序"
        android:textColor = "@android:color/black"
        android:textSize = "25dp" />
</LinearLayout>
```

(3) 打开 res\values 目录下的 strings.xml 布局文件,在文件中声明变量。代码为:

```
<?xml version = "1.0" encoding = "utf - 8"?>
<resources>
    <string name = "app_name">Eye_protection</string>
    <string name = "action_settings">Settings</string>
    <string name = "hello_world">Hello world!</string>
    <string name = "ticker_text">重要通知</string>
    <string name = "content_title">保护视力</string>
    <string name = "content_text">程序已经运行 1 分钟,请注意休息!</string>
```

```
</resources>
```

(4) 打开 src\fs.eye_protection 包下的 MainActivity.java 文件，在文件中创建 ServicesListActivity 类，继承了 Activity 类。并获取当前正在运行服务的列表。对于每个服务，获得其详细信息并在 Activity 中输出。代码为：

```java
package fs.eye_protection;
import java.util.Timer;
import java.util.TimerTask;
import android.app.Notification;
import android.app.NotificationManager;
import android.app.PendingIntent;
import android.app.Service;
import android.content.Context;
import android.content.Intent;
import android.os.IBinder;
public class MainActivity extends Service {
    private Timer timer;
    @Override
    public IBinder onBind(Intent intent) {
        return null;
    }
    @Override
    public void onCreate() {
        super.onCreate();
        timer = new Timer(true);          //创建 Timer 对象
    }
    @Override
    public void onStart(Intent intent, int startId) {
        super.onStart(intent, startId);
        timer.schedule(new TimerTask() {
            @Override
            public void run() {
                String ns = Context.NOTIFICATION_SERVICE;
                //获得通知管理器
                NotificationManager manager = (NotificationManager) getSystemService(ns);
                //创建通知
                Notification notification = new Notification(R.drawable.b9, getText(R.string.ticker_text), System.currentTimeMillis());
                CharSequence contentTitle = getText(R.string.content_title);      //定义通知的标题
                CharSequence contentText = getText(R.string.content_text);        //定义通知的内容
                Intent intent = new Intent(MainActivity.this, SecondActivity.class);  //创建 Intent 对象
                PendingIntent contentIntent = PendingIntent.getActivity(MainActivity.this, 0,
                        intent, Intent.FLAG_ACTIVITY_NEW_TASK);                   //创建 PendingIntent 对象
                //定义通知行为
                notification.setLatestEventInfo(MainActivity.this, contentTitle, contentText, contentIntent);
                manager.notify(0, notification);                                  //显示通知
                MainActivity.this.stopSelf();                                     //停止服务
            }
        }, 60000);
    }
}
```

(5) 在 src\fs.eye_protection 包下新建一个 SecondActivity.java 文件，在文件中实现调用 Service 服务程序。代码为：

```java
import android.app.Activity;
```

```
import android.content.Intent;
import android.os.Bundle;
public class SecondActivity extends Activity {
    @Override
    protected void onCreate(Bundle savedInstanceState) {
        super.onCreate(savedInstanceState);
        setContentView(R.layout.main);
        startService(new Intent(this, MainActivity.class));
    }
}
```

(6) 打开 AndroidManifest.xml 文件,增加 Activity 和 Service 配置。代码为:

```
<?xml version = "1.0" encoding = "utf-8"?>
<manifest xmlns:android = "http://schemas.android.com/apk/res/android"
    package = "fs.eye_protection"
    android:versionCode = "1"
    android:versionName = "1.0" >
    <uses-sdk
        android:minSdkVersion = "8"
        android:targetSdkVersion = "18" />
    <application
        android:allowBackup = "true"
        android:icon = "@drawable/ic_launcher"
        android:label = "@string/app_name"
        android:theme = "@style/AppTheme" >
        <activity
            android:name = "fs.eye_protection.SecondActivity"
            android:label = "@string/app_name" >
            <intent-filter >
                <action android:name = "android.intent.action.MAIN" />
                <category android:name = "android.intent.category.LAUNCHER" />
            </intent-filter >
        </activity>
        <!-- 添加服务配置 -->
        <service android:name = ".MainActivity"/>
    </application>
</manifest>
```

运行程序,界面如图 5-11(a)所示。在应用程序启动 1 分钟后会显示提示信息,如图 5-11(b)所示。

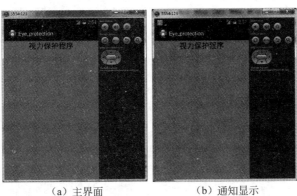

(a) 主界面　　　　(b) 通知显示

图 5-11　视力保护程序

5.11 天气预报实例

随着智能手机和平板电脑等移动终端设备的迅速发展,现在的 Internet 已经不再只是传统的有线互联网,还包括移动互联网。同有线互联网一样,移动互联网也可以使用 HTTP 访问网络。

本实例通过 Android 手机实时查询天气预报,通过本实例演示了手机网络的具体用法。本实例的具体实现步骤如下。

(1) 在 Eclipse 中创建一个 Android 应用项目,命名为 Weather_forecast。

(2) 打开 res\layout 目录下的 main.xml 布局文件,在文件中声明 6 个 Button 控件及一个 WebView 控件。代码为:

```xml
<?xml version = "1.0" encoding = "utf - 8"?>
<LinearLayout xmlns:android = "http://schemas.android.com/apk/res/android"
    android:orientation = "vertical"
    android:gravity = "center_horizontal"
    android:layout_width = "fill_parent"
    android:layout_height = "fill_parent"
    android:background = "#0cc">
    <LinearLayout
        android:orientation = "horizontal"
        android:layout_width = "wrap_content"
        android:layout_height = "wrap_content"
        android:background = "#0cc" >
        <Button
            android:id = "@ + id/tj"
            android:layout_width = "wrap_content"
            android:layout_height = "wrap_content"
            android:text = "天津" />
        <Button
            android:id = "@ + id/bj"
            android:layout_width = "wrap_content"
            android:layout_height = "wrap_content"
            android:text = "北京" />
        <Button
            android:id = "@ + id/gzh"
            android:layout_width = "wrap_content"
            android:layout_height = "wrap_content"
            android:text = "广州" />
        <Button
            android:id = "@ + id/nj"
            android:layout_width = "wrap_content"
            android:layout_height = "wrap_content"
            android:text = "南京" />
        <Button
            android:id = "@ + id/hzh"
            android:layout_width = "wrap_content"
            android:layout_height = "wrap_content"
            android:text = "杭州" />
        <Button
            android:id = "@ + id/wh"
```

```
                android:layout_width = "wrap_content"
                android:layout_height = "wrap_content"
                android:text = "武汉" />
    </LinearLayout >
    < WebView android:id = "@ + id/webView1"
        android:layout_width = "wrap_content"
        android:layout_height = "0dip"
         android:focusable = "false"
         android:layout_weight = "1"/>
</LinearLayout >
```

(3) 打开 src\fs. weather_forecast 包下的 MainActivity. java 文件,在文件中首先获取布局管理器中添加的 WebView 组件,然后设置该组件允许使用 JavaScript,并处理 JavaScript 对话框和请求事件,再为 WebView 组件指定要加载的天气预报信息,最后将网页内容放大 4 倍,当单击界面中的城市按钮时,即弹出相应的天气预报。代码为:

```
package fs.weather_forecast;
import android.app.Activity;
import android.os.Bundle;
import android.view.View;
import android.view.View.OnClickListener;
import android.webkit.WebChromeClient;
import android.webkit.WebView;
import android.webkit.WebViewClient;
import android.widget.Button;
public class MainActivity extends Activity implements OnClickListener {
    private WebView webView;                                    //声明 WebView 组件的对象
    @Override
    protected void onCreate(Bundle savedInstanceState) {
        super.onCreate(savedInstanceState);
        setContentView(R.layout.main);
        webView = (WebView)findViewById(R.id.webView1);         //获取 WebView 组件
        webView.getSettings().setJavaScriptEnabled(true);       //设置 JavaScript 可用
        webView.setWebChromeClient(new WebChromeClient());      //处理 JavaScript 对话框
//处理各种通知和请求事件,如果不使用该句代码,将使用内置浏览器访问网页
        webView.setWebViewClient(new WebViewClient());
//设置默认显示的天气预报信息
        webView.loadUrl("http://m.weather.com.cn/m/pn12/weather.htm ");webView.setInitialScale
(57 * 4);                                                       //将网页内容放大 4 倍
    Button bj = (Button)findViewById(R.id.bj);                  //获取布局管理器中添加的"北京"按钮
        bj.setOnClickListener(this);
Button sh = (Button)findViewById(R.id.gzh);                     //获取布局管理器中添加的"广州"按钮
        sh.setOnClickListener(this);
Button heb = (Button)findViewById(R.id.hzh);                    //获取布局管理器中添加的"杭州"按钮
        heb.setOnClickListener(this);
Button cc = (Button)findViewById(R.id.tj);                      //获取布局管理器中添加的"天津"按钮
//获取布局管理器中添加的"南京"按钮
        cc.setOnClickListener(this);Button sy = (Button)findViewById(R.id.nj);
            sy.setOnClickListener(this);
Button gz = (Button)findViewById(R.id.wh);                      //获取布局管理器中添加的"武汉"按钮
        gz.setOnClickListener(this);
    }
    @Override
    public void onClick(View view){
        switch(view.getId()){
```

```
            case R.id.bj:                              //单击的是"北京"按钮
                openUrl("101280101T");
                break;
            case R.id.gzh:                             //单击的是"广州"按钮
                openUrl("101020100T");
                break;
            case R.id.hzh:                             //单击的是"杭州"按钮
                openUrl("101210101T");
                break;
            case R.id.tj:                              //单击的是"天津"按钮
                openUrl("101030100T");
                break;
            case R.id.nj:                              //单击的是"南京"按钮
                openUrl("101190101T");
                break;
            case R.id.wh:                              //单击的是"武汉"按钮
                openUrl("101070101T");
                break;
        }
    }
    //打开网页的方法
    private void openUrl(String id){
        webView.loadUrl("http://m.weather.com.cn/m/pn12/weather.htm?id=" + id + " ");
                                                       //获取并显示天气预报信息
    }
}
```

(4) 打开 AndroidManifest.xml 文件，在文件中指定允许访问网络资源的权限。代码为：

```xml
<?xml version = "1.0" encoding = "utf - 8"?>
<manifest xmlns:android = "http://schemas.android.com/apk/res/android"
    package = "fs.weather_forecast"
    android:versionCode = "1"
    android:versionName = "1.0" >
    <uses - sdk
        android:minSdkVersion = "8"
        android:targetSdkVersion = "18" />
    <!-- 设置允许网络资源权限 -->
    <uses - permission android:name = "android.permission.INTERNET"/>
    <application
        android:allowBackup = "true"
        android:icon = "@drawable/ic_launcher"
        android:label = "@string/app_name"
        android:theme = "@style/AppTheme" >
        <activity
            android:name = "fs.weather_forecast.MainActivity"
            android:label = "@string/app_name" >
            <intent - filter>
                <action android:name = "android.intent.action.MAIN" />
                <category android:name = "android.intent.category.LAUNCHER" />
            </intent - filter>
        </activity>
    </application>
</manifest>
```

运行程序，单击需要查询的城市按钮，效果如图 5-12 所示。

图 5-12　天气预报

5.12　通信组件实例

在实际开发中,经常遇到在 Service 中下载文件完成或者其他长时间的操作完成后,需要通知用户知晓当前的状态。除了 Service 直接向 Activity 传输数据外,还可以通过 Service 发送广播,Activity 负责以接收的方式来完成消息的更新。

本实例用于实现多组件之间的通信,通过本实例演示了 Android 程序组件间的应用。本实例的具体实现步骤如下。

(1) 在 Eclipse 中创建一个 Android 应用项目,命名为 Communicate_test。

(2) 打开 res\layout 目录下的 main.xml 布局文件,在文件中声明一个 TextView 控件。代码为:

```
<RelativeLayout xmlns:android = "http://schemas.android.com/apk/res/android"
    xmlns:tools = "http://schemas.android.com/tools"
    android:layout_width = "match_parent"
    android:layout_height = "match_parent"
    android:paddingBottom = "@dimen/activity_vertical_margin"
    android:paddingLeft = "@dimen/activity_horizontal_margin"
    android:paddingRight = "@dimen/activity_horizontal_margin"
    android:paddingTop = "@dimen/activity_vertical_margin"
    tools:context = ".MainActivity"
    android:background = "#0aa">
    <TextView
        android:id = "@+id/tv"
        android:layout_width = "wrap_content"
        android:layout_height = "wrap_content"
        android:layout_centerHorizontal = "true"
        android:layout_centerVertical = "true"
        android:text = "通信组件" />
```

```
</RelativeLayout>
```

（3）打开 src\fs.communicate_test 包下的 MainActivity.java 文件，在文件中实现广播的注册及启动服务。代码为：

```java
package fs.communicate_test;
import android.os.Bundle;
import android.app.Activity;
import android.content.BroadcastReceiver;
import android.content.Context;
import android.content.Intent;
import android.content.IntentFilter;
import android.util.Log;
import android.view.Menu;
import android.widget.TextView;
public class MainActivity extends Activity {
    private TextView tv;
    ServiceA mService;
    MyReceiver receiver;
    @Override
    protected void onCreate(Bundle savedInstanceState) {
        super.onCreate(savedInstanceState);
        setContentView(R.layout.main);
        tv = (TextView)findViewById(R.id.tv);
        //注册广播
        receiver = new MyReceiver();
        IntentFilter intentFilter = new IntentFilter(ServiceA.SELF_ACTION);
        this.registerReceiver(receiver, intentFilter);
        //启动服务
        Intent intent = new Intent(MainActivity.this, ServiceA.class);
        startService(intent);
    }
    @Override
    public boolean onCreateOptionsMenu(Menu menu) {
        getMenuInflater().inflate(R.menu.main, menu);
        return true;
    }
    public class MyReceiver extends BroadcastReceiver {
        @Override
        public void onReceive(Context context, Intent intent) {
            //TODO 自动存根法
            Log.i(ServiceA.TAG,"OnReceiver");
            Bundle bundle = intent.getExtras();
            int i = bundle.getInt("i");
            //处理接收到的内容
            tv.setText("已经过去了 " + i + " 秒");
        }
        public MyReceiver() {
            Log.i(ServiceA.TAG,"MyReceiver");
        }
    }
}
```

(4) 打开 src\fs.communicate_test 包下的 ServiceA.java 文件，在文件中实现新建线程及销毁线程。代码为：

```java
package fs.communicate_test;
import android.app.Service;
import android.content.Intent;
import android.os.IBinder;
import android.util.Log;
public class ServiceA extends Service {
    static final String TAG = "Sample_7_8";
    static final String SELF_ACTION = "sample.intent.action.test";
    boolean isstop = false;
    @Override
    public IBinder onBind(Intent intent) {
        //TODO 自动存根法
        return null;
    }
    public void onCreate() {
        Log.i(TAG, "Services onCreate");
        super.onCreate();
    }
    public void onStart(Intent intent, int startId) {
        Log.i(TAG, "Services onStart");
        super.onStart(intent, startId);
        new Thread() {//新建线程，每隔1秒发送一次广播，同时把i放进intent传出
            public void run() {
                int i = 0;
                while (!isstop) {
                    Intent intent = new Intent();
                    intent.putExtra("i", i);
                    i++;
                    intent.setAction(SELF_ACTION);      //action与接收器相同
                    sendBroadcast(intent);
                    Log.i(TAG, String.valueOf(i));
                    try {
                        sleep(1000);
                    } catch (InterruptedException e) {
                        //TODO 自动存根法
                        e.printStackTrace();
                    }
                }
            }
        }.start();
    }
    @Override
    public void onDestroy() {
        Log.i("TAG", "Services onDestory");
        //即使Service销毁线程也不会停止，所以这里通过设置isstop来停止线程
        isstop = true;
        super.onDestroy();
    }
}
```

(5) 打开 AndroidManifest.xml 文件，在文件中添加 Service 声明。代码为：

```xml
<?xml version="1.0" encoding="utf-8"?>
<manifest xmlns:android="http://schemas.android.com/apk/res/android"
    package="fs.communicate_test"
    android:versionCode="1"
    android:versionName="1.0" >
    <uses-sdk
        android:minSdkVersion="8"
        android:targetSdkVersion="18" />
    <application
        android:allowBackup="true"
        android:icon="@drawable/ic_launcher"
        android:label="@string/app_name"
        android:theme="@style/AppTheme" >
        <activity
            android:name="fs.communicate_test.MainActivity"
            android:label="@string/app_name" >
            <intent-filter>
                <action android:name="android.intent.action.MAIN" />
                <category android:name="android.intent.category.LAUNCHER" />
            </intent-filter>
        </activity>
        <service android:name=".ServiceA"></service>
    </application>
</manifest>
```

运行程序，效果如图 5-13(a)、5-13(b)所示，广播打印输出 LogCat 的效果如图 5-14 所示。由图 5-14 可以明显地看出，广播的注册、服务的启动以及广播接收器接收的数据。

（a）广播效果1

（b）广播效果2

图 5-13 传播广播

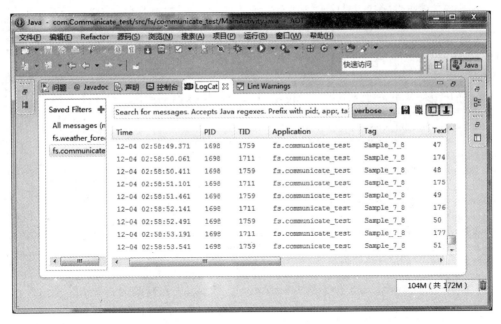

图 5-14 LogCat 输出

5.13 Wi-Fi 实例

Wi-Fi 是一种可以将个人计算机、手持设备(如 PDA、手机)等终端以无线方式互相连接的技术。Wi-Fi 是一个无线网路通信技术的品牌,由 Wi-Fi 联盟(Wi-Fi Alliance)所持有。目的是改善基于 IEEE 802.11 标准的无线网路产品之间的互通性。现时很多人会把 Wi-Fi 及 IEEE 802.11 混为一谈。甚至把 Wi-Fi 等同于无线网际网路。

Wi-Fi 是一种短程无线传输技术,能够在数百英尺范围内支持互联网接入的无线电信号。随着技术的发展,IEEE 802.11a 及 IEEE 802.11g 等标准的出现,现在 IEEE 802.11 这个标准已被统称作 Wi-Fi。从应用层面来说,要使用 Wi-Fi,用户首先要有 Wi-Fi 兼容的用户端装置。

本实例用于实现 Wi-Fi 的打开与关闭功能。通过本实例演示了 Wi-Fi 在 Android 上的使用。其具体实现步骤如下。

(1) 在 Eclipse 环境下建立一个 Android 应用项目,命名为 WiFi_test。

(2) 打开 res/Layout 目录下的 main.xml 文件,代码修改为:

```
<?xml version = "1.0" encoding = "utf-8"?>
<LinearLayout
    xmlns:android = "http://schemas.android.com/apk/res/android"
android:orientation = "vertical"
    android:layout_width = "fill_parent"
    android:layout_height = "fill_parent">
    <TextView
        android:id = "@+id/myTextView1"
        android:layout_width = "fill_parent"
        android:layout_height = "wrap_content"
        android:text = "@string/hello_world"/>
    <CheckBox
```

```
    android:id = "@ + id/myCheckBox1"
    android:layout_width = "wrap_content"
    android:layout_height = "wrap_content"
    android:text = "@string/str_checked"/>
</LinearLayout>
```

(3) 打开 res/Value 目录下的 strings.xml 文件,添加如下代码:

```
<string name = "str_checked">打开 Wi-Fi</string>
<string name = "str_uncheck">关闭 Wi-Fi</string>
<string name = "str_start_wifi_failed">打开失败</string>
<string name = "str_start_wifi_done">打开成功</string>
<string name = "str_stop_wifi_failed">关闭失败</string>
<string name = "str_stop_wifi_done">关闭成功</string>
<string name = "str_wifi_enabling">正在启动……</string>
<string name = "str_wifi_disabling">正在关闭……</string>
<string name = "str_wifi_disabled">已关闭</string>
<string name = "str_wifi_unknow">未知……</string>
```

(4) 打开 src/fs.wifi_test 包下的 MainActivity 文件,代码修改为:

```
public class MainActivity extends Activity
{
    private TextView mTextView1;
    private CheckBox mCheckBox1;
    /* 创建 WiFiManager 对象 */
    private WifiManager mWiFiManager1;
//定义了 mTextView1 和 mCheckBox1,分别用于显示提示文本和获取复选框的选择状态
    /** 第一次调用 Activity 活动 */
    @Override
    public void onCreate(Bundle savedInstanceState)
    {
        super.onCreate(savedInstanceState);
        setContentView(R.layout.main);
        mTextView1 = (TextView) findViewById(R.id.myTextView1);
        mCheckBox1 = (CheckBox) findViewById(R.id.myCheckBox1);
        /* 以 getSystemServices 取得 WIFI_SERVICE,然后通过 if 语句来判断动
         * 作程序后的 Wi-Fi 状态是否打开或打开中,这样即可显示对应的提示信息.
         */
        mWiFiManager1 = (WifiManager) this.getSystemService(Context.WIFI_SERVICE);
        /* 判断运行程序后的 Wi-Fi 状态是否打开或打开中 */
        if(mWiFiManager1.isWifiEnabled())
        {
            /* 判断 Wi-Fi 状态是否"已打开" */
            if(mWiFiManager1.getWifiState() == WifiManager.WIFI_STATE_ENABLED)
            {
                /* 若 Wi-Fi 已打开,将选取项勾选 */
                mCheckBox1.setChecked(true);
                /* 更改选取项文字为关闭 Wi-Fi */
                mCheckBox1.setText(R.string.str_uncheck);
            }
            else
            {
                /* 若 Wi-Fi 未打开,将选取项取消 */
                mCheckBox1.setChecked(false);
                /* 更改选取项文字为打开 Wi-Fi */
```

```java
            mCheckBox1.setText(R.string.str_checked);
        }
    }
    else
    {
        mCheckBox1.setChecked(false);
        mCheckBox1.setText(R.string.str_checked);
    }
    /* 捕捉 CheckBox 的单击事件 */
    mCheckBox1.setOnClickListener(
    new CheckBox.OnClickListener()
    {
        @Override
        public void onClick(View v)
        {
            //TODO 自动存根法
            /* 当选取项为取消选取状态 */
            if(mCheckBox1.isChecked() == false)
            {
                /* 尝试关闭 Wi-Fi 服务 */
                try
                {
                    /* 判断 Wi-Fi 状态是否为已打开 */
                    if(mWiFiManager1.isWifiEnabled())
                    {
                        /* 关闭 Wi-Fi */
                        if(mWiFiManager1.setWifiEnabled(false))
                        {
                            mTextView1.setText(R.string.str_stop_wifi_done);
                        }
                        else
                        {
                            mTextView1.setText(R.string.str_stop_wifi_failed);
                        }
                    }
                    else
                    {
                        /* Wi-Fi 状态不是已打开状态时 */
                        switch(mWiFiManager1.getWifiState())
                        {
                            /* Wi-Fi 正在打开过程中,导致无法关闭…… */
                            case WifiManager.WIFI_STATE_ENABLING:
                                mTextView1.setText
                                (
                                    getResources().getText
                                    (R.string.str_stop_wifi_failed) + ":" +
                                    getResources().getText
                                    (R.string.str_wifi_enabling)
                                );
                                break;
                            /* Wi-Fi 正在关闭过程中,导致无法关闭…… */
                            case WifiManager.WIFI_STATE_DISABLING:
                                mTextView1.setText
                                (
                                    getResources().getText
```

```java
                    (R.string.str_stop_wifi_failed) + ":" +
                    getResources().getText
                    (R.string.str_wifi_disabling)
                    );
                    break;
                  /* Wi-Fi 已经关闭 */
                  case WifiManager.WIFI_STATE_DISABLED:
                    mTextView1.setText
                    (
                      getResources().getText
                      (R.string.str_stop_wifi_failed) + ":" +
                      getResources().getText
                      (R.string.str_wifi_disabled)
                    );
                    break;
                  /* 无法取得或辨识 Wi-Fi 状态 */
                  case WifiManager.WIFI_STATE_UNKNOWN:
                  default:
                    mTextView1.setText
                    (
                      getResources().getText
                      (R.string.str_stop_wifi_failed) + ":" +
                      getResources().getText
                      (R.string.str_wifi_unknow)
                    );
                    break;
                }
                mCheckBox1.setText(R.string.str_checked);
            }
        }
        catch (Exception e)
        {
            Log.i("HIPPO", e.toString());
            e.printStackTrace();
        }
    }
    else if(mCheckBox1.isChecked() == true)
    {
        /* 尝试打开 Wi-Fi 服务 */
        try
        {
            /* 确认 Wi-Fi 服务是关闭且不在打开作业中 */
            if(!mWiFiManager1.isWifiEnabled() &&
                mWiFiManager1.getWifiState()!=
                WifiManager.WIFI_STATE_ENABLING )
            {
                if(mWiFiManager1.setWifiEnabled(true))
                {
                    switch(mWiFiManager1.getWifiState())
                    {
                        /* Wi-Fi 正在打开过程中,导致无法打开…… */
                        case WifiManager.WIFI_STATE_ENABLING:
                            mTextView1.setText
                            (
                                getResources().getText
```

```java
          (R.string.str_wifi_enabling)
        );
        break;
      /* Wi-Fi已经为打开,无法再次打开…… */
      case WifiManager.WIFI_STATE_ENABLED:
        mTextView1.setText
        (
          getResources().getText
          (R.string.str_start_wifi_done)
        );
        break;
      /* 其他未知的错误 */
      default:
        mTextView1.setText
        (
          getResources().getText
          (R.string.str_start_wifi_failed) + ":" +
          getResources().getText
          (R.string.str_wifi_unknow)
        );
        break;
    }
  }
  else
  {
    mTextView1.setText(R.string.str_start_wifi_failed);
  }
}
else
{
  switch(mWiFiManager1.getWifiState())
  {
    /* Wi-Fi正在打开过程中,导致无法打开…… */
    case WifiManager.WIFI_STATE_ENABLING:
      mTextView1.setText
      (
        getResources().getText
        (R.string.str_start_wifi_failed) + ":" +
        getResources().getText
        (R.string.str_wifi_enabling)
      );
      break;
    /* Wi-Fi正在关闭过程中,导致无法打开…… */
    case WifiManager.WIFI_STATE_DISABLING:
      mTextView1.setText
      (
        getResources().getText
        (R.string.str_start_wifi_failed) + ":" +
        getResources().getText
        (R.string.str_wifi_disabling)
      );
      break;
    /* Wi-Fi已经关闭 */
    case WifiManager.WIFI_STATE_DISABLED:
      mTextView1.setText
```

```java
                            (
                                getResources().getText
                                (R.string.str_start_wifi_failed) + ":" +
                                getResources().getText
                                (R.string.str_wifi_disabled)
                            );
                            break;
                        /* 无法取得或识别 Wi-Fi 状态 */
                        case WifiManager.WIFI_STATE_UNKNOWN:
                        default:
                            mTextView1.setText
                            (
                                getResources().getText
                                (R.string.str_start_wifi_failed) + ":" +
                                getResources().getText
                                (R.string.str_wifi_unknow)
                            );
                            break;
                    }
                }
                mCheckBox1.setText(R.string.str_uncheck);
            }
            catch (Exception e)
            {
                Log.i("HIPPO", e.toString());
                e.printStackTrace();
            }
        }
    });
}
//定义 mMakeTextToast()方法来根据当前操作显示对应的提示性信息
public void mMakeTextToast(String str, boolean isLong)
{
    if(isLong == true)
    {
        Toast.makeText(MainActivity.this, str, Toast.LENGTH_LONG).show();
    }
    else
    {
        Toast.makeText(MainActivity.this, str, Toast.LENGTH_SHORT).show();
    }
}
@Override
protected void onResume()
{
    /* 在 onResume 重写事件时,取得打开程序当前 Wi-Fi 的状态 */
    try
    {
        switch(mWiFiManager1.getWifiState())
        {
            /* Wi-Fi 已经在打开状态…… */
            case WifiManager.WIFI_STATE_ENABLED:
                mTextView1.setText
                (
```

```java
            getResources().getText(R.string.str_wifi_enabling)
          );
          break;
        /* Wi-Fi 正在打开过程中状态…… */
        case WifiManager.WIFI_STATE_ENABLING:
          mTextView1.setText
          (
            getResources().getText(R.string.str_wifi_enabling)
          );
          break;
        /* Wi-Fi 正在关闭过程中…… */
        case WifiManager.WIFI_STATE_DISABLING:
          mTextView1.setText
          (
            getResources().getText(R.string.str_wifi_disabling)
          );
          break;
        /* Wi-Fi 已经关闭 */
        case WifiManager.WIFI_STATE_DISABLED:
          mTextView1.setText
          (
            getResources().getText(R.string.str_wifi_disabled)
          );
          break;
        /* 无法取得或识别 Wi-Fi 状态 */
        case WifiManager.WIFI_STATE_UNKNOWN:
        default:
          mTextView1.setText
          (
            getResources().getText(R.string.str_wifi_unknow)
          );
          break;
      }
    }
    catch(Exception e)
    {
      mTextView1.setText(e.toString());
      e.getStackTrace();
    }
    super.onResume();
  }
  @Override
  protected void onPause()
  {
    super.onPause();
  }
}
```

(5) 授予权限。打开 AndroidManifest.xml 文件,代码如下代码:

```xml
…
</application>
<!-- 声明 Wi-Fi 以及网络等相关权限 -->
<uses-permission android:name="android.permission.CHANGE_NETWORK_STATE"/>
<uses-permission android:name="android.permission.CHANGE_WIFI_STATE"/> <uses-permission android:name="android.permission.ACCESS_NETWORK_STATE"/> <uses-permission android:name="
```

android.permission.ACCESS_WIFI_STATE"/> < uses - permission android:name = "android.permission.INTERNET"/>
< uses - permission android:name = "android.permission.WAKE_LOCK"/>
</manifest >

运行程序,效果如图 5-15 所示。当选择复选框后会执行对应的操作,并显示对应的提示信息。

图 5-15　Wi-Fi 页面

5.14　查看手机信息实例

每一部手机都有其详细的信息,这些信息包括了手机号码、电信网络国别等。

TelephonyManager 是一个管理手机通话状态和电话网络信息的服务类。以下实现中获取 TelephonyManager 十分简单,主要通过 getSystemService 获取 TelephonyManager 对象,接着通过 TelephonyManager 的方法来获取和电信有关的网络信息。然后通过 Android.provider.Setting.System.getString()获取手机的相关设置信息,最后以 setListAdapter 内的信息显示在 ListView 中。

下面通过一个案例来实现在界面上有一个按钮,当单击按钮时,即可获取手机上的相关信息,信息包括手机号码、电信网络国别、电信公司名称、手机 SIM 码、手机通信类型、手机网络类型、是否漫游以及蓝牙和 Wi-Fi 的状态等。当单击其中的一条信息时,会有 Toast 弹出,会给出所单击的信息。其具体实现步骤如下。

(1) 在 Eclipse 中创建一个 Android 应用项目,命名为 Tele_Phone。

(2) 在 res\layout 目录下的 main.xml 布局文件中定义一个 RelativeLayout 布局、一个 Button 控件及一个 LinearLayout 布局,在 LinearLayout 布局中声明一个 ListView,用于显示手机的相关信息。代码为:

```
<RelativeLayout
    xmlns:android = "http://schemas.android.com/apk/res/android"
    xmlns:tools = "http://schemas.android.com/tools"
    android:layout_width = "match_parent"
    android:layout_height = "match_parent"
    android:background = "#ffcc66">
    <Button
        android:text = "获取手机信息"
        android:id = "@ + id/Button1"
        android:layout_width = "fill_parent"
        android:layout_height = "wrap_content"/>
    <LinearLayout
        android:orientation = "vertical"
        android:layout_width = "fill_parent"
        android:layout_height = "wrap_content"
        android:background = "#ffcc66"
        android:paddingLeft = "0dip"
```

```xml
            android:paddingRight = "5dip"
            android:paddingTop = "5dip">
    <ListView
            android:id = "@ + id/ListView1"
            android:layout_width = "wrap_content"
            android:layout_height = "wrap_content"
            android:cacheColorHint = "#ffcc66"/>
    </LinearLayout>
</RelativeLayout>
```

（3）打开 src\fs.tele_phone 包下的 MainActivity.java 文件，在文件中实现获取手机的相关信息并将其显示在 ListView 控件中。代码为：

```java
package fs.tele_phone;
import java.util.ArrayList;
import java.util.List;
import android.app.Activity;
import android.content.ContentResolver;
import android.graphics.Color;
import android.os.Bundle;
import android.telephony.TelephonyManager;
import android.view.Gravity;
import android.view.View;
import android.view.ViewGroup;
import android.view.View.OnClickListener;
import android.widget.AdapterView;
import android.widget.BaseAdapter;
import android.widget.Button;
import android.widget.LinearLayout;
import android.widget.ListView;
import android.widget.TextView;
import android.widget.Toast;
import android.widget.AdapterView.OnItemClickListener;
public class MainActivity extends Activity {
    private ListView lv;
    private TelephonyManager tm;
    private ContentResolver cr;
    private List<String> list = new ArrayList<String>();
    private List<String> name = new ArrayList<String>();
    private Button bCheck;
    @Override
    public void onCreate(Bundle savedInstanceState) {
        super.onCreate(savedInstanceState);
        setContentView(R.layout.main);
        lv = (ListView)this.findViewById(R.id.ListView1);
        tm = (TelephonyManager)getSystemService(TELEPHONY_SERVICE);
        cr = MainActivity.this.getContentResolver();
        bCheck = (Button)this.findViewById(R.id.Button1);
        String str = null;                                //记录 cr 获取的信息
        name.add("手机号码：");
        name.add("电信网络国别：");
        name.add("电信公司代码：");
        name.add("电信公司名称：");
        name.add("SIM 码：");
        name.add("手机通信类型：");
        name.add("手机网络类型 ：");
        name.add("手机是否漫游：");
        name.add("蓝牙状态：");
```

```java
        name.add("Wi-Fi 状态: ");
        if(tm.getLine1Number()!= null)//手机号码
        {
            list.add(tm.getLine1Number());
        }else
        {
            list.add("无法取得您的电话号码");
        }
        if(!tm.getNetworkCountryIso().equals(""))         //电信网络国别
        {
            list.add(tm.getNetworkCountryIso());
        }else
        {
            list.add("无法取得您的电信网络国别");
        }
        if(!tm.getNetworkOperator().equals(""))           //电信公司代码
        {
            list.add(tm.getNetworkOperator());
        }else
        {
            list.add("无法获取电信公司代码");
        }
        if(!tm.getNetworkOperatorName().equals(""))       //电信公司名称
        {
            list.add(tm.getNetworkOperatorName());
        }else
        {
            list.add("无法获取电信公司名称");
        }
        if(tm.getSimSerialNumber()!= null)                //手机 SIM 码
        {
            list.add(tm.getSimSerialNumber());
        }else
        {
            list.add("无法获取手机 SIM 码");
        }
if(tm.getPhoneType() == TelephonyManager.PHONE_TYPE_GSM)  //手机行动通信类型
        {
            list.add("GSM");
        }
        else
        {
            list.add("无法获取手机通信类型");
        }
    //获取手机网络类型
        if(tm.getNetworkType() == TelephonyManager.NETWORK_TYPE_EDGE)
        {
            list.add("EDGE");
        }else if(tm.getNetworkType() == TelephonyManager.NETWORK_TYPE_GPRS)
        {
            list.add("GPRS");
        }else if(tm.getNetworkType() == TelephonyManager.NETWORK_TYPE_UMTS)
        {
            list.add("UMTS");
        }
        else
        {
            list.add("无法获取手机网络类型");
```

```java
}
if(tm.isNetworkRoaming())                               //手机是否漫游
{
    list.add("手机漫游中");
}else
{
    list.add("手机无漫游");
}
str = android.provider.Settings.System.getString(
        cr,android.provider.Settings.System.BLUETOOTH_ON
        );
if(str.equals("1"))
{
    list.add("蓝牙已打开");
}else
{
    list.add("蓝牙未打开");
}
str = android.provider.Settings.System.getString(cr,android.provider.Settings.System.WIFI_ON);
if(str.equals("1"))
{
    list.add("Wi-Fi已打开");
}else
{
    list.add("Wi-Fi未打开");
}
bCheck.setOnClickListener
(
        new OnClickListener()
        {
            public void onClick(View v) {
                BaseAdapter ba = new BaseAdapter() //创建适配器
                {
                    public int getCount() {
                        return list.size();
                    }
                    public Object getItem(int position) {
                        return null;
                    }
                    public long getItemId(int position) {
                        return 0;
                    }
                    public View getView(int arg0, View arg1, ViewGroup arg2) {
                        LinearLayout ll = new LinearLayout(MainActivity.this);
                            ll.setOrientation(LinearLayout.HORIZONTAL);
                            ll.setPadding(5, 5, 5, 5);
                        TextView tv = new TextView(MainActivity.this);  //初始化TextView
                            tv.setTextColor(Color.BLACK);    //设置字体颜色
                            tv.setPadding(5,5,5,5);
                            tv.setText(name.get(arg0));      //添加任务名字
                            tv.setGravity(Gravity.LEFT);     //左对齐
                            tv.setTextSize(18);              //字体大小
                            ll.addView(tv);                  //LinearLayout添加TextView
                        TextView tvv = new TextView(MainActivity.this); //初始化TextView
                            tvv.setTextColor(Color.BLACK);   //设置字体颜色
                            tvv.setPadding(5,5,5,5);
                            tvv.setText(list.get(arg0));     //添加任务名字
```

```
                        tvv.setGravity(Gravity.LEFT);          //左对齐
                        tvv.setTextSize(18);                   //字体大小
                        ll.addView(tvv);                       //LinearLayout 添加 TextView
                        return ll;
                    }
                };
                lv.setAdapter(ba);                             //设置适配器
                lv.setOnItemClickListener                      //设置选中菜单的监听器
                (
                        new OnItemClickListener()
                        {
                    public void onItemClick(AdapterView<?> arg0, View arg1,
                                int arg2, long arg3) {
                        Toast.makeText(MainActivity.this, name.get(arg2) + "" + list.get(arg2), Toast.LENGTH_SHORT).show();
                    }
                        }
                );
            }
        }
```

(4) 打开 AndroidManifest.xml 文件,为程序设置权限。添加代码为:

…
        </activity>
    </application>
<!-- 添加权限 -->
<uses-permission android:name = "android.permission.READ_PHONE_STATE"></uses-permission>
</manifest>

运行程序,效果如图 5-16(a)所示,单击界面中的"查看手机信息"按钮,效果如图 5-16(b)所示,单击任何一项信息,即显示对应的 Toast 提示,效果如图 5-16(c)所示。

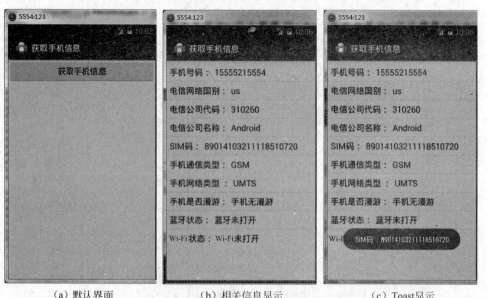

(a) 默认界面　　　　　　(b) 相关信息显示　　　　　　(c) Toast显示

图 5-16　显示手机相关信息

5.15 读取 SIM 卡参数实例

SIM 卡是(Subscriber Identity Module,客户识别模块)的缩写,也称为用户身份识别卡和智能卡,GSM 数字移动电话机必须装上此卡方能使用。它在一计算机芯片上存储了数字移动电话客户的信息、加密的密钥以及用户的电话簿等内容,可供 GSM 网络客户身份进行鉴别,并对客户通话时的语音信息进行加密。

手机的 SIM 卡是手机的一个重要的组成部分,没有 SIM 卡也不能正常拨打电话。本节将介绍怎样获取 SIM 卡的信息。

该案例通过使用 TelephonyManager 获取手机 SIM 卡的相关信息,并将获得的 SIM 卡的状态、卡号、SIM 卡供应商、SIM 卡供应商名称和 SIM 卡国别以 ListView 形式呈现在界面上。使用 TelephonyManager 获取手机 SIM 卡的信息需要为手机添加权限声明,权限代码为:
<uses-permission android:name="android.permission.READ_PHONE_STATE"/>。

读取 SIM 卡参数的具体步骤如下。

(1) 在 Eclipse 中创建一个 Android 应用项目,命名为 SIM_Message。

(2) 在 res\layout 目录下的 main.xml 布局文件中定义一个 RelativeLayout 布局、一个 Button 控件及一个 LinearLayout 布局,在 LinearLayout 布局中声明一个 ListView,用于显示手机的相关信息。代码为:

```xml
<RelativeLayout
    xmlns:android="http://schemas.android.com/apk/res/android"
    xmlns:tools="http://schemas.android.com/tools"
    android:layout_width="match_parent"
    android:layout_height="match_parent"
    android:background="#ffcc66">
    <Button
        android:text="获取 SIM 卡信息"
        android:id="@+id/Button1"
        android:layout_width="fill_parent"
        android:layout_height="wrap_content"/>
    <LinearLayout
        android:orientation="vertical"
        android:layout_width="fill_parent"
        android:layout_height="wrap_content"
        android:background="#ffcc66"
        android:paddingLeft="0dip"
        android:paddingRight="5dip"
        android:paddingTop="5dip">
        <ListView
            android:id="@+id/ListView1"
            android:layout_width="wrap_content"
            android:layout_height="wrap_content"
            android:cacheColorHint="#ffcc66"/>
    </LinearLayout>
</RelativeLayout>
```

(3) 打开 src\fs.sim_message 包下的 MainActivity.java 文件,在文件中实现获取 SIM 卡的相关信息将其显示在 ListView 控件中。代码为:

```java
package fs.sim_message;
import java.util.ArrayList;
import java.util.List;
import android.app.Activity;
import android.graphics.Color;
import android.os.Bundle;
import android.telephony.TelephonyManager;
import android.view.Gravity;
import android.view.View;
import android.view.ViewGroup;
import android.view.View.OnClickListener;
import android.widget.AdapterView;
import android.widget.BaseAdapter;
import android.widget.Button;
import android.widget.LinearLayout;
import android.widget.ListView;
import android.widget.TextView;
import android.widget.Toast;
import android.widget.AdapterView.OnItemClickListener;
public class MainActivity extends Activity {
    private ListView lv;
    private TelephonyManager tm;
    private List<String> list = new ArrayList<String>();
    private List<String> name = new ArrayList<String>();
    private Button bCheck;
    @Override
    public void onCreate(Bundle savedInstanceState) {
        super.onCreate(savedInstanceState);
        setContentView(R.layout.main);
        lv = (ListView)this.findViewById(R.id.ListView1);
        tm = (TelephonyManager)getSystemService(TELEPHONY_SERVICE);
        bCheck = (Button)this.findViewById(R.id.Button1);
        name.add("SIM 卡的状态: ");
        name.add("SIM 卡号: ");
        name.add("SIM 卡供应商号: ");
        name.add("SIM 卡供应商名称: ");
        name.add("SIM 卡国别: ");
        //通过 TelephonyManager 对象判断 SIM 卡状态、SIM 卡卡号、SIM 卡供应商
        //代号、供应商名称以及 SIM 卡国别,最后将获取的信息添加到 list 列表
        if(tm.getSimState() == TelephonyManager.SIM_STATE_READY)//SIM 卡状态
        {
            list.add("状态良好");
        }else if(tm.getSimState() == TelephonyManager.SIM_STATE_ABSENT)
        {
            list.add("您目前没有 SIM 卡");
        }else if(tm.getSimState() == TelephonyManager.SIM_STATE_UNKNOWN)
        {
            list.add("SIM 卡处于未知状态");
        }
        if(tm.getSimSerialNumber()!= null)                    //SIM 卡卡号
        {
            list.add(tm.getSimSerialNumber());
        }else
        {
            list.add("没有 SIM 卡卡号");
```

```java
        }
        if(!tm.getSimOperator().equals(""))                    //SIM卡供应商代号
        {
            list.add(tm.getSimOperator());
        }else
        {
            list.add("没有 SIM 卡供应商代号");
        }
        if(!tm.getSimOperatorName().equals(""))                //SIM卡供应商名称
        {
            list.add(tm.getSimOperatorName());
        }else
        {
            list.add("没有 SIM 卡供应商名称");
        }
        if(!tm.getSimCountryIso().equals(""))
        {
            list.add(tm.getSimCountryIso());
        }else
        {
            list.add("无法获取 SIM 国别");
        }
        bCheck.setOnClickListener
        (
                new OnClickListener()
                {
                    public void onClick(View v) {
                        BaseAdapter ba = new BaseAdapter()            //创建适配器
                        {
                            public int getCount() {
                                return list.size();
                            }
                            public Object getItem(int position) {
                                return null;
                            }
                            public long getItemId(int position) {
                                return 0;
                            }
                            public View getView(int arg0, View arg1, ViewGroup arg2) {
                                LinearLayout ll = new LinearLayout(MainActivity.this);
                                ll.setOrientation(LinearLayout.HORIZONTAL);
                                ll.setPadding(5, 5, 5, 5);
                                TextView tv = new TextView(MainActivity.this);    //初始化 TextView
                                tv.setTextColor(Color.BLACK);             //设置字体颜色
                                tv.setPadding(5,5,5,5);
                                tv.setText(name.get(arg0));               //添加任务名字
                                tv.setGravity(Gravity.LEFT);              //左对齐
                                tv.setTextSize(18);                       //字体大小
                                ll.addView(tv);                           //LinearLayout 添加 TextViewTextView
                                tvv = new TextView(MainActivity.this);    //初始化 TextView
                                tvv.setTextColor(Color.BLACK);            //设置字体颜色
                                tvv.setPadding(5,5,5,5);
                                tvv.setText(list.get(arg0));              //添加任务名字
                                tvv.setGravity(Gravity.LEFT);             //左对齐
                                tvv.setTextSize(18);                      //字体大小
```

```
                        ll.addView(tvv);              //LinearLayout 添加 TextView
                        return ll;
                    }
                };
                lv.setAdapter(ba);                    //设置适配器
                lv.setOnItemClickListener            //设置选中菜单的监听器
                (
                    new OnItemClickListener()
                    {
            public void onItemClick(AdapterView<?> arg0, View arg1,
                        int arg2, long arg3) {
                        Toast.makeText(MainActivity.this, name.get(arg2) + "" +
                        list.get(arg2), Toast.LENGTH_SHORT).show();
                        }
                    }
                );
            }
        }
    );
    }
}
```

(4) 打开 AndroidManifest.xml 文件,设置 SIM 卡权限。添加代码为:

…
　　</application>
<!-- 添加权限 -->
<uses-permission android:name = "android.permission.READ_PHONE_STATE"/>
</manifest>

运行程序,效果如图 5-17(a)所示,当单击界面中的"获取 SIM 卡信息"按钮时,效果如图 5-17(b)所示,当单击图 5-17(b)中任何一项时即弹出相应的 Toast 说明,如 5-17(c)所示。

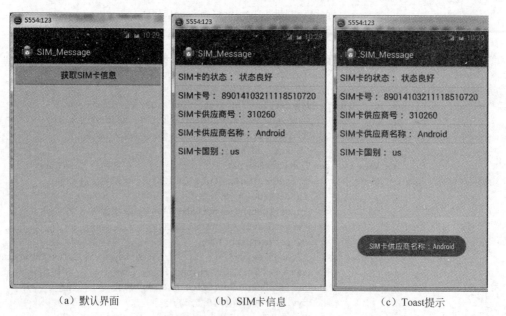

　　(a)默认界面　　　　　　　(b) SIM 卡信息　　　　　　(c) Toast 提示

图 5-17　显示 SIM 相关信息

5.16 查询电池剩余量实例

在使用过程中,最担心的是手机没电,所以及时显示电池容量功能是非常必要的。

可以使用 Android API 中的 BroadcastReseiver 类和 Button 的 Listener 类,当 Reseiver 被注册后会在后台等待被其他程序调用。当指定要捕捉的 Action 发生时,Reseiver 就会被调用,并运行 onReseiver 来实现里面的程序。

在实例中,将利用 BroadcastReseiver 的特性来获取手机电池的容量。通过注册 BroadcastReseiver 时设置的 IntentFiler 来获取系统发出的 Intent.ACTION_BATTERY_CHANGED,然后以此来获取电池的容量、温度及电压等。其具体操作步骤如下。

（1）在 Eclipse 中创建一个 Android 应用项目,命名为 Souces_Message。

（2）打开 res\layout 目录下的 main.xml 布局文件,在文件中声明一个 TextView 控件。用于显示电池情况。代码为:

```
<RelativeLayout xmlns:android = "http://schemas.android.com/apk/res/android"
    xmlns:tools = "http://schemas.android.com/tools"
    android:layout_width = "match_parent"
    android:layout_height = "match_parent"
    android:paddingBottom = "@dimen/activity_vertical_margin"
    android:paddingLeft = "@dimen/activity_horizontal_margin"
    android:paddingRight = "@dimen/activity_horizontal_margin"
    android:paddingTop = "@dimen/activity_vertical_margin"
    tools:context = ".MainActivity"
    android:background = "#ffcc66">
    <TextView
        android:id = "@+id/TV"
        android:layout_width = "wrap_content"
        android:layout_height = "wrap_content"/>
</RelativeLayout>
```

（3）打开 src\fs.souces_message 包下的 MainActivity.java 文件,在文件中实现显示电池的电量、温度及电压。代码为:

```
package fs.souces_message;
import android.app.Activity;
import android.content.BroadcastReceiver;
import android.content.Context;
import android.content.Intent;
import android.content.IntentFilter;
import android.os.BatteryManager;
import android.os.Bundle;
import android.widget.TextView;
public class MainActivity extends Activity
{
    private int BatteryN;                              //目前电量
    private int BatteryV;                              //电池电压
    private double BatteryT;                           //电池温度
    private String BatteryStatus;                      //电池状态
    private String BatteryTemp;                        //电池使用情况
    public TextView TV;
```

```java
@Override
public void onCreate(Bundle savedInstanceState)
{
    super.onCreate(savedInstanceState);
    setContentView(R.layout.main);
    //注册一个系统 BroadcastReceiver,作为访问电池计量之用这个不能直接在
    //AndroidManifest.xml 中注册
    registerReceiver(mBatInfoReceiver, new IntentFilter(Intent.ACTION_BATTERY_CHANGED));
    TV = (TextView)findViewById(R.id.TV);
}
/* 创建广播接收器 */
private BroadcastReceiver mBatInfoReceiver = new BroadcastReceiver()
{
    public void onReceive(Context context, Intent intent)
    {
        String action = intent.getAction();
        /*
         * 如果捕捉到的 action 是 ACTION_BATTERY_CHANGED, 则运行 onBatteryInfoReceiver()
         */
        if (Intent.ACTION_BATTERY_CHANGED.equals(action))
        {
            BatteryN = intent.getIntExtra("level", 0);       //目前电量
            BatteryV = intent.getIntExtra("voltage", 0);     //电池电压
            BatteryT = intent.getIntExtra("temperature", 0);//电池温度
            switch (intent.getIntExtra("status", BatteryManager.BATTERY_STATUS_UNKNOWN))
            {
            case BatteryManager.BATTERY_STATUS_CHARGING:
                BatteryStatus = "充电状态";
                break;
            case BatteryManager.BATTERY_STATUS_DISCHARGING:
                BatteryStatus = "放电状态";
                break;
            case BatteryManager.BATTERY_STATUS_NOT_CHARGING:
                BatteryStatus = "未充电";
                break;
            case BatteryManager.BATTERY_STATUS_FULL:
                BatteryStatus = "充满电";
                break;
            case BatteryManager.BATTERY_STATUS_UNKNOWN:
                BatteryStatus = "未知状态";
                break;
            }
            switch (intent.getIntExtra("health", BatteryManager.BATTERY_HEALTH_UNKNOWN))
            {
            case BatteryManager.BATTERY_HEALTH_UNKNOWN:
                BatteryTemp = "未知错误";
                break;
            case BatteryManager.BATTERY_HEALTH_GOOD:
                BatteryTemp = "状态良好";
                break;
            case BatteryManager.BATTERY_HEALTH_DEAD:
                BatteryTemp = "电池没有电";
                break;
            case BatteryManager.BATTERY_HEALTH_OVER_VOLTAGE:
                BatteryTemp = "电池电压过高";
```

```
                    break;
                case BatteryManager.BATTERY_HEALTH_OVERHEAT:
                    BatteryTemp = "电池过热";
                    break;
                }
            TV.setText("目前电量为" + BatteryN + "% —— " + BatteryStatus + "\n" + "电压为"
                + BatteryV + "mV ——" + BatteryTemp + "\n" + "温度为" + (BatteryT * 0.1) + "℃");
            }
        }
    };
}
```

运行程序,效果如图 5-18 所示。

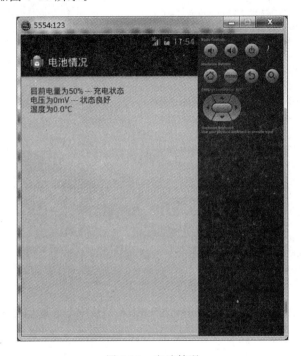

图 5-18 电池情况

5.17 通讯录实例

通讯录是 Android 中最常用的一种应用。个人通讯录主要包括联系人列表和联系人详细信息等界面。

本实例用于利用 ContentResolver 存储系统实现个人通讯录,通过本实例演示了 ContentResolver 的具体用法。本实例的具体实现步骤如下。

(1) 在 Eclipse 中创建一个 Android 应用项目,命名为 Person_Contacts。

(2) 打开 res\layout 目录下的 main.xml 布局文件,在文件中主要实现标题及 Logo 图像以及一个负责显示联系人的列表。代码为:

```
<?xml version = "1.0" encoding = "utf - 8"?>
< LinearLayout xmlns:android = "http://schemas.android.com/apk/res/android"
```

```xml
    android:orientation = "vertical"
    android:layout_width = "fill_parent"
    android:layout_height = "fill_parent"
    android:background = "#a00">
    <LinearLayout
        android:orientation = "horizontal"
        android:layout_width = "wrap_content"
        android:layout_height = "wrap_content"
        android:layout_gravity = "center_horizontal">
        <TextView
            android:layout_width = "wrap_content"
            android:layout_height = "wrap_content"
            android:layout_gravity = "center_horizontal"
            android:text = "@string/title"
            android:textSize = "24px"
            android:textColor = "@color/text"/>
        <ImageView
            android:layout_width = "wrap_content"
            android:layout_height = "wrap_content"
            android:src = "@drawable/td" />
    </LinearLayout>
    <ScrollView
        android:layout_width = "fill_parent"
        android:layout_height = "fill_parent"
        android:fillViewport = "true">
        <ListView
            android:id = "@+id/lv"
            android:layout_width = "fill_parent"
            android:layout_height = "fill_parent"
            android:choiceMode = "singleChoice"/>
    </ScrollView>
</LinearLayout>
```

（3）在 res\layout 目录下创建一个 detail.xml 布局文件，主要用于显示联系人的详细信息。代码为：

```xml
<?xml version = "1.0" encoding = "utf-8"?>
<LinearLayout
    xmlns:android = "http://schemas.android.com/apk/res/android"
    android:orientation = "vertical"
    android:layout_width = "fill_parent"
    android:layout_height = "wrap_content"
    android:background = "@drawable/kb">
    <!-- 显示联系人姓名线性布局 -->
    <LinearLayout
        android:orientation = "horizontal"
        android:layout_width = "fill_parent"
        android:layout_height = "wrap_content"
        android:layout_gravity = "center_horizontal">
        <TextView
            android:layout_width = "100px"
            android:layout_height = "wrap_content"
            android:textSize = "18px"
            android:textColor = "@color/text"
            android:layout_gravity = "left|center_vertical"
```

```xml
            android:text = "@string/tvName"/>
    </LinearLayout>
        <EditText
            android:id = "@ + id/etName"
            android:layout_width = "fill_parent"
            android:layout_height = "wrap_content"/>
    </LinearLayout>
<!-- 显示联系人固定电话的线性布局 -->
    <LinearLayout
        android:orientation = "horizontal"
        android:layout_width = "fill_parent"
        android:layout_height = "wrap_content">
        <TextView
            android:layout_width = "100px"
            android:layout_height = "wrap_content"
            android:textSize = "18px"
            android:textColor = "@color/text"
            android:layout_gravity = "center_vertical|left"
            android:text = "@string/tvPhone"/>
        <EditText
            android:id = "@ + id/etPhone"
            android:layout_width = "fill_parent"
            android:layout_height = "wrap_content"
            android:phoneNumber = "true"/>
    </LinearLayout>
<!-- 显示联系人手机号码的线性布局 -->
    <LinearLayout
        android:orientation = "horizontal"
        android:layout_width = "fill_parent"
        android:layout_height = "wrap_content">
        <TextView
            android:layout_width = "100px"
            android:layout_height = "wrap_content"
            android:textSize = "18px"
            android:textColor = "@color/text"
            android:layout_gravity = "left|center_vertical"
            android:text = "@string/tvMobile"/>
        <EditText
            android:id = "@ + id/etMobile"
            android:layout_width = "fill_parent"
            android:layout_height = "wrap_content"
            android:phoneNumber = "true"/>
    </LinearLayout>
<!-- 显示联系人电子邮件的线性布局 -->
    <LinearLayout
        android:orientation = "horizontal"
        android:layout_width = "fill_parent"
        android:layout_height = "wrap_content">
        <TextView
            android:layout_width = "100px"
            android:layout_height = "wrap_content"
            android:textSize = "18px"
            android:textColor = "@color/text"
            android:layout_gravity = "left|center_vertical"
            android:text = "@string/tvEmail"/>
        <EditText
```

```xml
            android:id = "@ + id/etEmail"
            android:layout_width = "fill_parent"
            android:layout_height = "wrap_content"/>
    </LinearLayout>
    <!-- 显示联系人邮编的线性布局 -->
    <LinearLayout
        android:orientation = "horizontal"
        android:layout_width = "fill_parent"
        android:layout_height = "wrap_content">
        <TextView
            android:layout_width = "100px"
            android:layout_height = "wrap_content"
            android:textSize = "18px"
            android:textColor = "@color/text"
            android:layout_gravity = "left|center_vertical"
            android:text = "@string/tvPost"/>
        <EditText
            android:id = "@ + id/etPost"
            android:layout_width = "fill_parent"
            android:layout_height = "wrap_content"/>
    </LinearLayout>
    <!-- 显示联系人通信地址的线性布局 -->
    <LinearLayout
        android:orientation = "horizontal"
        android:layout_width = "fill_parent"
        android:layout_height = "wrap_content">
        <TextView
            android:layout_width = "100px"
            android:layout_height = "wrap_content"
            android:textSize = "18px"
            android:textColor = "@color/text"
            android:layout_gravity = "left|center_vertical"
            android:text = "@string/tvAddr"/>
        <EditText
            android:id = "@ + id/etAddr"
            android:layout_width = "fill_parent"
            android:layout_height = "wrap_content"/>
    </LinearLayout>
    <!-- 显示联系人公司的线性布局 -->
    <LinearLayout
        android:orientation = "horizontal"
        android:layout_width = "fill_parent"
        android:layout_height = "wrap_content">
        <TextView
            android:layout_width = "100px"
            android:layout_height = "wrap_content"
            android:textSize = "18px"
            android:textColor = "@color/text"
            android:layout_gravity = "left|center_vertical"
            android:text = "@string/tvComp"/>
        <EditText
            android:id = "@ + id/etComp"
            android:layout_width = "fill_parent"
            android:layout_height = "wrap_content"/>
    </LinearLayout>
```

```xml
        <ImageButton
            android:id = "@ + id/ibSave"
            android:layout_width = "fill_parent"
            android:layout_height = "wrap_content"
            android:src = "@drawable/sd"/>
</LinearLayout>
```

(4) 打开 res\values 目录下的 strings.xml 文件,在文件中为变量赋值。代码为:

```xml
<?xml version = "1.0" encoding = "utf - 8"?>
<resources>
    <string name = "app_name">个人通讯录</string>
    <string name = "action_settings">Settings</string>
    <string name = "hello_world">Hello world!</string>
    <string name = "tvName">姓名:</string>
    <string name = "tvPhone">固定电话:</string>
    <string name = "tvMobile">移动电话:</string>
    <string name = "tvEmail">电子邮件:</string>
    <string name = "tvPost">邮政编码:</string>
    <string name = "tvAddr">通信地址:</string>
    <string name = "tvComp">公司名称:</string>
    <string name = "menu_add">添加</string>
    <string name = "menu_modify">修改</string>
    <string name = "menu_delete">删除</string>
    <string name = "menu_save">保存</string>
    <string name = "dialog_message">确认删除此人吗?</string>
    <string name = "ok">确定</string>
    <string name = "cancel">取消</string>
    <string name = "title">联系人列表</string>
</resources>
```

(5) 在 res\values 目录下新建一个 color.xml 布局文件,用于实现颜色资源。代码为:

```xml
<?xml version = "1.0" encoding = "utf - 8"?>
<resources>
    <color name = "text">#000000</color>
</resources>
```

(6) 在 src\fs.person_contacts 包下新建一个 AOpenHelper.java 文件,用于实现开发数据库的辅助类。代码为:

```java
package fs.person_contacts;
import android.content.Context;
import android.database.sqlite.SQLiteDatabase;
import android.database.sqlite.SQLiteOpenHelper;
import android.database.sqlite.SQLiteDatabase.CursorFactory;
public class AOpenHelper extends SQLiteOpenHelper{
    public static final String DB_NAME = "personal_contacts";    //数据库文件名称
    public static final String TABLE_NAME = "contacts";          //表名
    public static final String ID = "_id";                        //ID
    public static final String NAME = "name";                     //名称
    public static final String PHONE = "phone";                   //固定电话
    public static final String MOBILE = "mobile";                 //手机号码
    public static final String EMAIL = "email";                   //电子邮件地址
    public static final String POST = "post";                     //邮政编码
    public static final String ADDR = "addr";                     //通信地址
```

```java
        public static final String COMP = "comp";                          //公司
        public AOpenHelper(Context context, String name, CursorFactory factory,
                int version) {
            super(context, name, factory, version);                        //调用父类构造器
        }
        @Override
        public void onCreate(SQLiteDatabase db) {                          //重写 onCreate 方法
            //调用 execSQL 方法创建表
            db.execSQL("create table if not exists " + TABLE_NAME + " (" + ID + " integer primary key,"
                    + NAME + " varchar,"
                    + PHONE + " varchar,"
                    + MOBILE + " varchar,"
                    + EMAIL + " varchar,"
                    + POST + " varchar,"
                    + ADDR + " varchar,"
                    + COMP + " varchar)");
        }
        @Override
        public void onUpgrade(SQLiteDatabase db, int oldVersion, int newVersion) {
            //重写 onUpgrade 方法
        }
}
```

（7）打开 src\fs.person_contacts 包下的 MainActivity.java 文件，显示联系人的列表并删除指定项的功能。代码为：

```java
package fs.person_contacts;
import static fs.person_contacts.AOpenHelper.*;
import android.app.Activity;
import android.app.AlertDialog;
import android.app.Dialog;
import android.app.AlertDialog.Builder;
import android.content.ContentValues;
import android.content.DialogInterface;
import android.content.Intent;
import android.content.DialogInterface.OnClickListener;
import android.database.Cursor;
import android.database.sqlite.SQLiteDatabase;
import android.graphics.Color;
import android.os.Bundle;
import android.view.Gravity;
import android.view.Menu;
import android.view.MenuItem;
import android.view.View;
import android.view.ViewGroup;
import android.view.ViewGroup.LayoutParams;
import android.widget.AdapterView;
import android.widget.BaseAdapter;
import android.widget.LinearLayout;
import android.widget.ListView;
import android.widget.TextView;
import android.widget.AdapterView.OnItemClickListener;
public class MainActivity extends Activity {
    AOpenHelper myHelper;                                                  //声明 MyOpenHelper 对象
```

```java
        String [] contactsName;                                   //声明用于存放联系人姓名的数组
        String [] contactsPhone;                                  //声明用于存放联系人电话的数组
        int [] contactsId;                                        //声明用于存放联系人 id 的数组
        final int MENU_ADD = Menu.FIRST;                          //声明菜单选行的 id
        final int MENU_DELETE = Menu.FIRST + 1;                   //声明菜单项的编号
        final int DIALOG_DELETE = 0;                              //确认删除对话框的 id
        ListView lv;                                              //声明 ListView 对象
        BaseAdapter myAdapter = new BaseAdapter(){
            @Override
            public int getCount() {
                if(contactsName != null){                         //如果姓名数组不为空
                    return contactsName.length;
                }
                else {
                    return 0;                                     //如果姓名数组为空则返回 0
                }
            }
            @Override
            public Object getItem(int arg0) {
                return null;
            }
            @Override
            public long getItemId(int arg0) {
                return 0;
            }
            @Override
            public View getView(int position, View convertView, ViewGroup parent) {
                LinearLayout ll = new LinearLayout(MainActivity.this);
                ll.setOrientation(LinearLayout.HORIZONTAL);
                TextView tv = new TextView(MainActivity.this);
                tv.setText(contactsName[position]);
                tv.setTextSize(32);
                tv.setTextColor(Color.BLACK);
                tv.setLayoutParams(new LayoutParams(LayoutParams.WRAP_CONTENT, LayoutParams.WRAP_
                    CONTENT));
                tv.setGravity(Gravity.CENTER_VERTICAL);
                TextView tv2 = new TextView(MainActivity.this);
                tv2.setText("[" + contactsPhone[position] + "]");
                tv2.setTextSize(28);
                tv2.setTextColor(Color.BLACK);
                tv2.setLayoutParams(new LayoutParams(LayoutParams.WRAP_CONTENT, LayoutParams.WRAP
                    _CONTENT));
    //设置 TextView 控件在父容器中的位置
                tv2.setGravity(Gravity.BOTTOM|Gravity.RIGHT);
                ll.addView(tv);
                ll.addView(tv2);
                return ll;
            }
        };
        @Override
        public void onCreate(Bundle savedInstanceState) {
            super.onCreate(savedInstanceState);
```

```java
        setContentView(R.layout.main);
        myHelper = new AOpenHelper(this, DB_NAME, null, 1);
        lv = (ListView)findViewById(R.id.lv);
        lv.setAdapter(myAdapter);
        lv.setOnItemClickListener(new OnItemClickListener() {
            @Override
            public void onItemClick(AdapterView<?> arg0, View view, int position,
                    long id) {
                Intent intent = new Intent(MainActivity.this,DetailActivity.class);
                intent.putExtra("cmd", 0);                  //0 代表查询联系人,1 代表添加联系人
                intent.putExtra("id", contactsId[position]);
                startActivity(intent);
            }
        });
    }
    @Override
    protected void onResume() {
        getBasicInfo(myHelper);
        myAdapter.notifyDataSetChanged();
        super.onResume();
    }
    //获取所有联系人的姓名
    public void getBasicInfo(AOpenHelper helper){
        SQLiteDatabase db = helper.getWritableDatabase();  //获取数据库连接
        Cursor c = db.query(TABLE_NAME, new String[]{ID,NAME,PHONE}, null, null, null, null,
            ID);
        int idIndex = c.getColumnIndex(ID);
        int nameIndex = c.getColumnIndex(NAME);             //获得姓名列的列号
        int phoneIndex = c.getColumnIndex(PHONE);           //获得电话列的序号
        contactsName = new String[c.getCount()];            //创建存放姓名的 String 数组对象
        contactsId = new int[c.getCount()];                 //创建存放 id 的 int 数组对象
        contactsPhone = new String[c.getCount()];           //创建存放 phone 的数组对象
        int i = 0;                                          //声明一个计数器
        for(c.moveToFirst();!(c.isAfterLast());c.moveToNext()){
            contactsName[i] = c.getString(nameIndex);       //将姓名添加到 String 数组中
            contactsId[i] = c.getInt(idIndex);
            contactsPhone[i] = c.getString(phoneIndex);         //将固定电话添加到 String 数组中
            i++;
        }
        c.close();                                          //关闭 Cursor 对象
        db.close();                                         //关闭 SQLiteDatabase 对象
    }
    @Override
    public boolean onCreateOptionsMenu(Menu menu) {
        menu.add(0, MENU_ADD, 0, R.string.menu_add)
            .setIcon(R.drawable.ad);                        //添加"添加"菜单选项
        menu.add(0, MENU_DELETE, 0, R.string.menu_delete)
            .setIcon(R.drawable.delete);                    //添加"删除"菜单选项
        return super.onCreateOptionsMenu(menu);
    }
    @Override
    public boolean onOptionsItemSelected(MenuItem item) {
```

```java
            switch(item.getItemId()){                              //判断按下的菜单选项
            case MENU_ADD:                                         //按下添加按钮
                Intent intent = new Intent(MainActivity.this,DetailActivity.class);
                intent.putExtra("cmd", 1);
                startActivity(intent);
                break;
            case MENU_DELETE:                                      //按下了删除选项
                showDialog(DIALOG_DELETE);                         //显示确认删除对话框
                break;
            }
            return super.onOptionsItemSelected(item);
    }
    @Override
    protected Dialog onCreateDialog(int id) {
        Dialog dialog = null;
        switch(id){                                                //对对话框id进行判断
        case DIALOG_DELETE:                                        //创建删除确认对话框
            Builder b = new AlertDialog.Builder(this);
            b.setIcon(R.drawable.hd);                              //设置对话框图标
            b.setTitle("提示");                                    //设置对话框标题
            b.setMessage(R.string.dialog_message);                 //设置对话框内容
            b.setPositiveButton(
                    R.string.ok,
                    new OnClickListener() {                        //单击确认删除按钮
                        @Override
                        public void onClick(DialogInterface dialog, int which) {
                            int position = MainActivity.this.lv.getSelectedItemPosition();
                            deleteContact(contactsId[position]);
                            getBasicInfo(myHelper);
                            myAdapter.notifyDataSetChanged();
                        }
                    });
            b.setNegativeButton(
                    R.string.cancel,
                    new OnClickListener() {
                        @Override
                        public void onClick(DialogInterface dialog, int which) {}
                    });
            dialog = b.create();
            break;
        }
        return dialog;
    }
    //删除指定联系人
    public void deleteContact(int id){
        SQLiteDatabase db = myHelper.getWritableDatabase();//获得数据库对象
        db.delete(TABLE_NAME, ID + " = ?", new String[]{id + ""});
        db.close();
    }
```

(8) 在 src\fs.person_contacts 包下新建 DetailActivity.java 文件，用于显示联系人详细信息界面，并对该类进行开发。代码为：

```java
package fs.person_contacts;
import static fs.person_contacts.AOpenHelper.*;
import android.annotation.SuppressLint;
import android.app.Activity;
import android.content.ContentValues;
import android.content.Intent;
import android.database.Cursor;
import android.database.sqlite.SQLiteDatabase;
import android.os.Bundle;
import android.view.Menu;
import android.view.View;
import android.widget.EditText;
import android.widget.ImageButton;
import android.widget.Toast;
@SuppressLint("NewApi")
public class DetailActivity extends Activity{
    AOpenHelper myHelper;                             //声明一个 MyOpenHelper 对象
    final int MENU_ADD = Menu.FIRST;                  //声明菜单项的编号
    final int MENU_MODIFY = Menu.FIRST + 1;           //声明菜单项的编号
    final int MENU_DELETE = Menu.FIRST + 2;           //声明菜单项的编号
    final int MENU_SAVE = Menu.FIRST + 3;             //声明菜单项的编号
    int id = -1;                                      //记录当前显示的联系人 id
    int [] textIds = {
        R.id.etName,
        R.id.etPhone,
        R.id.etMobile,
        R.id.etEmail,
        R.id.etPost,
        R.id.etAddr,
        R.id.etComp
    };
    EditText [] textArray;                            //存放界面中的 EditText 控件的数组
    ImageButton ibSave;                               //保存按钮
    int status = -1;              //0 表示查看信息,1 表示添加联系人,2 表示修改联系人
    View.OnClickListener myListener = new View.OnClickListener() {
        @Override
        public void onClick(View v) {                 //保存按钮按下触发该事件
            String [] strArray = new String[textArray.length];
            for(int i = 0;i < strArray.length;i++){
                strArray[i] = textArray[i].getText().toString().trim();   //获得用户输入的信息数组
            }
            if(strArray[0].equals("") || strArray[1].equals("")){
                Toast.makeText(DetailActivity.this, "对不起,姓名和电话必须填写完整!",
                    Toast.LENGTH_LONG).show();
            }
            switch(status){                           //判断当前的状态
            case 0:                                   //查询联系人详细信息时按下保存
                updateContact(strArray);              //更新联系人信息
                break;
```

```java
                case 1:                                     //新建联系人时按下保存按钮
                    insertContact(strArray);                //插入联系人信息
                    break;
            }
        }
    };
    @Override
    protected void onCreate(Bundle savedInstanceState) {
        super.onCreate(savedInstanceState);
        setContentView(R.layout.detail);
        textArray = new EditText[textIds.length];
        for(int i = 0; i < textIds.length; i++){
            textArray[i] = (EditText)findViewById(textIds[i]);
        }
        ibSave = (ImageButton)findViewById(R.id.ibSave);
        ibSave.setOnClickListener(myListener);
        myHelper = new AOpenHelper(this, AOpenHelper.DB_NAME, null, 1);
        Intent intent = getIntent();
        status = intent.getExtras().getInt("cmd");           //读命令类型
        switch(status){
        case 0:                                              //查看联系人的详细信息
            id = intent.getExtras().getInt("id");            //获得要显示的联系人的id
            SQLiteDatabase db = myHelper.getWritableDatabase();
            Cursor c = db.query(AOpenHelper.TABLE_NAME, new String[]{NAME,PHONE,MOBILE,
            EMAIL,POST,ADDR,COMP}, ID + " = ?", new String[]{id + ""}, null, null, null);
            if(c.getCount() == 0){                           //没有查询到指定的人
                Toast.makeText(this, "对不起,没有找到对应的联系人!", Toast.LENGTH_LONG).
                show();
            }
            else{                                            //查询到了这个人
                c.moveToFirst();                             //移动到第一条记录
                textArray[0].setText(c.getString(0));        //设置姓名框中的内容
                textArray[1].setText(c.getString(1));        //设置固话框中的内容
                textArray[2].setText(c.getString(2));        //设置手机号码框中的内容
                textArray[3].setText(c.getString(3));        //设置电子邮件框中的内容
                textArray[4].setText(c.getString(4));        //设置邮政编码框中的内容
                textArray[5].setText(c.getString(5));        //设置通信地址框中的内容
                textArray[6].setText(c.getString(6));        //设置公司名称框中的内容
            }
            c.close();
            db.close();
            break;
        case 1:                                              //新建详细人信息
            for(EditText et:textArray){
                et.getEditableText().clear();                //清空各个EditText控件中内容
            }
            break;
        }
    }
    //添加指定联系人
    public void insertContact(String [] strArray){
        SQLiteDatabase db = myHelper.getWritableDatabase();//获得数据库对象
```

```java
        ContentValues values = new ContentValues();
        values.put(NAME, strArray[0]);
        values.put(PHONE, strArray[1]);
        values.put(MOBILE, strArray[2]);
        values.put(EMAIL, strArray[3]);
        values.put(POST, strArray[4]);
        values.put(ADDR, strArray[5]);
        values.put(COMP, strArray[6]);
        long count = db.insert(TABLE_NAME, ID, values);        //插入数据
        db.close();
        if(count == -1){
            Toast.makeText(this, "添加联系人失败!", Toast.LENGTH_LONG).show();
        }
        else{
            Toast.makeText(this, "添加联系人成功!", Toast.LENGTH_LONG).show();
        }
    }
    //更新某个联系人信息
    public void updateContact(String [] strArray){
        SQLiteDatabase db = myHelper.getWritableDatabase();               //获得数据库对象
        ContentValues values = new ContentValues();
        values.put(NAME, strArray[0]);
        values.put(PHONE, strArray[1]);
        values.put(MOBILE, strArray[2]);
        values.put(EMAIL, strArray[3]);
        values.put(POST, strArray[4]);
        values.put(ADDR, strArray[5]);
        values.put(COMP, strArray[6]);
        //更新数据库
        int count = db.update(TABLE_NAME, values, ID + " = ?", new String[]{id + ""});
        db.close();
        if(count == 1){
            Toast.makeText(this, "修改联系人成功!", Toast.LENGTH_LONG).show();
        }
        else{
            Toast.makeText(this, "修改联系人失败!", Toast.LENGTH_LONG).show();
        }
    }
}
```

(9) 打开 AndroidManifest.xml 文件，用于声明 DetailActivity 类。代码为：

```
        ...
            </intent-filter>
        </activity>
            <activity android:name=".DetailActivity"/>
    </application>
</manifest>
```

运行程序，默认效果如图 5-19(a)所示，单击界面中的"MENU"按钮或右上角的菜单项，即弹出"添加"与"删除"子菜单，如图 5-19(b)所示，选择"添加"子菜单，即实现通讯录的添加功能，如图 5-19(c)所示，在该界面中填写相应的信息如图 5-19(d)所示并单击界面中的"保存"

图标按钮,即实现通讯录添加,如图 5-19(e)所示,返回联系人列表即效果如图 5-19(f)所示。

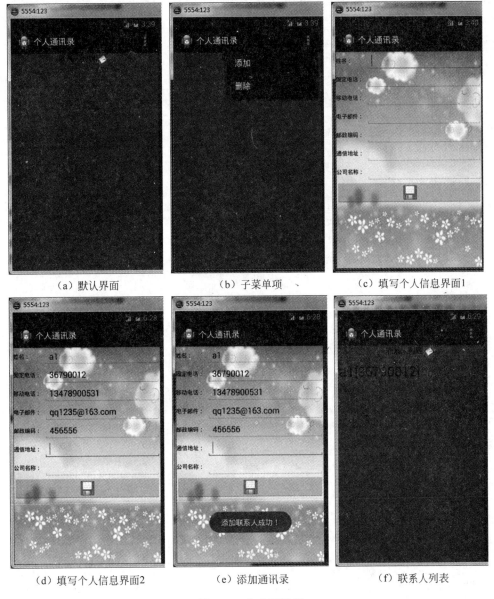

图 5-19 个人通讯录

5.18 在线查询实例

现在的人都喜欢当遇到不认识的单词或不理解的词语时,第一时间都会借助于网络来进行查询。也就是说在线查询也成为人们生活中不可缺少的助手。本实例实现对词语的翻译以及理解的实例。通过本实例演示了 Android 网络通信的具体应用。

本实例的具体实现步骤为:

(1) 在 Eclipse 中创建一个 Android 应用项目,命名为 Online_inquiry。

(2) 打开 res\layout 目录下的 main.xml 布局文件，在文件中声明了文本输入框、单选框、按钮组及网页控件，并且分别为其指定 id。代码为：

```xml
<RelativeLayout xmlns:android="http://schemas.android.com/apk/res/android"
    xmlns:tools="http://schemas.android.com/tools"
    android:layout_width="match_parent"
    android:layout_height="match_parent"
    android:paddingBottom="@dimen/activity_vertical_margin"
    android:paddingLeft="@dimen/activity_horizontal_margin"
    android:paddingRight="@dimen/activity_horizontal_margin"
    android:paddingTop="@dimen/activity_vertical_margin"
    tools:context=".MainActivity"
    android:background="#a0a">
    <TextView
        android:id="@+id/textView1"
        android:layout_width="wrap_content"
        android:layout_height="wrap_content"
        android:layout_centerHorizontal="true"
        android:textSize="15sp"
        android:text="在线查询" />
    <EditText
        android:id="@+id/tinput"
        android:layout_width="fill_parent"
        android:layout_height="wrap_content"
        android:layout_alignParentLeft="true"
        android:layout_below="@+id/textView1"
        android:ems="10"
        android:hint="输入要查询的词" />
    <RadioGroup
        android:id="@+id/myRadioGroup"
        android:layout_width="wrap_content"
        android:layout_height="wrap_content"
        android:layout_alignParentLeft="true"
        android:layout_alignParentRight="true"
        android:layout_below="@+id/tinput"
        android:orientation="horizontal" >
        <RadioButton
            android:id="@+id/myRadioButton1"
            android:layout_height="wrap_content"
            android:text="翻译" />
        <RadioButton
            android:id="@+id/myRadioButton2"
            android:layout_height="wrap_content"
            android:text="百科" />
    </RadioGroup>
    <Button
        android:id="@+id/submit"
        android:layout_width="wrap_content"
        android:layout_height="wrap_content"
        android:layout_alignParentRight="true"
        android:layout_below="@+id/myRadioGroup"
        android:text="查询" />
    <TextView
        android:id="@+id/tips"
        android:layout_width="fill_parent"
```

```xml
        android:layout_height = "wrap_content"
        android:layout_alignParentLeft = "true"
        android:layout_below = "@ + id/submit"
        android:text = "查询结果如下："
        android:textSize = "14sp"
        android:visibility = "invisible"
        android:typeface = "sans" />
    <WebView
        android:id = "@ + id/toutput"
        android:layout_width = "fill_parent"
        android:layout_height = "270px"
        android:layout_alignParentBottom = "true"
        android:layout_alignParentLeft = "true"
        android:layout_below = "@ + id/tips"
        android:visibility = "invisible" />
</RelativeLayout>
```

（3）打开 src\ fs.online_inquiry 包下的 MainActivity.java 文件，在文件中实现在线查询单词。代码为：

```java
package fs.online_inquiry;
import android.os.Bundle;
import android.os.Handler;
import android.app.Activity;
import android.view.Menu;
import android.view.View;
import android.view.View.OnClickListener;
import android.webkit.WebSettings;
import android.webkit.WebView;
import android.webkit.WebViewClient;
import android.widget.Button;
import android.widget.EditText;
import android.widget.RadioButton;
import android.widget.RadioGroup;
import android.widget.TextView;
import android.widget.Toast;
public class MainActivity extends Activity {
    //定义控件
    private TextView tips;
    private EditText editText;
    private WebView webView;
    private Button submit;
    RadioButton rb1, rb2;
    RadioGroup rGroup;
    private Handler tHandler = new Handler();                    //定义 Handler
    //第一次调用 Activity 活动
    @Override
    protected void onCreate(Bundle savedInstanceState) {         //重写 onCreate()方法
        super.onCreate(savedInstanceState);
        setContentView(R.layout.main);
        webView = (WebView) findViewById(R.id.toutput);
        submit = (Button) findViewById(R.id.submit);
        editText = (EditText) findViewById(R.id.tinput);
        tips = (TextView) findViewById(R.id.tips);
        rb1 = (RadioButton) findViewById(R.id.myRadioButton1);
```

```java
            rb2 = (RadioButton) findViewById(R.id.myRadioButton2);
            rGroup = (RadioGroup) findViewById(R.id.myRadioGroup);
            rGroup.check(R.id.myRadioButton1);
            WebSettings webSettings = webView.getSettings();              //获取 WebSetting
            webSettings.setSaveFormData(false);
            webSettings.setSavePassword(false);
            webSettings.setSupportZoom(false);
            webView.setWebViewClient(new WebViewClient() {                //设置 WebViewClient
                @Override
                public boolean shouldOverrideUrlLoading(
                        WebView view, String url) {
                    //使用自定义的 WebView 加载
                    view.loadUrl(url);
                    return true;
                }
            });
            submit.setOnClickListener(new OnClickListener() {             //设置按钮单击事件
                @Override
                public void onClick(View v) {
                    if (editText.getText().toString().equals("")) {      //判断输入是否为空
                        Toast.makeText(MainActivity.this, "请输入查询的词",
                                Toast.LENGTH_LONG);
                        return;
                    }
                    tips.setVisibility(TextView.VISIBLE);
                    webView.setVisibility(WebView.VISIBLE);
                    tHandler.post(new Runnable() {
                        public void run() {
                            if (rGroup.getCheckedRadioButtonId() == R.id.myRadioButton1) {
                                webView.loadUrl("http://3g.dict.cn/s.php?q="
                                        + editText.getText().toString());    //加载翻译
                            } else {
                                webView.loadUrl("http://www.baike.com/wiki/"
                                        + editText.getText().toString());    //加载百科
                            }
                        }
                    });
                }
            });
        }
        @Override
        public boolean onCreateOptionsMenu(Menu menu) {
            //Inflate the menu; this adds items to the action bar if it is present.
            getMenuInflater().inflate(R.menu.main, menu);
            return true;
        }
    }
```

(4) 打开 AndroidManifest.xml 文件,在文件中添加在线查询权限。代码为:

```
…
    <uses-sdk
        android:minSdkVersion = "8"
        android:targetSdkVersion = "18" />
<!-- 添加在线查询权限 -->
<uses-permission android:name = "android.permission.INTERNET" />
```

```
<application
    android:allowBackup = "true"
    android:icon = "@drawable/ic_launcher"
    android:label = "@string/app_name"
    android:theme = "@style/AppTheme" >
...
```

运行程序,默认效果如图 5-20(a)所示,在编辑框中输入需要查询的词,选择"翻译"项,并单击"查询"按钮,效果如图 5-20(b)所示,选择"百科"项,并单击"查询"按钮,效果如图 5-20(c)所示。

(a)默认界面　　　　　　(b)翻译结果　　　　　　(c)百科结果

图 5-20　在线查询

5.19　自动朗读实例

Android 提供了自动朗读支持。自动朗读支持可以对指定文本内容进行朗读,从而发生声音;不仅如此,Android 的自动朗读支持还允许把文本对应的音频录制成音频文件,方便以后播放。这种自动朗读支持的英文名称为 TextToSpeech,简称 TTS。借助于 TTS 的支持,可以在应用程序中动态地增加音频输出,从而改善用户体验。

本实例利用 Android 提供的 TextToSpeech 方法来实现声音的朗读与记录声音功能。通过本实例演示了 TextToSpeech 方法的具体用法。

本实例的具体实现步骤如下。

(1) 在 Eclipse 中新建一个 Android 应用项目,命名为 TTS_test。

(2) 打开 res\layout 目录下的 main.xml 布局文件,在文件中声明按钮控件和编辑框控件。代码为:

```
<?xml version = "1.0" encoding = "utf-8"?>
<LinearLayout xmlns:android = "http://schemas.android.com/apk/res/android"
    android:orientation = "vertical"
```

```xml
            android:layout_width = "fill_parent"
            android:layout_height = "fill_parent"
            android:gravity = "center_horizontal"
            android:background = "#0a0">
    <EditText
            android:id = "@+id/txt"
            android:layout_width = "fill_parent"
            android:layout_height = "wrap_content"
            android:lines = "5"/>
    <LinearLayout
            android:orientation = "horizontal"
            android:layout_width = "fill_parent"
            android:layout_height = "wrap_content"
            android:gravity = "center_horizontal">
        <Button
            android:id = "@+id/speech"
            android:layout_width = "wrap_content"
            android:layout_height = "wrap_content"
            android:text = "@string/speech"/>
        <Button
            android:id = "@+id/record"
            android:layout_width = "wrap_content"
            android:layout_height = "wrap_content"
            android:text = "@string/record"/>
    </LinearLayout>
</LinearLayout>
```

(3) 打开 res\values 目录下的 strings.xml 文件,在文件中为变量赋值。代码为:

```xml
<?xml version = "1.0" encoding = "utf-8"?>
<resources>
    <string name = "app_name">自动朗读</string>
    <string name = "action_settings">Settings</string>
    <string name = "hello_world">Hello world!</string>
    <string name = "speech">朗读</string>
    <string name = "record">记录声音</string>
</resources>
```

(4) 打开 src\fs.tts_test 包下的 MainActivity.java 文件,在文件中实现当在编辑框中输入对应的语句时,并单击按钮时即可实现自动朗读功能。代码为:

```java
package fs.tts_test;
import java.util.Locale;
import android.app.Activity;
import android.os.Bundle;
import android.speech.tts.TextToSpeech;
import android.speech.tts.TextToSpeech.OnInitListener;
import android.view.View;
import android.view.View.OnClickListener;
import android.widget.Button;
import android.widget.EditText;
import android.widget.Toast;
public class MainActivity extends Activity {
    //定义控件
    TextToSpeech tts;
    EditText editText;
```

```java
        Button speech;
        Button record;
        @Override
        public void onCreate(Bundle savedInstanceState) {
            super.onCreate(savedInstanceState);
            setContentView(R.layout.main);
            //初始化 TextToSpeech 对象
            tts = new TextToSpeech(this, new OnInitListener() {
                public void onInit(int status) {
                    //TODO 自动存根法
                    //如果装载 TTS 引擎成功
                    if(status == TextToSpeech.SUCCESS){
                        //设置使用美式英语朗读
                        int result = tts.setLanguage(Locale.US);
                        //如果不支持所设置的语言
                        if(result != TextToSpeech.LANG_COUNTRY_AVAILABLE
                            && result != TextToSpeech.LANG_AVAILABLE){
Toast.makeText(MainActivity.this, "TTS暂时不支持这种语言的朗读.", 50000).show();
                        }
                    }
                }
            });
            editText = (EditText)findViewById(R.id.txt);
            speech = (Button) findViewById(R.id.speech);
            record = (Button) findViewById(R.id.record);
            speech.setOnClickListener(new OnClickListener() {
                public void onClick(View v) {
                    //TODO 自动存根法
                    //执行朗读
                    tts.speak(editText.getText().toString(), TextToSpeech.QUEUE_ADD, null);
                }
            });
            record.setOnClickListener(new OnClickListener() {
                public void onClick(View v) {
                    //TODO 自动存根法
                    //将朗读文本的音频记录到指定文件
      tts.synthesizeToFile(editText.getText().toString(), null, "/mnt/sdcard/sound.wav");
                    Toast.makeText(MainActivity.this, "声音记录成功! ", 50000).show();
                }
            });
        }
        @Override
        protected void onDestroy() {
                //关闭 TextToSpeech 对象
            if(tts != null){
                tts.shutdown();
            }
            super.onDestroy();
        }
    }
```

运行程序，在编辑框中输入对应的语句时，并单击"朗读"按钮时，即可实现朗读功能，效果如图 5-21 所示，单击"记录声音"按钮时，即可将自动朗读的声音记录下来。

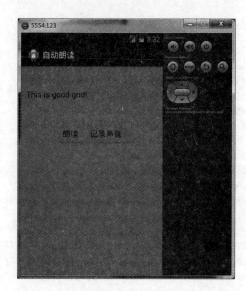

图 5-21 自动朗读

第 6 章

Android 手机功能实例

现在几乎每人手中都有一台智能手机,而 Android 作为最大的智能手机系统,其本身提供了相应完善的功能,因此本章节利用实例形式来对 Android 手机的服务功能进行介绍。

6.1 振动器实例

有些时候,程序需要启动系统振动器,例如手机静音时使用振动提示用户;再例如玩游戏时,当系统碰撞、爆炸时使用振动带给用户更逼真的体验等。总之,振动是除视频、声音之外的另一种"多媒体",充分利用系统的振动器会带给用户更好的体验。

系统获取振动器是调用 Context 的 getSystemService()方法即可,接着就可调用振动器的方法来控制手机振动了。

本实例利用了 Android 的 onTouchEvent 方法,当用户用触碰手机触摸屏时将会启动手机振动。通过本实例演示了 onTouchEvent 方法的具体用法。

本实例的具体实现步骤如下。

(1) 在 Eclipse 中创建一个 Android 应用项目,命名为 Vibrator_test。

(2) 打开 res\layout 目录下的 main.xml 布局文件,在文件中声明两个 ToggleButton 控件和两个 TextView 控件。代码为:

```
<?xml version = "1.0" encoding = "utf - 8"?>
< LinearLayout xmlns:android = "http://schemas.android.com/apk/res/android"
    android:orientation = "vertical"
    android:layout_width = "fill_parent"
    android:layout_height = "fill_parent"
    android:background = "#0c0">
    < LinearLayout
        android:orientation = "horizontal"
        android:layout_width = "fill_parent"
        android:layout_height = "wrap_content">
        < ToggleButton
            android:id = "@ + id/tb1"
            android:textOn = "关闭振动"
            android:textOff = "启动振动"
            android:checked = "false"
```

```xml
            android:layout_width = "wrap_content"
            android:layout_height = "wrap_content" />
        <TextView
            android:id = "@ + id/tv1"
            android:text = "振动已关闭"
            android:layout_width = "wrap_content"
            android:layout_height = "wrap_content" />
    </LinearLayout>
    <LinearLayout
        android:orientation = "horizontal"
        android:layout_width = "fill_parent"
        android:layout_height = "wrap_content">
        <ToggleButton
            android:id = "@ + id/tb2"
            android:textOn = "关闭振动"
            android:textOff = "启动振动"
            android:checked = "false"
            android:layout_width = "wrap_content"
            android:layout_height = "wrap_content" />
        <TextView
            android:id = "@ + id/tv2"
            android:text = "振动已关闭"
            android:layout_width = "wrap_content"
            android:layout_height = "wrap_content" />
    </LinearLayout>
</LinearLayout>
```

（3）打开 src\fs.vibrator_test 包下的 MainActivity.java 文件，在文件中实现手机的振动功能。代码为：

```java
package fs.vibrator_test;
import android.app.Activity;
import android.app.Service;
import android.os.Bundle;
import android.os.Vibrator;
import android.widget.CompoundButton;
import android.widget.TextView;
import android.widget.ToggleButton;
import android.widget.CompoundButton.OnCheckedChangeListener;
public class MainActivity extends Activity {
    //定义控件
    private Vibrator vibrator = null;
    private ToggleButton tb1 = null, tb2 = null;
    private TextView tv1 = null, tv2 = null;
    @Override
    public void onCreate(Bundle savedInstanceState) {
        super.onCreate(savedInstanceState);
        setContentView(R.layout.main);
        //注意模拟器是模拟不了振动的,需要真机测试
        //创建 Vibrator 对象
        vibrator = (Vibrator)getSystemService(Service.VIBRATOR_SERVICE);
        tv1 = (TextView)findViewById(R.id.tv1);
        tv2 = (TextView)findViewById(R.id.tv2);
        tb1 = (ToggleButton)findViewById(R.id.tb1);
```

```
            tb2 = (ToggleButton)findViewById(R.id.tb2);
            tb1.setOnCheckedChangeListener(listener);
            tb2.setOnCheckedChangeListener(listener);
        }
        OnCheckedChangeListener listener = new OnCheckedChangeListener(){
            public void onCheckedChanged(CompoundButton buttonView, boolean isChecked) {
                ToggleButton toggleButton = (ToggleButton)buttonView;
                switch (toggleButton.getId()) {
                case R.id.tb1:
                    if(isChecked){
/** 根据指定的模式进行振动
 * 第1个参数:该数组中第1个元素是等待多长的时间才启动振动,
 * 之后将会是开启和关闭振动的持续时间,单位为毫秒
 * 第2个参数:重复振动时在pattern中的索引,如果设置为-1则表示不重复振动
 */
                        vibrator.vibrate(new long[]{1000,50,50,100,50}, -1);
                        tv1.setText("振动已启动");
                    }else {
                        //关闭振动
                        vibrator.cancel();
                        tv1.setText("振动已关闭");
                    }
                    break;
                case R.id.tb2:
                    if(isChecked){
                        //启动振动,并持续指定的时间
                        vibrator.vibrate(3500);
                        tv2.setText("振动已启动");
                    }else {
                        //关闭启动
                        vibrator.cancel();
                        tv2.setText("振动已关闭");
                    }
                    break;
                }
            }
        };
}
```

(4) 打开 AndroidManifest.xml 文件,为手机振动设置权限。代码为:

```
…
            <intent-filter>
                <action android:name = "android.intent.action.MAIN" />
                <category android:name = "android.intent.category.LAUNCHER" />
            </intent-filter>
        </activity>
    </application>
<!-- 设置手机振动权限 -->
    <uses-permission android:name = "android.permission.VIBRATE" />
</manifest>
```

运行程序,效果如图 6-1 所示。

图 6-1 振动实例

6.2 闹钟实例

闹钟是现在每台手机都具有的功能,在 Android 中提供了 AlarmManager 用于开发手机闹钟,但实际上它的作用不止于此。它的本质是一个全局的定时器,AlarmManager 可在指定时间或指定周期启动其他组件。

本实例用于实现一个按钮让用户来设置闹铃时间,当用户设置好闹铃时间后,即使退出该程序,到了预设时间 AlarmManager 会启动指定组件——这是因为 AlarmManager 是一个全局定时器。

本实例的具体实现步骤如下。

(1) 在 Eclipse 中创建一个 Android 应用项目,命名为 AlarmClock_test。

(2) 打开 res\layout 目录下的 main.xml 布局文件,在文件中声明一个 Button 控件,用于实现设置闹钟按钮。代码为:

```xml
<?xml version = "1.0" encoding = "utf-8"?>
<LinearLayout xmlns:android = "http://schemas.android.com/apk/res/android"
    android:orientation = "vertical"
    android:layout_width = "fill_parent"
    android:layout_height = "fill_parent"
    android:background = "#0c0">
    <Button
        android:id = "@+id/btn"
        android:text = "设置闹钟"
        android:layout_width = "fill_parent"
        android:layout_height = "wrap_content" />
</LinearLayout>
```

(3) 在 src\fs.alarmclock_test 包下新建一个 AlarmReceiver.java 类,用于调用 Intent 机制,并启动 Activity 活动。代码为:

```java
package fs.alarmclock_test;
import android.content.BroadcastReceiver;
import android.content.Context;
import android.content.Intent;
public class AlarmReceiver extends BroadcastReceiver{
    @Override
    public void onReceive(Context context, Intent intent) {
        Intent i = new Intent(context, AlarmActivity.class);
        i.addFlags(Intent.FLAG_ACTIVITY_NEW_TASK);
        context.startActivity(i);
    }
}
```

(4) 在 src\fs.alarmclock_test 包下新建一个 AlarmActivity.java 类,用于调用时间对话框,并返回闹钟时间到对话框。代码为:

```java
package fs.alarmclock_test;
import android.app.Activity;
import android.app.AlertDialog;
import android.content.DialogInterface;
import android.content.DialogInterface.OnClickListener;
import android.os.Bundle;
public class AlarmActivity extends Activity {
    @Override
    public void onCreate(Bundle savedInstanceState) {
        super.onCreate(savedInstanceState);
        //显示对话框
        new AlertDialog.Builder(AlarmActivity.this).setTitle("闹钟").      //设置标题
        setMessage("时间到了!").                                            //设置内容
        setPositiveButton("知道了", new OnClickListener(){                  //设置按钮
            public void onClick(DialogInterface dialog, int which) {
                AlarmActivity.this.finish();                                //关闭 Activity
            }
        }).create().show();
    }
}
```

(5) 打开 src\fs.alarmclock_test 包下的 MainActivity.java 文件,在文件中实现闹钟设置事件。代码为:

```java
package fs.alarmclock_test;
import java.util.Calendar;
import android.app.Activity;
import android.app.AlarmManager;
import android.app.Dialog;
import android.app.PendingIntent;
import android.app.TimePickerDialog;
import android.content.Context;
import android.content.Intent;
import android.os.Bundle;
import android.view.View;
import android.widget.Button;
import android.widget.TimePicker;
import android.widget.Toast;
public class MainActivity extends Activity{
    private Button btn = null;
```

```java
        private AlarmManager alarmManager = null;
        Calendar cal = Calendar.getInstance();
        final int DIALOG_TIME = 0;                    //设置对话框id
        @Override
        protected void onCreate(Bundle savedInstanceState) {
            super.onCreate(savedInstanceState);
            setContentView(R.layout.main);
            alarmManager = (AlarmManager)getSystemService(Context.ALARM_SERVICE);
            btn = (Button)findViewById(R.id.btn);
            btn.setOnClickListener(new View.OnClickListener(){
                public void onClick(View view) {
                    showDialog(DIALOG_TIME);          //显示时间选择对话框
                }
            });
        }
        @Override
        protected Dialog onCreateDialog(int id) {
            Dialog dialog = null;
            switch (id) {
            case DIALOG_TIME:
                dialog = new TimePickerDialog(
                        this,new TimePickerDialog.OnTimeSetListener(){
            public void onTimeSet(TimePicker timePicker, int hourOfDay,int minute) {
                Calendar c = Calendar.getInstance();    //获取日期对象
                c.setTimeInMillis(System.currentTimeMillis());          //设置Calendar对象
                c.set(Calendar.HOUR, hourOfDay);        //设置闹钟的小时数
                c.set(Calendar.MINUTE, minute);         //设置闹钟的分钟数
                c.set(Calendar.SECOND, 0);              //设置闹钟的秒数
                c.set(Calendar.MILLISECOND, 0);         //设置闹钟的毫秒数
    Intent intent = new Intent(MainActivity.this, AlarmReceiver.class);    //创建Intent对象
    //创建PendingIntent
    PendingIntent pi = PendingIntent.getBroadcast(MainActivity.this, 0, intent, 0);
        //设置闹钟,满足当前时间则唤醒
    alarmManager.set(AlarmManager.RTC_WAKEUP, System.currentTimeMillis(), pi);
    Toast.makeText(MainActivity.this, "闹钟设置成功", Toast.LENGTH_LONG).show();  //提示用户
                }
            },
            cal.get(Calendar.HOUR_OF_DAY),
            cal.get(Calendar.MINUTE),
            false);
                break;
            }
            return dialog;
        }
    }
```

(6) 打开AndroidManifest.xml文件,设置闹钟权限。代码为:

```xml
...
            </intent-filter>
        </activity>
    <activity android:name=".AlarmActivity" />
<!-- android:process=":remote": 新开一个进程 -->
    <receiver android:name=".AlarmReceiver" android:process=":remote"/>
    </application>
</manifest>
```

运行程序，默认界面如图 6-2(a)所示。若单击界面中的"设置闹钟"按钮，则弹出设置闹钟时间对话框，效果如图 6-2(b)所示。选择相应时间并单击"Done"按钮，当时间到时即实现闹钟提醒，效果如图 6-2(c)所示。

　　（a）默认界面　　　　　　（b）设置闹钟时间界面　　　　　（c）闹钟提醒界面

图 6-2　闹钟实例

6.3　计算器实例

计算器在人们生活中扮演着不可缺少的角色，因此计算器是当今所有手机上都集成拥有的功能，Android 手机也不例外。Android 手机的计算器功能越来越多，使用越来越方便。

下面通过一个实例来实现计算器在 Android 手机中的集成，本实例的具体实现步骤如下。

(1) 在 Eclipse 中创建一个 Android 应用项目，命名为 Calculators_test。

(2) 打开 res\layout 目录下的 main.xml 文件，在文件中实现相应的控件声明和计算器界面。代码为：

```xml
<?xml version = "1.0" encoding = "utf-8"?>
<LinearLayout xmlns:android = "http://schemas.android.com/apk/res/android"
    android:layout_width = "match_parent"
    android:layout_height = "match_parent"
    android:orientation = "vertical"
    android:background = "#aaaccc">
    <LinearLayout android:layout_width = "fill_parent"
    android:layout_height = "wrap_content">
<TextView
    android:id = "@ + id/tvResult"
    android:layout_width = "fill_parent"
    android:layout_height = "wrap_content"
    android:height = "50dp"
    android:text = "" />
</LinearLayout>
```

```xml
<LinearLayout android:layout_width = "fill_parent"
    android:layout_height = "wrap_content">
    <Button
        android:id = "@ + id/btnBackspace"
        android:layout_width = "wrap_content"
        android:layout_height = "wrap_content"
        android:width = "150dp"
        android:layout_marginLeft = "10dp"
            android:text = "backspace"/>
    <Button
        android:id = "@ + id/btnCE"
        android:layout_width = "wrap_content"
        android:layout_height = "wrap_content"
         android:width = "150dp"
            android:text = "CE"/>
</LinearLayout>
<LinearLayout android:layout_width = "fill_parent"
    android:layout_height = "wrap_content">
    <Button
        android:id = "@ + id/btn7"
        android:layout_width = "wrap_content"
        android:layout_height = "wrap_content"
        android:layout_marginLeft = "10dp"
         android:width = "75dp"
         android:text = "7"/>
     <Button
        android:id = "@ + id/btn8"
        android:layout_width = "wrap_content"
        android:layout_height = "wrap_content"
        android:width = "75dp"
        android:text = "8"/>
    <Button
        android:id = "@ + id/btn9"
        android:layout_width = "wrap_content"
        android:layout_height = "wrap_content"
        android:width = "75dp"
        android:text = "9"/>
    <Button
        android:id = "@ + id/btnDiv"
        android:layout_width = "wrap_content"
        android:layout_height = "wrap_content"
        android:width = "75dp"
        android:text = "/"/>
</LinearLayout>
<LinearLayout android:layout_width = "fill_parent"
    android:layout_height = "wrap_content">
    <Button
        android:id = "@ + id/btn4"
        android:layout_width = "wrap_content"
        android:layout_height = "wrap_content"
        android:layout_marginLeft = "10dp"
        android:width = "75dp"
        android:text = "4"/>
    <Button
        android:id = "@ + id/btn5"
```

```xml
            android:layout_width = "wrap_content"
            android:layout_height = "wrap_content"
            android:width = "75dp"
            android:text = "5"/>
        <Button
            android:id = "@ + id/btn6"
            android:layout_width = "wrap_content"
            android:layout_height = "wrap_content"
            android:width = "75dp"
            android:text = "6"/>
        <Button
            android:id = "@ + id/btnMul"
            android:layout_width = "wrap_content"
            android:layout_height = "wrap_content"
            android:width = "75dp"
            android:text = " * 1"/>
</LinearLayout>
<LinearLayout android:layout_width = "fill_parent"
    android:layout_height = "wrap_content">
        <Button
            android:id = "@ + id/btn1"
            android:layout_width = "wrap_content"
            android:layout_height = "wrap_content"
            android:layout_marginLeft = "10dp"
            android:width = "75dp"
            android:text = "1"/>
        <Button
            android:id = "@ + id/btn2"
            android:layout_width = "wrap_content"
            android:layout_height = "wrap_content"
            android:width = "75dp"
            android:text = "2"/>
        <Button
            android:id = "@ + id/btn3"
            android:layout_width = "wrap_content"
            android:layout_height = "wrap_content"
            android:width = "75dp"
            android:text = "3"/>
        <Button
            android:id = "@ + id/btnAdd"
            android:layout_width = "wrap_content"
            android:layout_height = "wrap_content"
            android:width = "75dp"
            android:text = " + "/>
</LinearLayout>
<LinearLayout android:layout_width = "fill_parent"
    android:layout_height = "wrap_content">
        <Button
            android:id = "@ + id/btn0"
            android:layout_width = "wrap_content"
            android:layout_height = "wrap_content"
            android:layout_marginLeft = "10dp"
            android:width = "75dp"
            android:text = "0"/>
        <Button
```

```xml
            android:id = "@ + id/btnC"
            android:layout_width = "wrap_content"
            android:layout_height = "wrap_content"
            android:width = "75dp"
            android:text = "C"/>
        <Button
            android:id = "@ + id/btnEqu"
            android:layout_width = "wrap_content"
            android:layout_height = "wrap_content"
            android:width = "75dp"
            android:text = " = "/>
        <Button
            android:id = "@ + id/btnSub"
            android:layout_width = "wrap_content"
            android:layout_height = "wrap_content"
            android:width = "75dp"
            android:text = " - "/>
    </LinearLayout>
</LinearLayout>
```

(3) 打开 src\fs.calculators_test 包下的 MainActivity.java 文件,在文件中实现数据相应运算。代码为:

```java
0package fs.li9_8calculators;
import android.os.Bundle;
import android.view.View;
import android.view.View.OnClickListener;
import android.widget.Button;
import android.widget.TextView;
import android.app.Activity;
public class MainActivity extends Activity implements OnClickListener{
    //声明一些控件
    Button btn0 = null;
    Button btn1 = null;
    Button btn2 = null;
    Button btn3 = null;
    Button btn4 = null;
    Button btn5 = null;
    Button btn6 = null;
    Button btn7 = null;
    Button btn8 = null;
    Button btn9 = null;
    Button btnBackspace = null;
    Button btnCE = null;
    Button btnC = null;
    Button btnAdd = null;
    Button btnSub = null;
    Button btnMul = null;
    Button btnDiv = null;
    Button btnEqu = null;
    TextView tvResult = null;
    //声明两个参数,接收 tvResult 前后的值
    double num1 = 0, num2 = 0;
    double Result = 0;                              //计算结果
    int op = 0;                                     //判断操作数
    boolean isClickEqu = false;                     //判断是否单击了" = "按钮
```

```java
@Override
protected void onCreate(Bundle savedInstanceState) {
    super.onCreate(savedInstanceState);
    setContentView(R.layout.main);
    //从布局文件中获取控件
    btn0 = (Button)findViewById(R.id.btn0);
    btn1 = (Button)findViewById(R.id.btn1);
    btn2 = (Button)findViewById(R.id.btn2);
    btn3 = (Button)findViewById(R.id.btn3);
    btn4 = (Button)findViewById(R.id.btn4);
    btn5 = (Button)findViewById(R.id.btn5);
    btn6 = (Button)findViewById(R.id.btn6);
    btn7 = (Button)findViewById(R.id.btn7);
    btn8 = (Button)findViewById(R.id.btn8);
    btn9 = (Button)findViewById(R.id.btn9);
    btnBackspace = (Button)findViewById(R.id.btnBackspace);
    btnCE = (Button)findViewById(R.id.btnCE);
    btnC = (Button)findViewById(R.id.btnC);
    btnEqu = (Button)findViewById(R.id.btnEqu);
    btnAdd = (Button)findViewById(R.id.btnAdd);
    btnSub = (Button)findViewById(R.id.btnSub);
    btnMul = (Button)findViewById(R.id.btnMul);
    btnDiv = (Button)findViewById(R.id.btnDiv);
    tvResult = (TextView)findViewById(R.id.tvResult);
    //添加监听
    btnBackspace.setOnClickListener(this);
    btnCE.setOnClickListener(this);
    btn0.setOnClickListener(this);
    btn1.setOnClickListener(this);
    btn2.setOnClickListener(this);
    btn3.setOnClickListener(this);
    btn4.setOnClickListener(this);
    btn5.setOnClickListener(this);
    btn6.setOnClickListener(this);
    btn7.setOnClickListener(this);
    btn8.setOnClickListener(this);
    btn9.setOnClickListener(this);
    btnAdd.setOnClickListener(this);
    btnSub.setOnClickListener(this);
    btnMul.setOnClickListener(this);
    btnDiv.setOnClickListener(this);
    btnEqu.setOnClickListener(this);
}
@Override
public void onClick(View v) {
    switch (v.getId()) {
    //btnBackspace 和 CE
    case R.id.btnBackspace:
        String myStr = tvResult.getText().toString();
        try {
            tvResult.setText(myStr.substring(0, myStr.length() - 1));
        } catch (Exception e) {
            tvResult.setText("");
        }
        break;
```

```java
        case R.id.btnCE:
            tvResult.setText(null);
            break;
            //btn0 到 btn 9
        case R.id.btn0:
            if(isClickEqu)
            {
                tvResult.setText(null);
                isClickEqu = false;
            }
            String myString = tvResult.getText().toString();
            myString += "0";
            tvResult.setText(myString);
            break;
        case R.id.btn1:
            if(isClickEqu)
            {
                tvResult.setText(null);
                isClickEqu = false;
            }
            String myString1 = tvResult.getText().toString();
            myString1 += "1";
            tvResult.setText(myString1);
            break;
        case R.id.btn2:
            if(isClickEqu)
            {
                tvResult.setText(null);
                isClickEqu = false;
            }
            String myString2 = tvResult.getText().toString();
            myString2 += "2";
            tvResult.setText(myString2);
            break;
        case R.id.btn3:
            if(isClickEqu)
            {
                tvResult.setText(null);
                isClickEqu = false;
            }
            String myString3 = tvResult.getText().toString();
            myString3 += "3";
            tvResult.setText(myString3);
            break;
        case R.id.btn4:
            if(isClickEqu)
            {
                tvResult.setText(null);
                isClickEqu = false;
            }
            String myString4 = tvResult.getText().toString();
            myString4 += "4";
            tvResult.setText(myString4);
            break;
        case R.id.btn5:
```

```java
            if(isClickEqu)
            {
                tvResult.setText(null);
                isClickEqu = false;
            }
            String myString5 = tvResult.getText().toString();
            myString5 += "5";
            tvResult.setText(myString5);
            break;
        case R.id.btn6:
            if(isClickEqu)
            {
                tvResult.setText(null);
                isClickEqu = false;
            }
            String myString6 = tvResult.getText().toString();
            myString6 += "6";
            tvResult.setText(myString6);
            break;
        case R.id.btn7:
            if(isClickEqu)
            {
                tvResult.setText(null);
                isClickEqu = false;
            }
            String myString7 = tvResult.getText().toString();
            myString7 += "7";
            tvResult.setText(myString7);
            break;
        case R.id.btn8:
            if(isClickEqu)
            {
                tvResult.setText(null);
                isClickEqu = false;
            }
            String myString8 = tvResult.getText().toString();
            myString8 += "8";
            tvResult.setText(myString8);
            break;
        case R.id.btn9:
            if(isClickEqu)
            {
                tvResult.setText(null);
                isClickEqu = false;
            }
            String myString9 = tvResult.getText().toString();
            myString9 += "9";
            tvResult.setText(myString9);
            break;
            //btn +- * / =
        case R.id.btnAdd:
            String myStringAdd = tvResult.getText().toString();
            if(myStringAdd.equals(null))
            {
                return;
```

```
        }
        num1 = Double.valueOf(myStringAdd);
        tvResult.setText(null);
        op = 1;
        isClickEqu = false;
        break;
    case R.id.btnSub:
        String myStringSub = tvResult.getText().toString();
        if(myStringSub.equals(null))
        {
            return;
        }
        num1 = Double.valueOf(myStringSub);
        tvResult.setText(null);
        op = 2;
        isClickEqu = false;
        break;
    case R.id.btnMul:
        String myStringMul = tvResult.getText().toString();
        if(myStringMul.equals(null))
        {
            return;
        }
        num1 = Double.valueOf(myStringMul);
        tvResult.setText(null);
        op = 3;
        isClickEqu = false;
        break;
    case R.id.btnDiv:
        String myStringDiv = tvResult.getText().toString();
        if(myStringDiv.equals(null))
        {
            return;
        }
        num1 = Double.valueOf(myStringDiv);
        tvResult.setText(null);
        op = 4;
        isClickEqu = false;
        break;
    case R.id.btnEqu:
        String myStringEqu = tvResult.getText().toString();
        if(myStringEqu.equals(null))
        {
            return;
        }
        num2 = Double.valueOf(myStringEqu);
        tvResult.setText(null);
        switch (op) {
        case 0:
            Result = num2;
            break;
        case 1:
            Result = num1 + num2;
            break;
        case 2:
```

```
                    Result = num1 - num2;
                    break;
                case 3:
                    Result = num1 * num2;
                    break;
                case 4:
                    Result = num1/num2;
                    break;
                default:
                    Result = 0;
                    break;
                }
                tvResult.setText(String.valueOf(Result));
                isClickEqu = true;
                break;
            default:
                break;
            }
        }
    }
```

运行程序,效果如图 6-3 所示。

图 6-3　计算器

6.4　手电筒实例

手电筒手机功能现在是比较普遍的,Android 智能手机也一样,那么怎样在 Android 智能手机中实现手电筒呢？

本实例用于实现一个用闪光灯做的手电筒,其具体实现步骤如下。

(1) 在 Android 中创建一个 Android 应用项目,命名为 Flashlight_test。

(2) 打开 res\layout 目录下的 main.xml 布局文件,在文件中声明一个 TextView 控件。代码为:

```
<?xml version = "1.0" encoding = "utf-8"?>
```

```xml
<LinearLayout xmlns:android = "http://schemas.android.com/apk/res/android"
    android:layout_width = "match_parent"
    android:layout_height = "match_parent"
    android:orientation = "vertical" >
    <TextView
        android:id = "@+id/main_img"
        android:layout_width = "fill_parent"
        android:layout_height = "fill_parent"
        android:background = "@drawable/ic_launcher">
    </TextView>
</LinearLayout>
```

(3) 打开 src\ fs.flashlight_test 包下的 MainActivity.java 文件,在文件中实现手电筒功能。代码为:

```java
package fs.flashlight_test;
import android.app.Activity;
import android.hardware.Camera;
import android.hardware.Camera.Parameters;
import android.os.Bundle;
import android.view.View;
import android.widget.TextView;
import android.widget.Toast;
public class MainActivity extends Activity {
    private boolean isopent = false;
    private Camera camera;
    @Override
    protected void onCreate(Bundle savedInstanceState) {
        //TODO 自动存根法
        super.onCreate(savedInstanceState);
        View view = View.inflate(this, R.layout.main, null);
        setContentView(view);
        TextView img_but = (TextView) findViewById(R.id.main_img);
        img_but.setOnClickListener(new View.OnClickListener() {
            @Override
            public void onClick(View v) {
                //TODO 自动存根法
                if (!isopent) {
                //Toast 提示
                  Toast.makeText(getApplicationContext(), "您已经打开了手电筒", 0)
                        .show();
                    camera = Camera.open();
                    Parameters params = camera.getParameters();
                    params.setFlashMode(Parameters.FLASH_MODE_TORCH);
                    camera.setParameters(params);
                    camera.startPreview();          //开始亮灯
                    isopent = true;
                } else {
                    Toast.makeText(getApplicationContext(), "关闭了手电筒",
                        Toast.LENGTH_SHORT).show();
                    camera.stopPreview();           //关掉亮灯
                    camera.release();               //关掉照相机
                    isopent = false;
                }
            }
        });
    }
}
```

(4) 打开 AndroidManifest.xml 文件,为文件添加手电筒及拍照权限。代码为:

```xml
<?xml version = "1.0" encoding = "utf-8"?>
<manifest xmlns:android = "http://schemas.android.com/apk/res/android"
    package = "fs.flashlight_test"
    android:versionCode = "1"
    android:versionName = "1.0" >
    <uses-sdk
        android:minSdkVersion = "8"
        android:targetSdkVersion = "18" />
    <application
        android:allowBackup = "true"
        android:icon = "@drawable/ic_launcher"
        android:label = "@string/app_name"
        android:theme = "@style/AppTheme" >
        <activity
            android:name = "fs.flashlight_test.MainActivity"
            android:label = "@string/app_name" >
            <intent-filter >
                <action android:name = "android.intent.action.MAIN" />
                <category android:name = "android.intent.category.LAUNCHER" />
            </intent-filter>
        </activity>
    </application>
<!-- 摄像头、手电筒 -->
    <uses-permission android:name = "android.permission.CAMERA" />
    <uses-permission android:name = "android.permission.FLASHLIGHT" />
    <uses-feature android:name = "android.hardware.camera" />
    <uses-feature android:name = "android.hardware.camera.autofocus" />
    <uses-feature android:name = "android.hardware.camera.flash" />
</manifest>
```

运行程序,默认效果如图 6-4(a)所示。当单击界面中的图像按钮时,则打开手电筒,弹出 Toast,效果如图 6-4(b)所示。再单击图像按钮时,则关闭手电筒,也弹出 Toast 提示,效果如图 6-4(c)所示。

(a) 默认界面　　　　　　　(b) 打开手电筒　　　　　　(c) 关闭手电筒

图 6-4　手电筒实例

6.5 备忘录实例

现在手机的一个很重要的功能就是能够设置备忘录,并且在设定的时间提醒用户。Android 平台下的手机同样可以设定,并且可以自己制作备忘录,在设定好的时间提醒用户。

本实例用于制作 Android 手机备忘录,实现备忘录的制作。当运行软件时,在主界面可以设定备忘录内容,单击按钮设置提醒的时间。每次设置一个备忘录时,会在按钮下方记录所设定的备忘录,方便用户查看。当设定备忘录的时间到时,在主界面上自动弹出一个对话框,进行提示。

本实例的具体实现步骤如下。

(1) 在 Eclipse 中创建一个 Android 应用项目,命名为 Memo_test。

(2) 打开 res\layout 目录下的 main.xml 文件,用于实现主界面。代码为:

```xml
<LinearLayout xmlns:android="http://schemas.android.com/apk/res/android"
    android:layout_width="fill_parent"
    android:layout_height="fill_parent"
    android:orientation="vertical"
    android:background="#aaaccc">
    <LinearLayout
        android:id="@+id/LinearLayout1"
        android:layout_width="fill_parent"
        android:layout_height="wrap_content"
        android:orientation="horizontal">
        <TextView
            android:layout_width="100dip"
            android:layout_height="wrap_content"
            android:text="备忘录内容:"/>
        <EditText
            android:text=""
            android:id="@+id/EditText1"
            android:layout_width="fill_parent"
            android:layout_height="wrap_content"/>
    </LinearLayout>
    <TextView
        android:text=""
        android:id="@+id/TextView1"
        android:layout_width="fill_parent"
        android:layout_height="wrap_content"/>
    <Button
        android:id="@+id/Button1"
        android:layout_width="wrap_content"
        android:layout_height="wrap_content"
        android:layout_alignLeft="@+id/LinearLayout1"
        android:layout_below="@+id/LinearLayout1"
        android:layout_marginTop="46dp"
        android:text="设置闹钟"/>
</LinearLayout>
```

(3) 在 res\layout 目录下创建一个名为 dialog.xml 的文件,用于实现闹钟设置界面。代码为:

```xml
<?xml version = "1.0" encoding = "utf-8"?>
<LinearLayout xmlns:android = "http://schemas.android.com/apk/res/android"
    android:orientation = "vertical"
    android:layout_width = "220dip"
    android:layout_height = "fill_parent"
    android:background = "#FFFFFF"
    android:paddingLeft = "10dip"
    android:paddingRight = "10dip"
    android:paddingTop = "10dip"
    android:paddingBottom = "10dip"
    android:gravity = "center">
    <TextView
    android:text = "备忘录时间到了,请注意!"
    android:layout_width = "fill_parent"
    android:layout_height = "wrap_content"
    android:textSize = "20dip"
    android:textColor = "#FFFFFF"
    android:gravity = "left"/>
    <Button
     android:text = "关闭"
     android:id = "@+id/mywktzOk"
     android:layout_width = "60dip"
     android:layout_height = "40dip"
     android:textSize = "18dip"
     android:gravity = "center"/>
</LinearLayout>
```

(4) 打开 src\fs.memo_test 包下的 MainActivity.java 文件,在文件中是继承 Activity 类的开发,在该类中主要完成的是备忘录的设置。代码为:

```java
public class MainActivity extends Activity {
    EditText et;                            //备忘录编辑框
    Button button;                          //设置按钮
    String msg;                             //备忘录信息
    Dialog dialog;                          //对话框
    private final int DIALOG = 0;
    TextView tv;                            //记录备忘录
    StringBuilder sb;
    int count;
    @Override
    public void onCreate(Bundle savedInstanceState) {
        super.onCreate(savedInstanceState);
        setContentView(R.layout.main);
        et = (EditText)this.findViewById(R.id.EditText1);
        button = (Button)this.findViewById(R.id.Button1);
        tv = (TextView)this.findViewById(R.id.TextView1);
        Bundle bundle = this.getIntent().getExtras();       //获得短信发来 bundle
        sb = new StringBuilder();
        if(bundle!= null)
        {
            showDialog(DIALOG);                             //显示对话框
        }
        final Calendar c = Calendar.getInstance();
        button.setOnClickListener                           //设置按钮监听器
        (
            new OnClickListener()
```

```java
                {
                    public void onClick(View v) {
                        msg = et.getText().toString().trim();              //获取备忘录信息
                        sb.append(count++);
                        sb.append(".备忘录内容为:");
                        sb.append(msg);
                        sb.append("\n");
                        tv.setText(sb.toString().trim());
                        c.setTimeInMillis(System.currentTimeMillis());     //将当前事件设置为默认时间
                        int hour = c.get(Calendar.HOUR_OF_DAY);            //小时
                        int minute = c.get(Calendar.MINUTE);               //分钟
                        new TimePickerDialog(
                                MainActivity.this,
                                new TimePickerDialog.OnTimeSetListener() {
                            public void onTimeSet(TimePicker view, int hourOfDay, int minute) {
                                    //TODO 自动存根法
                                c.setTimeInMillis(System.currentTimeMillis());     //设置当前时间
                                c.set(Calendar.HOUR_OF_DAY, hourOfDay);            //设置小时
                                c.set(Calendar.MINUTE, minute);                    //设置分钟
                                c.set(Calendar.SECOND, 0);                         //设置秒
                                c.set(Calendar.MILLISECOND, 0);                    //设置毫秒
                                Intent intent = new Intent(MainActivity.this, AlarmRece.
                                class);        //运行 AlarmReceiver 类
                                PendingIntent pi = PendingIntent.getBroadcast(
                                                                        //创建 PendingIntent 对象
                                        MainActivity.this, 0, intent, 0);
                                AlarmManager alarm = (AlarmManager)MainActivity.this.
                                        getSystemService(ALARM_SERVICE);
                                alarm.set(AlarmManager.RTC_WAKEUP, c.getTimeInMillis
                                        (), pi);   //设置闹钟提醒一次
alarm.setRepeating(AlarmManager.RTC_WAKEUP, c.getTimeInMillis(), 120000, pi);
                                                                        //每两分钟提醒一次
String tempHour = (hourOfDay + "").length()>1?hourOfDay + "":"0" + hourOfDay;
String tempMinute = (minute + "").length()>1?minute + "":"0" + minute;
                                Toast.makeText(
                                        MainActivity.this,
                            设置的时间为:" + tempHour + ":" + tempMinute,
                                        Toast.LENGTH_SHORT).show();
                                }
                            }, hour, minute, true).show();
                    }
                }
            );
        }
        @Override
        public Dialog onCreateDialog(int id)
        {
            Dialog result = null;
            switch(id)
            {
            case DIALOG://初始化
                AlertDialog.Builder mywktzb = new AlertDialog.Builder(this);
                mywktzb.setItems(
                    null,
                    null);
```

```
                dialog = mywktzb.create();
                result = dialog;
            break;
        }
        return result;
    }
    @Override
    public void onPrepareDialog(int id, final Dialog dialog)
    {
        switch(id)
        {
        case DIALOG://对话框
            dialog.setContentView(R.layout.dialog);
            Button mywktzBOk = (Button)dialog.findViewById(R.id.mywktzOk);
            mywktzBOk.setOnClickListener
            (
                    new OnClickListener()
                    {
                        public void onClick(View arg0) {
                            dialog.cancel();
                            MainActivity.this.finish();
                        }
                    }
            );
            dialog.setCancelable(true);              //设置在dialog界面可以单击返回键
        break;
        }
    }
}
```

(5) 在 src\fs.memo_test 包下新建一个 AlarmRece.java 文件,在文件中主要实现 Intent 对象的创建及启动 Activity 活动。代码为:

```
package fs.li9_9memo;
import android.content.BroadcastReceiver;
import android.content.Context;
import android.content.Intent;
import android.os.Bundle;
public class AlarmRece extends BroadcastReceiver{
    @Override
    public void onReceive(Context context, Intent intent) {
        Intent tempIntent = new Intent(context,MainActivity.class);        //创建 Intent 对象
        Bundle myBundle = new Bundle();
        myBundle.putString("msg", "msg");
        tempIntent.putExtras(myBundle);
        tempIntent.addFlags(Intent.FLAG_ACTIVITY_NEW_TASK);                //设置新的 task
        context.startActivity(tempIntent);                                  //启动 Actvity
    }
}
```

运行程序,效果如图 6-5(a)所示。当在备忘录内容右侧的文本框中填写对应的备忘内容,单击主界面中的"设置闹钟"按钮时,则可弹出闹钟设置界面,如图 6-5(b)所示,设置完成时间后,效果如图 6-5(c)所示。

(a) 备忘录主界面　　　　　(b) 闹钟设置界面　　　　(c) 完成备忘录设置界面

图 6-5　设置备忘录

6.6　手机状态提醒实例

手机的状态可以分为待机状态、来电状态以及电话接通状态,怎样来判断手机的当前状态,本节将通过实例来介绍。

本实例主要实现根据当前是否有电话拨入,判断手机的状态,如果没有则为待机状态。当有电话拨入时,提醒用户有来电以及来电人的电话号码,并且手机开始振动。当接听该电话时,提醒用户电话已接通,手机停止振动。如果没有接听该电话,手机又重新回到待机状态。

本实例的具体实现步骤如下。

(1) 在 Eclipse 中创建一个 Android 应用项目,命名为 Phonestatus_test。

(2) 打开 res\layout 目录下的 main.xml 布局文件,在文件中声明一个 TextView 控件。代码为:

```
<?xml version = "1.0" encoding = "utf-8"?>
<LinearLayout xmlns:android = "http://schemas.android.com/apk/res/android"
    android:orientation = "vertical"
    android:layout_width = "fill_parent"
    android:layout_height = "fill_parent"
    android:background = "#a00">
    <TextView
        android:text = "1111111"
        android:id = "@+id/TextView01"
        android:layout_width = "wrap_content"
        android:layout_height = "wrap_content">
    </TextView>
</LinearLayout>
```

(3) 打开 src\fs.phonestatus_test 包下的 MainActivity.java 文件,在文件中主要实现 3 种手机的状态。代码为:

```java
package fs.phonestatus_test;
import android.app.Activity;
import android.app.Service;
import android.content.Context;
import android.os.Bundle;
import android.os.Vibrator;
import android.telephony.PhoneStateListener;
import android.telephony.TelephonyManager;
import android.widget.TextView;
import android.widget.Toast;
public class MainActivity extends Activity {
    TextView tv;
    Vibrator vibrator;                                   //手机振动引用
    @Override
    public void onCreate(Bundle savedInstanceState) {
        super.onCreate(savedInstanceState);
        setContentView(R.layout.main);
        tv = (TextView)this.findViewById(R.id.TextView01);
//创建 Vibrator 对象
     vibrator = (Vibrator)getApplication().getSystemService(Service.VIBRATOR_SERVICE);
        myPhoneStateListener mPSL = new myPhoneStateListener();           //创建对象
        TelephonyManager
//创建 TelephonyManager 对象
tm = (TelephonyManager) this.getSystemService(Context.TELEPHONY_SERVICE); tm.listen(mPSL,
PhoneStateListener.LISTEN_CALL_STATE);           //添加监听器
    }
    public class myPhoneStateListener extends PhoneStateListener
    {
        public void onCallStateChanged(int state,String incomingNumber)
        {
            switch(state)
            {
                case TelephonyManager.CALL_STATE_IDLE:               //手机待机状态
                    Toast.makeText(
                            MainActivity.this,
                            "手机处于待机状态!",
                            Toast.LENGTH_SHORT).show();
                    tv.setText("手机处于待机状态!");
                    break;
                case TelephonyManager.CALL_STATE_OFFHOOK:             //接电话状态
                    Toast.makeText(
                            MainActivity.this,
                            "电话已接通!",
                            Toast.LENGTH_SHORT).show();
                    vibrator.cancel();          //取消振动
                    Toast.makeText(
                            MainActivity.this,
                            "振动取消!",
                            Toast.LENGTH_SHORT).show();
                    tv.setText("电话已接通!");
                    break;
                case TelephonyManager.CALL_STATE_RINGING:              //电话响铃状态
                    Toast.makeText(
                            MainActivity.this,
                         "有来电,电话号码: " + incomingNumber + ",请接听!",
```

```
                        Toast.LENGTH_SHORT).show();
    vibrator.vibrate(new long[]{100,10,100,1000},0);        //设置手机震动,0表示持续振动
                    Toast.makeText(
                        MainActivity.this,
                        "手机正在振动!",
                        Toast.LENGTH_SHORT).show();
            tv.setText("有来电,电话号码: " + incomingNumber + ",请接听!");
            break;
        }
        super.onCallStateChanged(state, incomingNumber);
    }
  }
}
```

(4) 打开 AndroidManifest.xml 布局文件,为文件添加权限。代码为:

```
            ...
                <intent-filter>
                    <action android:name="android.intent.action.MAIN"/>
                    <category android:name="android.intent.category.LAUNCHER"/>
                </intent-filter>
            </activity>
        </application>
        <!-- 权限设置 -->
    <uses-permission android:name="android.permission.VIBRATE"/>
    <uses-permission android:name="android.permission.READ_PHONE_STATE"/>
</manifest>
```

运行程序,得到手机待机状态界面如图 6-6 所示。

图 6-6 手机状态

6.7 来电自动短信回复实例

在前面介绍了怎样获取手机当前的状态,有了手机的状态就可以完成一些其他任务,例如不方便接电话时,可挂掉电话,给来电用户发送一条短信,或通知来电用户现在正在忙碌中,稍

后回复等信息。

本实例实现首先获取当前状态,并将状态显示在主界面上,当有来电时,提醒用户有来电,并自动发送一条短信通知来电用户请等待。如果手机用户挂掉电话,则发送短信给来电用户,通知来电用户稍后回复。

本实例的具体实现步骤如下。

(1) 在 Eclipse 中创建一个 Android 应用项目,命名为 Autocalls_test。

(2) 打开 res\layout 目录下的 main.xml 布局文件,在文件中定义一个 TextView 控件。代码为:

```xml
<?xml version = "1.0" encoding = "utf - 8"?>
<LinearLayout xmlns:android = "http://schemas.android.com/apk/res/android"
    android:orientation = "vertical"
    android:layout_width = "fill_parent"
    android:layout_height = "fill_parent"
    android:background = "#0d0">
<TextView
    android:layout_width = "fill_parent"
    android:layout_height = "wrap_content"
    android:id = "@+id/TextView01"/>
</LinearLayout>
```

(3) 打开 src\fs.autocalls_test 包下的 MainActivity.java 文件,在文件中实现当有来电时,手机自动回复短信。代码为:

```java
package fs.autocalls_test;
import android.app.Activity;
import android.app.PendingIntent;
import android.content.Context;
import android.content.Intent;
import android.os.Bundle;
import android.os.Handler;
import android.os.Message;
import android.telephony.PhoneStateListener;
import android.telephony.TelephonyManager;
import android.widget.TextView;
import android.widget.Toast;
import android.telephony.gsm.SmsManager;
public class MainActivity extends Activity {
    TextView tv;
    int count = 0;
    Handler hd = new Handler()
    {
        public void handleMessage(Message msg)
        {
            switch(msg.what)
            {
                case 0:
                    Bundle b = msg.getData();
                    String incomingNumber = (String) b.get("number");
                    SmsManager smsManager = SmsManager.getDefault();
            PendingIntent pi = PendingIntent.getBroadcast(            //创建 PendingIntent 对象
```

```java
                    MainActivity.this,0,new Intent(),0);
                String tempMsg = "请等待用户接听.";
                smsManager.sendTextMessage(
                    incomingNumber,
                    null,
                    tempMsg,
                    pi,
                    null);                      //发送短信
                Toast.makeText(
                    MainActivity.this,
                    "短信已发送到" + incomingNumber,
                    Toast.LENGTH_SHORT).show();
            break;
            case 1:
                Bundle bb = msg.getData();
                String incomingNumber2 = (String) bb.get("number");
                SmsManager smsManager2 = SmsManager.getDefault();
    PendingIntent pi2 = PendingIntent.getBroadcast(              //创建 PendingIntent 对象
                    MainActivity.this,0,new Intent(),0);
                smsManager2.sendTextMessage(
                    incomingNumber2,
                    null,
                    "我现在不方便接电话,稍后打给你",
                    pi2,
                    null);                      //发送短信
                Toast.makeText(
                    MainActivity.this,
                    "短信已发送到" + incomingNumber2,
                    Toast.LENGTH_SHORT).show();
            break;
        }
    }
};
@Override
public void onCreate(Bundle savedInstanceState) {
    super.onCreate(savedInstanceState);
    setContentView(R.layout.main);
    tv = (TextView)this.findViewById(R.id.TextView01);
    myPhoneStateListener mPSL = new myPhoneStateListener();       //创建对象
    TelephonyManager tm = (TelephonyManager)
 //创建 TelephonyManager 对象
    this.getSystemService(Context.TELEPHONY_SERVICE);
    tm.listen(mPSL, PhoneStateListener.LISTEN_CALL_STATE);        //添加监听器
}
public class myPhoneStateListener extends PhoneStateListener
{
    public void onCallStateChanged(int state,String incomingNumber)
    {
        switch(state)
        {
            case TelephonyManager.CALL_STATE_IDLE://手机待机状态
                Toast.makeText(
```

```java
                    MainActivity.this,
                    "手机处于待机状态!",
                    Toast.LENGTH_SHORT).show();
            tv.setText("手机处于待机状态!");
            if(count == 1)
            {
                Bundle bundle = new Bundle();
                bundle.putString("number", incomingNumber);
                Message m = new Message();
                m.what = 1;
                m.setData(bundle);
                hd.sendMessage(m);
                count = 0;
            }
            break;
        case TelephonyManager.CALL_STATE_OFFHOOK:           //接电话状态
            Toast.makeText(
                    MainActivity.this,
                    "电话已接通!",
                    Toast.LENGTH_SHORT).show();
            tv.setText("电话已接通!");
            count = 0;
            break;
        case TelephonyManager.CALL_STATE_RINGING:           //电话响铃状态
            Toast.makeText(
                    MainActivity.this,
                    "有来电,电话号码: " + incomingNumber + ",请接听!",
                    Toast.LENGTH_SHORT).show();
            tv.setText("有来电,电话号码: " + incomingNumber + ",请接听!");
            Bundle bundle = new Bundle();
            bundle.putString("number", incomingNumber);
            Message m = new Message();
            m.what = 0;
            m.setData(bundle);
            hd.sendMessage(m);
            count = 1;
            break;
        }
        super.onCallStateChanged(state, incomingNumber);
    }
  }
}
```

(4) 打开 AndroidManifest.xml 布局文件,在文件中添加权限。代码为:

```xml
    ...
            </intent-filter>
        </activity>
    </application>
    <!-- 设置权限 -->
    <uses-permission android:name="android.permission.READ_PHONE_STATE"/>
    <uses-permission android:name="android.permission.SEND_SMS"/>
</manifest>
```

运行程序,效果如图 6-7 所示。

图 6-7　手机自动回复短信

6.8　万年历实例

　　Android 平台上有几种有用的日历控件。日历控件在 Web 开发中有很多的解决方案,而且很容易实现,但是在 Android 平台上的解决方案较少且不容易实现。

　　在 Android 平台 3.0 中新增了日历视图控件,可以显示网格状的日历内容,那么对于 Android 3.0 以下的版本要使用日历控件只能借助第三方,目前用的最多的是 CalendarView。

　　本实例用于在 Android 中实现万能历的设置。其具体操作步骤如下。

　　(1) 在 Eclipse 中创建一个 Android 应用项目,命名为 Calendar_Manager。

　　(2) 打开 res\layout 目录下的 main.xml 布局文件,在文件中声明一个 TextView 控件用来显示日历界面以及声明两个 Button 控件,用于实现翻页。代码为:

```xml
<?xml version = "1.0" encoding = "utf-8"?>
<LinearLayout xmlns:android = "http://schemas.android.com/apk/res/android"
    android:layout_width = "fill_parent"
    android:layout_height = "fill_parent"
    android:gravity = "center_horizontal"
    android:orientation = "vertical" >
    <RelativeLayout
        android:id = "@+id/relativeLayout1"
        android:layout_width = "match_parent"
        android:layout_height = "40dip"
        android:background = "#EDE8DD" >
        <TextView
            android:id = "@+id/Top_Date"
            android:layout_width = "150dip"
            android:layout_height = "wrap_content"
            android:layout_centerInParent = "true"
            android:gravity = "center_horizontal|center"
```

```xml
            android:textColor = " # 424139"
            android:textSize = "19sp"
            android:textStyle = "bold" />
    <Button
            android:id = "@ + id/btn_pre_month"
            android:layout_width = "wrap_content"
            android:layout_height = "wrap_content"
            android:layout_alignParentLeft = "true"
            android:layout_centerVertical = "true"
            android:layout_marginLeft = "30dp"
            android:background = "@drawable/bt11" />
    <Button
            android:id = "@ + id/btn_next_month"
            android:layout_width = "wrap_content"
            android:layout_height = "wrap_content"
            android:layout_alignParentRight = "true"
            android:layout_centerVertical = "true"
            android:layout_marginRight = "30dp"
            android:background = "@drawable/bt12" />
    </RelativeLayout>
</LinearLayout>
```

(3) 在 res\values 目录下新建一个颜色渐变文件,命名为 colors.xml 文件。代码为:

```xml
<?xml version = "1.0" encoding = "utf-8"?>
<resources>
    <!-- 主应用程序 -->
    <color name = "title_icon"> # ff000000 </color>
    <color name = "text_minor"> # ff666666 </color>
    <color name = "application_backcolor"> # fff6eade </color>
    <color name = "login_font"> # 3D3E40 </color>
    <!-- 日历 -->
    <color name = "calendar_background"> # 000000 </color>
    <color name = "recordremind_background"> # FFFFFF </color>
    <color name = "border_color"> # FFFFFF </color>
    <color name = "weekname_color"> # FFFFFF </color>
    <color name = "day_color"> # FFFFFF </color>
    <color name = "inner_grid_color"> # FFFFFF </color>
    <color name = "prev_next_month_day_color"> # 999999 </color>
    <color name = "text_color"> # FFFF00 </color>
    <color name = "recordremindtext_color"> # 000000 </color>
    <color name = "current_day_color"> # FF0000 </color>
    <color name = "today_background_color"> # FF0000 </color>
    <color name = "today_color"> # FFFFFF </color>
    <color name = "sunday_saturday_color"> # FF0000 </color>
    <color name = "sunday_saturday_prev_next_month_day_color"> # 990000 </color>
    <color name = "transparent"> # 50000000 </color>
    <color name = "white"> # ffffff </color>
    <color name = "black"> # 000000 </color>
    <color name = "bgcolor"> # ffffff </color>
    <color name = "txtcolor"> # 000000 </color>
    <color name = "red"> # ff0000 </color>
    <color name = "Calendar_WeekBgColor"> # EFE7DE </color>
    <color name = "Calendar_WeekFontColor"> # 8C8A8C </color>
    <color name = "Calendar_DayBgColor"> # F7F7F7 </color>
    <color name = "isHoliday_BgColor"> # E0E0E0 </color>
```

```xml
<color name = "unPresentMonth_FontColor">#8C8A8C</color>
<color name = "isPresentMonth_FontColor">#000000</color>
<color name = "isToday_BgColor">#EBDCC1</color>
<color name = "specialReminder">#FF0000</color>
<color name = "commonReminder">#BDBAB5</color>
<color name = "gray">#808080</color>
</resources>
```

(4) 打开 src\calendar_manager 包下的 MainActivity.java 文件,在文件中实现日历生成、外层容器、当前日期操作、页面控件和数据源等功能。代码为:

```java
...
public class MainActivity extends Activity{
    //生成日历和外层容器
    private LinearLayout layContent = null;
    private ArrayList<DateWidgetDayCell> days = new ArrayList<DateWidgetDayCell>();
    //日期变量
    public static Calendar calStartDate = Calendar.getInstance();
    private Calendar calToday = Calendar.getInstance();
    private Calendar calCalendar = Calendar.getInstance();
    private Calendar calSelected = Calendar.getInstance();
    //当前操作日期
    private int iMonthViewCurrentMonth = 0;
    private int iMonthViewCurrentYear = 0;
    private int iFirstDayOfWeek = Calendar.MONDAY;
    private int Calendar_Width = 0;
    private int Cell_Width = 0;
    //页面控件
    TextView Top_Date = null;
    Button btn_pre_month = null;
    Button btn_next_month = null;
    TextView arrange_text = null;
    LinearLayout mainLayout = null;
    LinearLayout arrange_layout = null;
    //数据源
    ArrayList<String> Calendar_Source = null;
    Hashtable<Integer, Integer> calendar_Hashtable = new Hashtable<Integer, Integer>();
    Boolean[] flag = null;
    Calendar startDate = null;
    Calendar endDate = null;
    int dayvalue = -1;
    public static int Calendar_WeekBgColor = 0;
    public static int Calendar_DayBgColor = 0;
    public static int isHoliday_BgColor = 0;
    public static int unPresentMonth_FontColor = 0;
    public static int isPresentMonth_FontColor = 0;
    public static int isToday_BgColor = 0;
    public static int special_Reminder = 0;
    public static int common_Reminder = 0;
    public static int Calendar_WeekFontColor = 0;
    String UserName = "";
    @Override
    public void onCreate(Bundle savedInstanceState) {
        super.onCreate(savedInstanceState);
        //获得屏幕宽和高,并计算出屏幕宽度分七等份的大小
        WindowManager windowManager = getWindowManager();
```

```java
            Display display = windowManager.getDefaultDisplay();
            int screenWidth = display.getWidth();
            Calendar_Width = screenWidth;
            Cell_Width = Calendar_Width / 7 + 1;
            //制定布局文件,并设置属性
            mainLayout = (LinearLayout) getLayoutInflater().inflate(R.layout.main, null);
            setContentView(mainLayout);
            //声明控件,并绑定事件
            Top_Date = (TextView) findViewById(R.id.Top_Date);
            btn_pre_month = (Button) findViewById(R.id.btn_pre_month);
            btn_next_month = (Button) findViewById(R.id.btn_next_month);
            btn_pre_month.setOnClickListener(new Pre_MonthOnClickListener());
            btn_next_month.setOnClickListener(new Next_MonthOnClickListener());
            //计算本月日历中的第一天(一般是上月的某天),并更新日历
            calStartDate = getCalendarStartDate();
            mainLayout.addView(generateCalendarMain());
            DateWidgetDayCell daySelected = updateCalendar();
            if (daySelected != null)
                daySelected.requestFocus();
            LinearLayout.LayoutParams Param1 = new LinearLayout.LayoutParams(
                    ViewGroup.LayoutParams.MATCH_PARENT,
                    ViewGroup.LayoutParams.MATCH_PARENT);
            ScrollView view = new ScrollView(this);
            arrange_layout = createLayout(LinearLayout.VERTICAL);
            arrange_layout.setPadding(5, 2, 0, 0);
            arrange_text = new TextView(this);
            mainLayout.setBackgroundColor(Color.WHITE);
            arrange_text.setTextColor(Color.BLACK);
            arrange_text.setTextSize(18);
            arrange_layout.addView(arrange_text);
            startDate = GetStartDate();
            calToday = GetTodayDate();
            endDate = GetEndDate(startDate);
            view.addView(arrange_layout, Param1);
            mainLayout.addView(view);
...
```

(5) 在 src\calendar_manager 包下创建一个 DateWidgetDayCell.java 文件,该文件用于实现日期控件字串。代码为:

```java
...
public class DateWidgetDayCell extends View {
    //字体大小
    private static final int fTextSize = 28;
    //基本元素
    private OnItemClick itemClick = null;
    private Paint pt = new Paint();
    private RectF rect = new RectF();
    private String sDate = "";
    //当前日期
    private int iDateYear = 0;
    private int iDateMonth = 0;
    private int iDateDay = 0;
    //布尔变量
    private boolean bSelected = false;
    private boolean bIsActiveMonth = false;
```

```
private boolean bToday = false;
private boolean bTouchedDown = false;
private boolean bHoliday = false;
private boolean hasRecord = false;
public static int ANIM_ALPHA_DURATION = 100;
public interface OnItemClick {
    public void OnClick(DateWidgetDayCell item);
}
//构造函数
public DateWidgetDayCell(Context context, int iWidth, int iHeight) {
    super(context);
    setFocusable(true);
    setLayoutParams(new LayoutParams(iWidth, iHeight));
}
//取变量值
public Calendar getDate() {
    Calendar calDate = Calendar.getInstance();
    calDate.clear();
    calDate.set(Calendar.YEAR, iDateYear);
    calDate.set(Calendar.MONTH, iDateMonth);
    calDate.set(Calendar.DAY_OF_MONTH, iDateDay);
    return calDate;
}
...
```

(6) 在 src\calendar_manager 包下新建一个 DateWidgetDayHeader.java 文件,该文件用于设置日期控件的字串头。代码为:

```
package fs.calendar_manager;
import android.content.Context;
import android.graphics.Canvas;
import android.graphics.Paint;
import android.graphics.RectF;
import android.view.View;
import android.widget.LinearLayout.LayoutParams;
public class DateWidgetDayHeader extends View {
    //字体大小
    private final static int fTextSize = 22;
    private Paint pt = new Paint();
    private RectF rect = new RectF();
    private int iWeekDay = -1;
    public DateWidgetDayHeader(Context context, int iWidth, int iHeight) {
        super(context);
        setLayoutParams(new LayoutParams(iWidth, iHeight));
    }
    @Override
    protected void onDraw(Canvas canvas) {
        super.onDraw(canvas);
        //设置矩形大小
        rect.set(0, 0, this.getWidth(), this.getHeight());
        rect.inset(1, 1);
        //绘制日历头部
        drawDayHeader(canvas);
    }
    private void drawDayHeader(Canvas canvas) {
        //画矩形,并设置矩形画笔的颜色
```

```
            pt.setColor(MainActivity.Calendar_WeekBgColor);
            canvas.drawRect(rect, pt);
            //写入日历头部,设置画笔参数
            pt.setTypeface(null);
            pt.setTextSize(fTextSize);
            pt.setAntiAlias(true);
            pt.setFakeBoldText(true);
            pt.setColor(MainActivity.Calendar_WeekFontColor);
            //显示日期
            final String sDayName = DayStyle.getWeekDayName(iWeekDay);
            final int iPosX = (int) rect.left + ((int) rect.width() >> 1)
                    - ((int) pt.measureText(sDayName) >> 1);
            final int iPosY = (int) (this.getHeight()
                    - (this.getHeight() - getTextHeight()) / 2 - pt
                    .getFontMetrics().bottom);
            canvas.drawText(sDayName, iPosX, iPosY, pt);
        }
        //得到字体高度
        private int getTextHeight() {
            return (int) (-pt.ascent() + pt.descent());
        }
        //得到一个星期的第几天的文本标记
        public void setData(int iWeekDay) {
            this.iWeekDay = iWeekDay;
        }
    }
```

(7) 在src\calendar_manager 包下新建一个 DayStyle.java 文件,该文件用于实现日期类型。代码为:

```
package fs.calendar_manager;
import java.util.Calendar;
public class DayStyle {
    private final static String[] vecStrWeekDayNames = getWeekDayNames();
    private static String[] getWeekDayNames() {
        String[] vec = new String[10];
        vec[Calendar.SUNDAY] = "周日";
        vec[Calendar.MONDAY] = "周一";
        vec[Calendar.TUESDAY] = "周二";
        vec[Calendar.WEDNESDAY] = "周三";
        vec[Calendar.THURSDAY] = "周四";
        vec[Calendar.FRIDAY] = "周五";
        vec[Calendar.SATURDAY] = "周六";
        return vec;
    }
    public static String getWeekDayName(int iDay) {
        return vecStrWeekDayNames[iDay];
    }
    public static int getWeekDay(int index, int iFirstDayOfWeek) {
        int iWeekDay = -1;
        if (iFirstDayOfWeek == Calendar.MONDAY) {
            iWeekDay = index + Calendar.MONDAY;
            if (iWeekDay > Calendar.SATURDAY)
                iWeekDay = Calendar.SUNDAY;
        }
        if (iFirstDayOfWeek == Calendar.SUNDAY) {
```

```
                iWeekDay = index + Calendar.SUNDAY;
            }
            return iWeekDay;
        }
    }
```

运行程序,效果如图 6-8 所示,单击界面中的数字可改变日期,单击界面中的"上一页"按钮可翻转到上一个月,单击界面中的"下一个"按钮可翻转到下一个月。

图 6-8 万年历界面

6.9 存储卡查询实例

每一个手机都会有一张存储卡,在存储卡中放着自己喜欢的相册、音乐或电影,在日常生活中为人们生活提供了很大的帮助。

本实例用于实现怎样获取手机中存储卡的容量,使用户知道存储卡使用了多少空间(容量),还剩余多少空间(容量)。

本实例的具体实现步骤如下。

(1) 在 Eclipse 中创建一个 Android 应用项目,命名为 Scard_test。

(2) 打开 res\layout 目录下的 main.xml 布局文件,在文件中声明 3 个 TextView 控件、3 个 EditText 控件及一个 Button 控件。代码为:

```xml
<?xml version = "1.0" encoding = "utf-8"?>
<LinearLayout xmlns:android = "http://schemas.android.com/apk/res/android"
    android:orientation = "vertical"
    android:layout_width = "fill_parent"
    android:layout_height = "fill_parent"
    android:background = "#aabbcc">
    <LinearLayout
        android:id = "@ + id/LinearLayout01"
        android:layout_width = "fill_parent"
        android:layout_height = "wrap_content"
```

```xml
        android:orientation = "horizontal">
        <TextView
            android:layout_width = "100dip"
            android:layout_height = "wrap_content"
            android:text = "SD卡总容量: "/>
        <EditText
            android:id = "@ + id/EditText01"
            android:layout_width = "fill_parent"
            android:layout_height = "wrap_content"
            android:editable = "false"/>
    </LinearLayout>
    <LinearLayout
        android:id = "@ + id/LinearLayout02"
        android:layout_width = "fill_parent"
        android:layout_height = "wrap_content"
        android:orientation = "horizontal">
        <TextView
            android:layout_width = "100dip"
            android:layout_height = "wrap_content"
            android:text = "SD卡已使用: "/>
        <EditText
            android:id = "@ + id/EditText02"
            android:layout_width = "fill_parent"
            android:layout_height = "wrap_content"
            android:editable = "false"/>
    </LinearLayout>
    <LinearLayout
        android:id = "@ + id/LinearLayout03"
        android:layout_width = "fill_parent"
        android:layout_height = "wrap_content"
        android:orientation = "horizontal">
        <TextView
            android:layout_width = "100dip"
            android:layout_height = "wrap_content"
            android:text = "SD卡可用量: "/>
        <EditText
            android:id = "@ + id/EditText03"
            android:layout_width = "fill_parent"
            android:layout_height = "wrap_content"
            android:editable = "false"/>
    </LinearLayout>
    <Button
        android:text = "查询SD卡容量"
        android:id = "@ + id/Button01"
        android:layout_width = "wrap_content"
        android:layout_height = "wrap_content">
    </Button>
</LinearLayout>
```

(3) 打开 src\fs.scard_test 包下的 MainActivity.java 文件，在文件中实现查询手机 SD 卡的容量。当单击界面中的"查询 SD 卡容量"按钮时，先判断手机中是否已安装 SD 卡，如果未安装则在界面上显示容量为 0，如果安装，则从系统中获取 SD 卡的容量。实现代码为：

```
package fs.scard_test;
import java.io.File;
```

```java
import java.text.DecimalFormat;
import android.app.Activity;
import android.os.Bundle;
import android.os.Environment;
import android.os.StatFs;
import android.view.View;
import android.view.View.OnClickListener;
import android.widget.Button;
import android.widget.EditText;
public class MainActivity extends Activity {
    //定义控件
    Button button;                                              //查询按钮
    EditText etTotal;                                           //总容量
    EditText etUsed;                                            //已使用量
    EditText etAvailable;                                       //未使用量
    @Override
    public void onCreate(Bundle savedInstanceState) {
        super.onCreate(savedInstanceState);
        setContentView(R.layout.main);
        button = (Button)this.findViewById(R.id.Button01);      //查询按钮
        etTotal = (EditText)this.findViewById(R.id.EditText01); //总容量
        etUsed = (EditText)this.findViewById(R.id.EditText02);  //已使用量
        etAvailable = (EditText)this.findViewById(R.id.EditText03); //未使用量
        //设置单击监听事件
        button.setOnClickListener
        (
            new OnClickListener()
            {
                public void onClick(View v) {
                    if(Environment.getExternalStorageState().equals(Environment.MEDIA_MOUNTED))
                    {                                                   //判断存储卡是否插入
                        File path = Environment.getExternalStorageDirectory();  //获取路径
                        StatFs sf = new StatFs(path.getPath());         //创建 StatFs 对象
                        long size = sf.getBlockSize();                  //SD卡单位大小
                        long total = sf.getBlockCount();                //总数
                        long available = sf.getAvailableBlocks();       //可使用的数量
                        DecimalFormat df = new DecimalFormat();         //创建对象
                        df.setGroupingSize(3);                          //每3位分为一组
                        String totalSize = (size * total)/1024 >= 1024? //总容量
                            df.format((((size * total)/1024)/1024)) + "MB":
                            df.format((size * total)/1024) + "KB";
                        String availableSize = (size * available)/1024 >= 1024? //未使用量
                            df.format((((size * available)/1024)/1024)) + "MB":
                            df.format((size * available)/1024) + "KB";
                        String usedSize = (size * (total-available))/1024 >= 1024? //已使用量
                            df.format((((size * (total-available))/1024)/1024)) + "MB":
                            df.format((size * (total-available))/1024) + "KB";
                        etTotal.setText(totalSize);                     //总容量
                        etUsed.setText(usedSize);                       //已使用量
                        etAvailable.setText(availableSize);             //未使用量
                    }else if(Environment.getExternalStorageState().equals(Environment.MEDIA_REMOVED))
                    {
                        //SD卡已移除
                        etTotal.setText(0);                             //总容量
```

```
                    etUsed.setText(0);            //已使用量
                    etAvailable.setText(0);       //未使用量
                }
            }
        }
    );
    }
}
```

运行程序,得到的界面如图 6-9 所示。

图 6-9 查询 SD 卡容量

6.10 RSS 阅读器实例

RSS 阅读器(Rss Reader)是一个全功能 Google Reader 阅读器。通过使用 Rss Reader 可以查看、添加和删除自己的订阅。Rss Reader 支持离线浏览和图片缓存,在浏览订阅后可以在图库里面查看喜欢的图片。Rss Reader 同样支持手势操作,在内容页面通过不同手势切换到不同条目和展示方式。

在日常生活中,经常需要浏览多个网站,从而获取自己需要的信息,但是这样不仅浪费了时间,而且不利于信息的查找。而 RSS 阅读器很好地解决了这些问题,可以管理多个网站,并且可以显示网站的更新信息。

本实例主要实现在 Android 中实现 RSS 阅读器的制作。通过本实例演示 RSS 阅读器的功能。

本实例的具体实现步骤如下。

(1) 在 Eclipse 中创建一个 Android 应用项目,命名为 RSS_test。

(2) 打开 res\layout 目录下的 main.xml 文件,在文件中声明一个 TextView 控件、一个 EditText 控件及一个 Button 控件。代码为:

```
<?xml version = "1.0" encoding = "utf - 8"?>
```

```xml
<LinearLayout
    android:id = "@ + id/LinearLayout01"
    android:layout_width = "fill_parent"
    android:layout_height = "fill_parent"
    android:orientation = "vertical"
    android:background = "#aabbcc"
    xmlns:android = "http://schemas.android.com/apk/res/android">
    <TextView
        android:text = "请输入 RSS 源地址："
        android:id = "@ + id/TextView"
        android:textSize = "18sp"
        android:textColor = "@drawable/white"
        android:background = "@drawable/black"
        android:layout_width = "fill_parent"
        android:layout_height = "wrap_content">
    </TextView>
    <EditText
        android:text = "http://www.baidu.com/"
        android:id = "@ + id/EditText"
        android:textColor = "@drawable/black"
        android:layout_width = "wrap_content"
        android:layout_height = "wrap_content">
    </EditText>
    <Button
        android:text = "开始阅读"
        android:id = "@ + id/Button"
        android:gravity = "center"
        android:textColor = "@drawable/black"
        android:layout_width = "wrap_content"
        android:layout_height = "wrap_content">
    </Button>
</LinearLayout>
```

（3）在 res\layout 目录下新建一个 rows.xml 布局文件，在文件中声明一个 ImageView 控件及一个 TextView 控件。代码为：

```xml
<?xml version = "1.0" encoding = "utf-8"?>
<LinearLayout
    xmlns:android = "http://schemas.android.com/apk/res/android"
    android:orientation = "horizontal"
    android:layout_width = "fill_parent"
    android:layout_height = "fill_parent"
    android:background = "#abc">
    <ImageView
        android:id = "@ + id/icon"
        android:layout_width = "25dip"
        android:layout_height = "25dip"
        android:src = "@drawable/ic_launcher">
    </ImageView>
    <TextView
        android:id = "@ + id/news"
        android:layout_gravity = "center_vertical"
        android:layout_width = "wrap_content"
        android:layout_height = "wrap_content"
        android:textColor = "@drawable/black">
    </TextView>
```

        </LinearLayout>

（4）在 res\layout 目录下新建一个 contents.xml 布局文件，在文件中声明一个 ScrollView 控件及 3 个 TextView 控件。代码为：

```xml
<?xml version = "1.0" encoding = "utf-8"?>
<LinearLayout
    android:id = "@+id/LinearLayout01"
    android:layout_width = "fill_parent"
    android:layout_height = "fill_parent"
    xmlns:android = "http://schemas.android.com/apk/res/android"
    android:orientation = "vertical"
    android:background = "#cba">
    <TextView
        android:id = "@+id/myTitle"
        android:textColor = "@drawable/white"
        android:textSize = "18sp"
        android:layout_width = "fill_parent"
        android:gravity = "center"
        android:background = "@drawable/black"
        android:layout_height = "wrap_content">
    </TextView>
    <ScrollView
        android:id = "@+id/ScrollView01"
        android:layout_width = "wrap_content"
        android:layout_height = "wrap_content">
        <TextView
            android:id = "@+id/myDesc"
            android:textColor = "@drawable/black"
            android:layout_width = "300px"
            android:layout_height = "200px">
        </TextView>
    </ScrollView>
    <TextView
        android:id = "@+id/myLink"
        android:textColor = "#0000FF"
        android:layout_width = "wrap_content"
        android:layout_height = "wrap_content">
    </TextView>
</LinearLayout>
```

（5）在 res\layout 目录下新建一个 newlist.xml 布局文件，在文件中声明一个 TextView 控件及一个 ListView 控件。代码为：

```xml
<?xml version = "1.0" encoding = "utf-8"?>
<LinearLayout
    xmlns:android = "http://schemas.android.com/apk/res/android"
    android:layout_width = "fill_parent"
    android:layout_height = "fill_parent"
    android:orientation = "vertical"
    android:background = "#bbaacc">
    <TextView
        android:id = "@+id/myText"
        android:layout_width = "fill_parent"
        android:layout_height = "wrap_content"
        android:padding = "5px"
```

```xml
        android:textSize = "18sp"
        android:textColor = "@drawable/white"
        android:background = "@drawable/black"
        android:gravity = "center"/>
    <ListView
        android:id = "@android:id/list"
        android:layout_width = "wrap_content"
        android:layout_height = "wrap_content">
    </ListView>
</LinearLayout>
```

(6) 在 res\values 目录下新建一个 colors.xml 布局文件,用于实现颜色文件,代码为:

```xml
<?xml version = "1.0" encoding = "utf-8"?>
<resources>
    <drawable name = "white">#FFFFFFFF</drawable>
    <drawable name = "black">#000000</drawable>
    <drawable name = "blue">#0000FF</drawable>
<drawable name = "green">#00FFCC</drawable>
<drawable name = "red">#FF0000</drawable>
</resources>
```

(7) 打开 src\fs.rss_test 包下的 MainActivity.java 文件,在文件中当单击界面中的"开始阅读"按钮时,则跳转到信息更新的界面,单击信息更新界面中的任意一项,进入详细说明界面。代码为:

```java
package fs.rss_test;
import android.app.Activity;
import android.app.AlertDialog;
import android.content.DialogInterface;
import android.content.Intent;
import android.os.Bundle;
import android.view.View;
import android.widget.Button;
import android.widget.EditText;
public class MainActivity extends Activity
{
    //成员变量
    private Button myButton;
    private EditText myEditText;
    //重写 onCreate 方法
    @Override
    public void onCreate(Bundle savedInstanceState)
    {
        super.onCreate(savedInstanceState);
        setContentView(R.layout.main);
        //获得 EditText 和 Button
        myEditText = (EditText) findViewById(R.id.EditText);
        myButton = (Button) findViewById(R.id.Button);
        //为 Button 设置监听器
        myButton.setOnClickListener(new Button.OnClickListener()
        {
            @Override
            public void onClick(View v)
            {
                //得到 path
```

```java
                String path = myEditText.getText().toString();
                //判断网址是否正确
                if(path.equals(""))
                {
                    //显示对话框,提示没有输入网址
                    showDialog("请输入网址!");
                }
                else
                {
                    //新建 intent 发送消息给 Reader_1
                    Intent intent = new Intent();
                    intent.setClass(MainActivity.this,Reader_1.class);
                    Bundle b = new Bundle();
                    //将 path 传入
                    b.putString("path",path);
                    intent.putExtras(b);
                    startActivityForResult(intent,0);
                }
            }
        });
    }
    //重写 onActivityResult 方法
    @Override
    protected void onActivityResult(int requestCode, int resultCode, Intent data)
    {
        switch (resultCode)
        {
            //错误处理代码
            case 8080:
                Bundle b = data.getExtras();
                String error = b.getString("error");
                showDialog(error);
            break;
            default:
            break;
        }
    }
    //捕捉异常处理
    private void showDialog(String mess)
    {
        new AlertDialog.Builder(MainActivity.this).setTitle("错误提示")
        .setMessage(mess)
        .setNegativeButton("返回", new DialogInterface.OnClickListener()
        {
            public void onClick(DialogInterface dialog, int which)
            {
            }
        }).show();
    }
}
```

(8) 在 src\fs.rss_test 包下新建一个 ABaseAdapter.java 文件,该文件主要实现构造器。代码为:

```java
package fs.rss_test;
import java.util.List;
```

```java
import android.content.Context;
import android.view.LayoutInflater;
import android.view.View;
import android.view.ViewGroup;
import android.widget.BaseAdapter;
import android.widget.TextView;
public class ABaseAdapter extends BaseAdapter
{
    //成员变量
    private LayoutInflater myInflater;
    private List<News> list;
    //构造器
    public ABaseAdapter(Context context,List<News> list)
    {
        myInflater = LayoutInflater.from(context);
        this.list = list;
    }
    @Override
    public int getCount()
    {
        return list.size();
    }
    @Override
    public Object getItem(int position)
    {
        return list.get(position);
    }
    @Override
    public long getItemId(int position)
    {
        return position;
    }
    @Override
    public View getView(int position,View convertView,ViewGroup par)
    {
        ViewHolder holder;
        //使用news_row 为layout
        if(convertView == null)
        {
            convertView = myInflater.inflate(R.layout.rows, null);
            holder = new ViewHolder();
            holder.text = (TextView) convertView.findViewById(R.id.news);
            convertView.setTag(holder);
        }
        else
        {
            holder = (ViewHolder) convertView.getTag();
        }
        News tmpNews = (News)list.get(position);
        holder.text.setText(tmpNews.getTitle());
        return convertView;
    }
    private class ViewHolder
    {
        TextView text;
```

　　　　}
　　}

(9) 在 src\fs.rss_test 包下新建一个 AHandler.java 文件，主要调用 Handler 机制及解析 XML 文件。代码为：

```java
package fs.rss_test;
import java.util.ArrayList;
import java.util.List;
import org.xml.sax.Attributes;
import org.xml.sax.SAXException;
import org.xml.sax.helpers.DefaultHandler;
//解析 xml 文件的类
public class AHandler extends DefaultHandler
{
    //成员变量
    private boolean item_flag = false;
    private boolean title_flag = false;
    private boolean link_flag = false;
    private boolean desc_flag = false;
    private boolean date_flag = false;
    private boolean mainTitle_flag = false;
    private List<News> list;
    private News news;
    private String title = "";
    private StringBuffer buf = new StringBuffer();
    public List<News> getParsedData()
    {
        return list;
    }
    public String getRssTitle()
    {
        return title;
    }
    //开始解析方法
    @Override
    public void startDocument() throws SAXException
    {
        list = new ArrayList<News>();
    }
    //结束解析方法
    @Override
    public void endDocument() throws SAXException
    {
    }
    //解析前面部分
    @Override
    public void startElement(String namespaceURI, String localName,
            String qName, Attributes atts) throws SAXException
            {
        if (localName.equals("item"))
        {
            this.item_flag = true;
            news = new News();
        }
        else if (localName.equals("title"))
```

```java
            {
                if(this.item_flag)
                {
                    this.title_flag = true;
                }
                else
                {
                    this.mainTitle_flag = true;
                }
            }
            else if (localName.equals("link"))
            {
                if(this.item_flag)
                {
                    this.link_flag = true;
                }
            }
            else if (localName.equals("description"))
            {
                if(this.item_flag)
                {
                    this.desc_flag = true;
                }
            }
            else if (localName.equals("pubDate"))
            {
                if(this.item_flag)
                {
                    this.date_flag = true;
                }
            }
        }
        //解析后面部分
        @Override
        public void endElement(String namespaceURI, String localName,
                        String qName) throws SAXException
                        {
            if (localName.equals("item"))
            {
                this.item_flag = false;
                list.add(news);
            }
            else if (localName.equals("title"))
            {
                if(this.item_flag)
                {
                    news.setTitle(buf.toString().trim());
                    buf.setLength(0);
                    this.title_flag = false;
                }
                else
                {
                    title = buf.toString().trim();
                    buf.setLength(0);
                    this.mainTitle_flag = false;
```

```java
            }
        }
        else if (localName.equals("link"))
        {
            if(this.item_flag)
            {
                news.setLink(buf.toString().trim());
                buf.setLength(0);
                this.link_flag = false;
            }
        }
        else if (localName.equals("description"))
        {
            if(item_flag)
            {
                news.setDesc(buf.toString().trim());
                buf.setLength(0);
                this.desc_flag = false;
            }
        }
        else if (localName.equals("pubDate"))
        {
            if(item_flag)
            {
                news.setDate(buf.toString().trim());
                buf.setLength(0);
                this.date_flag = false;
            }
        }
    }
    //字符串处理
    @Override
    public void characters(char ch[], int start, int length)
    {
        if(this.item_flag||this.mainTitle_flag)
        {
            buf.append(ch,start,length);
        }
    }
}
```

(10) 在 src\fs.rss_test 包下新建一个 News.java 文件，用于设置和获取对应的方法。代码为：

```java
package fs.rss_test;
public class News
{
    //成员变量
    private String _title = "";
    private String _link = "";
    private String _desc = "";
    private String _date = "";
    //设置和获取 title 方法
    public String getTitle()
    {
        return _title;
```

```
        }
        public void setTitle(String title)
        {
            _title = title;
        }
        //设置和获取 link 方法
        public String getLink()
        {
            return _link;
        }
        public void setLink(String link)
        {
            _link = link;
        }
        //设置和获取 desc 方法
        public String getDesc()
        {
            return _desc;
        }
        public void setDesc(String desc)
        {
            _desc = desc;
        }
        //设置和获取 date 方法
        public String getDate()
        {
            return _date;
        }
        public void setDate(String date)
        {
            _date = date;
        }
    }
```

(11) 在 src\fs.rss_test 包下新建一个 Read_1.java 文件, 在文件中实现线程间的消息传递, 并自定义适配器和机制。代码为:

```
package fs.rss_test;
import java.net.URL;
import java.util.ArrayList;
import java.util.List;
import javax.xml.parsers.SAXParser;
import javax.xml.parsers.SAXParserFactory;
import org.xml.sax.InputSource;
import org.xml.sax.XMLReader;
import android.app.ListActivity;
import android.content.Intent;
import android.os.Bundle;
import android.view.View;
import android.widget.ListView;
import android.widget.TextView;
public class Reader_1 extends ListActivity
{
    //成员变量
    private TextView myTextView;
    private String myTitle;
```

```java
    private List<News> list = null;
    //重写 onCreate 方法
    @Override
    public void onCreate(Bundle savedInstanceState)
    {
        super.onCreate((android.os.Bundle) savedInstanceState);
        setContentView(R.layout.newlist);
        //获得 TextView
        myTextView = (TextView)this.findViewById(R.id.myText);
        //新建 intent 来传递消息
        Intent intent = this.getIntent();
        Bundle b = intent.getExtras();
        String path = b.getString("path");
        //获得 list
        list = getRss(path);
        myTextView.setText(myTitle);
        //设置自定义的设配器
        setListAdapter(new ABaseAdapter(this,list));
    }
    //添加监听器
    @Override
    public void onListItemClick(ListView l, View v, int position, long id)
    {
        News news = (News)list.get(position);
        //新建 intent 来传递消息
        Intent intent = new Intent();
        intent.setClass(Reader_1.this, Reader_2.class);
        Bundle ba = new Bundle();
        //传入 title、desc 和 link 参数
        ba.putString("title", news.getTitle());
        ba.putString("desc", news.getDesc());
        ba.putString("link", news.getLink());
        intent.putExtras(ba);
        //开启 Activity
        startActivity(intent);
    }
    //获得 RSS 方法
    private List<News> getRss(String path)
    {
        List<News> result = new ArrayList<News>();
        //下载信息的 url
        URL url = null;
        try
        {
            //传入 path
            url = new URL(path);
            //利用 SAXParserFactory 对 XML 文件进行解析
            SAXParserFactory spf = SAXParserFactory.newInstance();
            SAXParser sp = spf.newSAXParser();
            XMLReader xr = sp.getXMLReader();
            AHandler myExampleHandler = new AHandler();
            //设置自定义的 Handler
            xr.setContentHandler(myExampleHandler);
            xr.parse(new InputSource(url.openStream()));
            //获取信息
```

```
                result = myExampleHandler.getParsedData();
                myTitle = myExampleHandler.getRssTitle();
            }
            catch(Exception e)
            {
                //错误处理
                Intent intent = new Intent();
                Bundle b = new Bundle();
                b.putString("error","" + e);
                intent.putExtras(b);
                Reader_1.this.setResult(8080, intent);
                Reader_1.this.finish();
            }
            //返回获取的信息
            return result;
        }
    }
```

(12) 在 src\fs.rss_test 包下新建一个 Read_2.java 文件,该文件用于新建机制及接收消息。代码为：

```
package fs.rss_test;
import android.app.Activity;
import android.content.Intent;
import android.os.Bundle;
import android.text.util.Linkify;
import android.widget.TextView;
public class Reader_2 extends Activity
{
    //成员变量
    private TextView myTitle;
    private TextView myDesc;
    private TextView myLink;
    //重写 onCreat 方法
    @Override
    public void onCreate(Bundle savedInstanceState)
    {
        super.onCreate(savedInstanceState);
        setContentView(R.layout.contents);
        //获取 title、desc 和 link
        myTitle = (TextView) findViewById(R.id.myTitle);
        myDesc = (TextView) findViewById(R.id.myDesc);
        myLink = (TextView) findViewById(R.id.myLink);
        //新建 intent 传递消息
        Intent intent = this.getIntent();
        Bundle b = intent.getExtras();
        //设置 title、desc 和 link
        myTitle.setText(b.getString("title"));
        myDesc.setText(b.getString("desc"));
        myLink.setText(b.getString("link"));
        //获取 link 地址
        Linkify.addLinks(myLink,Linkify.WEB_URLS);
    }
}
```

(13) 打开 AndroidManifest.xml 文件,在文件中设置对应的权限及导入定义的类。代码为：

```
    ……
    </activity>
    <!-- 添加文件 -->
    <activity android:name = "Reader_1"></activity>
    <activity android:name = "Reader_2"></activity>
</application>
<!-- 添加权限 -->
<uses-permission android:name = "android.permission.INTERNET" />
</manifest>
```

运行程序，效果如图6-10所示，当单击界面中的按钮时，则开始实现阅读。

图6-10　RSS阅读器实例

第 7 章

Android 媒体应用实例

随着 3G 时代的到来,多媒体在手机和平板电脑上广泛应用。Android 作为手机和平板电脑的一个操作系统,对于多媒体应用也提供了良好的支持。它不仅支持音频和视频的播放,而且还支持音频、视频和摄像头拍照。

7.1 MediaPlayer 播放音频实例

在 Android 中使用 MediaPlayer 播放音频是十分简单的,当程序控制 MediaPlayer 对象装载音频完成后,程序可以调用 MediaPlayer 的如下 3 个方法进行播放控制。

- start():开始或恢复播放。
- stop():停止播放。
- pause():暂停播放。

在 Android 中利用 MediaPlayer 播放音频可以实现从资源文件中播放音频、外部存储设备播放音频及网络中播放音频。下面通过实例来演示从不同地方播放音频。

1. 从资源文件中播放音频

本实例主要利用 MediaPlayer 实现从资源文件中播放音频,其具体实现步骤如下。

(1) 在 Eclipse 中创建一个 Android 应用项目,命名为 MediaPlayer_1。

(2) 打开 res\layout 目录下的 main.xml 布局文件,在文件中声明 3 个 ImageButton 控件及一个 TextView 控件。TextView 控件用于随机显示音频播放的状态。代码为:

```
<RelativeLayout xmlns:android = "http://schemas.android.com/apk/res/android"
    xmlns:tools = "http://schemas.android.com/tools"
    android:layout_width = "match_parent"
    android:layout_height = "match_parent"
    android:paddingBottom = "@dimen/activity_vertical_margin"
    android:paddingLeft = "@dimen/activity_horizontal_margin"
    android:paddingRight = "@dimen/activity_horizontal_margin"
    android:paddingTop = "@dimen/activity_vertical_margin"
    tools:context = ".MainActivity"
    android:background = "#abc">
    <TextView
        android:id = "@+id/myTextView1"
```

```xml
        android:layout_width = "wrap_content"
        android:layout_height = "wrap_content"
        android:text = "@string/hello_world"
        android:layout_alignParentTop = "true"
        android:layout_alignParentLeft = "true"/>
    <ImageButton
        android:id = "@+id/myButton1"
        android:layout_width = "wrap_content"
        android:layout_height = "wrap_content"
        android:src = "@drawable/b12"
        android:layout_below = "@+id/myTextView1"/>
    <ImageButton
        android:id = "@+id/myButton3"
        android:layout_width = "wrap_content"
        android:layout_height = "wrap_content"
        android:src = "@drawable/b13"
        android:layout_alignTop = "@+id/myButton1"
        android:layout_toRightOf = "@+id/myButton1"/>
    <ImageButton
        android:id = "@+id/myButton2"
        android:layout_width = "wrap_content"
        android:layout_height = "wrap_content"
        android:src = "@drawable/b11"
        android:layout_alignTop = "@+id/myButton1"
        android:layout_toRightOf = "@+id/myButton3"/>
</RelativeLayout>
```

(3) 在 res 中创建一个 raw 文件夹，在文件夹中放置一个 .mp3 文件。

(4) 打开 src\fs.mediaplayer_1 目录下的 MainActivity.java 文件，在文件中实现当单击界面中的"播放"图像按钮时即播放音频，单击"暂停"图像按钮时即暂停正在播放的音频，单击"停止"图像按钮时即停止正在播放的音频，并将音频播放的状态显示在文本框中。代码为：

```java
package fs.mediaplayer_1;
import android.app.Activity;
import android.media.MediaPlayer;
import android.os.Bundle;
import android.view.View;
import android.widget.ImageButton;
import android.widget.TextView;
public class MainActivity extends Activity {
    private ImageButton mb1,mb2,mb3;
    private TextView tv;
    private MediaPlayer mp;
    //声明一个变量,判断是否为暂停,默认为 false
    private boolean isPaused = false;
        public void onCreate(Bundle savedInstanceState) {
            super.onCreate(savedInstanceState);
            setContentView(R.layout.main);
            //通过 findViewById 找到资源
            mb1 = (ImageButton)findViewById(R.id.myButton1);
            mb2 = (ImageButton)findViewById(R.id.myButton2);
            mb3 = (ImageButton)findViewById(R.id.myButton3);
            tv = (TextView)findViewById(R.id.myTextView1);
            //创建 MediaPlayer 对象,将 onlyyou.mp3 文件放置在 raw 文件夹下
            mp = MediaPlayer.create(this,R.raw.onlyyou);
```

```java
//增加播放音乐按钮的事件
mb1.setOnClickListener(new ImageButton.OnClickListener(){
 @Override
 public void onClick(View v) {
  try {
   if(mp != null)
   {
    mp.stop();
   }
   mp.prepare();
   mp.start();
   tv.setText("音乐播放中……");
  } catch (Exception e) {
   tv.setText("播放发生异常……");
   e.printStackTrace();
  }
 }
});
mb2.setOnClickListener(new ImageButton.OnClickListener(){
 @Override
 public void onClick(View v) {
  try {
   if(mp != null)
   {
    mp.stop();
    tv.setText("音乐停止播放……");
   }
  } catch (Exception e) {
   tv.setText("音乐停止发生异常……");
   e.printStackTrace();
  }
 }
});
mb3.setOnClickListener(new ImageButton.OnClickListener(){
 @Override
 public void onClick(View v) {
  try {
   if(mp != null)
   {
    if(isPaused == false)
    {
     mp.pause();
     isPaused = true;
     tv.setText("暂停播放!");
    }
    else if(isPaused == true)
    {
     mp.start();
     isPaused = false;
     tv.setText("开始播发!");
    }
   }
  } catch (Exception e) {
   tv.setText("发生异常……");
   e.printStackTrace();
```

```
            }
          }
        });
    /* 当 MediaPlayer.OnCompletionLister 运行 Listener */
        mp.setOnCompletionListener(
          new MediaPlayer.OnCompletionListener()
            {
              //@Override
              /* 覆盖文件播出完毕事件 */
              public void onCompletion(MediaPlayer arg0)
              {
                try
                {
                  /* 解除资源与 MediaPlayer 的赋值关系
                   * 让资源可以为其他程序利用 */
                  mp.release();
                  /* 改变 TextView 为播放结束 */
                  tv.setText("音乐播发结束!");
                }
                catch (Exception e)
                {
                  tv.setText(e.toString());
                  e.printStackTrace();
                }
              }
        });
        /* 当 MediaPlayer.OnErrorListener 运行 Listener */
        mp.setOnErrorListener(new MediaPlayer.OnErrorListener()
        {
          @Override
          /* 覆盖错误处理事件 */
          public boolean onError(MediaPlayer arg0, int arg1, int arg2)
          {
            //TODO 自动存根法
            try
            {
              /* 发生错误时也解除资源与 MediaPlayer 的赋值 */
              mp.release();
              tv.setText("播放发生异常!");
            }
            catch (Exception e)
            {
              tv.setText(e.toString());
              e.printStackTrace();
            }
            return false;
          }
        });
      }
}
```

运行程序,默认效果如图 7-1(a)所示。当单击界面中的图像"播放"按钮时,即播放音频,并将音频播放状态显示在文本框中,效果如图 7-1(b)所示。单击图像"暂停"按钮时,效果如图 7-1(c)所示。单击图像"停止"按钮时,效果如图 7-1(d)所示。

（a）默认界面

（b）播放音频

（c）暂停播放

（d）停止播放

图 7-1　从资源文件中播放音频

2. 从外部文件中播放音频

在 MATLAB 中利用 MediaPlayer 对象加载外部 MP3 媒体也非常简单，主要通过 MediaPlayer.setDataSource()方法来实现，构建 setDataSource()的方法有很多，比较简单的方法就是直接传入 MP3 媒体文件的路径。

本实例主要利用 MediaPlayer 实现从外部文件中播放音频，其具体实现步骤如下。

（1）在 Eclipse 中创建一个 Android 应用项目，命名为 MediaPlayer_2。

（2）打开 res\layout 目录下的 main.xml 布局文件，在文件中声明一个 EditText 控件，用于运行程序时，输入对应的外部文件路径，声明 4 个 Button 控件，用于播放音频。代码为：

```
<LinearLayout xmlns:android = "http://schemas.android.com/apk/res/android"
```

```xml
        android:orientation = "vertical"
        android:layout_width = "fill_parent"
        android:layout_height = "fill_parent"
        android:background = "#0c0">
    <EditText
        android:id = "@+id/et_path"
        android:layout_width = "fill_parent"
        android:layout_height = "wrap_content"
        android:lines = "2" />
    <Button
        android:id = "@+id/btn_play"
        android:layout_width = "match_parent"
        android:layout_height = "wrap_content"
        android:text = "播放" />
    <Button
        android:id = "@+id/btn_pause"
        android:layout_width = "match_parent"
        android:layout_height = "wrap_content"
        android:text = "暂停" />
    <Button
        android:id = "@+id/btn_replay"
        android:layout_width = "match_parent"
        android:layout_height = "wrap_content"
        android:text = "重播" />
    <Button
        android:id = "@+id/btn_stop"
        android:layout_width = "match_parent"
        android:layout_height = "wrap_content"
        android:text = "停止" />
</LinearLayout>
```

(3) 打开 src\fs.mediaplayer_2 包下的 MainActivity.java 文件,在文件中实现播放外部文件,当单击界面中相应的按钮时,则实现对应的操作,并利用 Toast 实现提示。代码为:

```java
package fs.mediaplayer_2;
import java.io.File;
import android.media.AudioManager;
import android.media.MediaPlayer;
import android.media.MediaPlayer.OnCompletionListener;
import android.media.MediaPlayer.OnErrorListener;
import android.media.MediaPlayer.OnPreparedListener;
import android.os.Bundle;
import android.app.Activity;
import android.view.View;
import android.widget.Button;
import android.widget.EditText;
import android.widget.Toast;
public class MainActivity extends Activity {
    private EditText et_path;
    private Button btn_play, btn_pause, btn_replay, btn_stop;
    private MediaPlayer mediaPlayer;
    @Override
    protected void onCreate(Bundle savedInstanceState) {
        super.onCreate(savedInstanceState);
        setContentView(R.layout.main);
        et_path = (EditText) findViewById(R.id.et_path);
```

```java
            btn_play = (Button) findViewById(R.id.btn_play);
            btn_pause = (Button) findViewById(R.id.btn_pause);
            btn_replay = (Button) findViewById(R.id.btn_replay);
            btn_stop = (Button) findViewById(R.id.btn_stop);
            btn_play.setOnClickListener(click);
            btn_pause.setOnClickListener(click);
            btn_replay.setOnClickListener(click);
            btn_stop.setOnClickListener(click);
    }
    private View.OnClickListener click = new View.OnClickListener() {
        @Override
        public void onClick(View v) {
            switch (v.getId()) {
                case R.id.btn_play:
                    play();
                    break;
                case R.id.btn_pause:
                    pause();
                    break;
                case R.id.btn_replay:
                    replay();
                    break;
                case R.id.btn_stop:
                    stop();
                    break;
                default:
                    break;
            }
        }
    };
    /**
     * 播放音乐
     */
    protected void play() {
        String path = et_path.getText().toString().trim();
        File file = new File(path);
        if (file.exists() && file.length() > 0) {
            try {
                mediaPlayer = new MediaPlayer();
                //设置指定的流媒体地址
                mediaPlayer.setDataSource(path);
                //设置音频流的类型
                mediaPlayer.setAudioStreamType(AudioManager.STREAM_MUSIC);
                //通过异步的方式装载媒体资源
                mediaPlayer.prepareAsync();
                mediaPlayer.setOnPreparedListener(new OnPreparedListener() {
                    @Override
                    public void onPrepared(MediaPlayer mp) {
                        //装载完毕,开始播放流媒体
                        mediaPlayer.start();
                        Toast.makeText(MainActivity.this, "开始播放", 0).show();
                        //避免重复播放,把播放按钮设置为不可用
                        btn_play.setEnabled(false);
                    }
                });
```

```java
            //设置循环播放
            //mediaPlayer.setLooping(true);
            mediaPlayer.setOnCompletionListener(new OnCompletionListener() {
                @Override
                public void onCompletion(MediaPlayer mp) {
                    //在播放完毕被回调
                    btn_play.setEnabled(true);
                }
            });
            mediaPlayer.setOnErrorListener(new OnErrorListener() {
                @Override
                public boolean onError(MediaPlayer mp, int what, int extra) {
                    //如果发生错误,重新播放
                    replay();
                    return false;
                }
            });
        } catch (Exception e) {
            e.printStackTrace();
            Toast.makeText(this, "播放失败", 0).show();
        }
    } else {
        Toast.makeText(this, "文件不存在", 0).show();
    }
}
/**
 * 暂停
 */
protected void pause() {
    if (btn_pause.getText().toString().trim().equals("继续")) {
        btn_pause.setText("暂停");
        mediaPlayer.start();
        Toast.makeText(this, "继续播放", 0).show();
        return;
    }
    if (mediaPlayer != null && mediaPlayer.isPlaying()) {
        mediaPlayer.pause();
        btn_pause.setText("继续");
        Toast.makeText(this, "暂停播放", 0).show();
    }
}
/**
 * 重新播放
 */
protected void replay() {
    if (mediaPlayer != null && mediaPlayer.isPlaying()) {
        mediaPlayer.seekTo(0);
        Toast.makeText(this, "重新播放", 0).show();
        btn_pause.setText("暂停");
        return;
    }
    play();
}
/**
 * 停止播放
```

```
        */
        protected void stop() {
            if (mediaPlayer != null && mediaPlayer.isPlaying()) {
                mediaPlayer.stop();
                mediaPlayer.release();
                mediaPlayer = null;
                btn_play.setEnabled(true);
                Toast.makeText(this, "停止播放", 0).show();
            }
        }
        @Override
        protected void onDestroy() {
            //在Activity结束的时候回收资源
            if (mediaPlayer != null && mediaPlayer.isPlaying()) {
                mediaPlayer.stop();
                mediaPlayer.release();
                mediaPlayer = null;
            }
            super.onDestroy();
        }
    }
```

运行程序,效果如图7-2所示,在编辑框中输入对应的外部文件路径,即可实现音频的播放效果。

图7-2 从外部文件中播放音频

3. 从网络中播放音频

随着3G技术的逐渐成熟,移动互联网时代已经到来,网络资费的不断降低,使直接利用网络资源已经不再是问题,在Android中通过MediaPlayer.setDataSource()方法可直接传入网络媒体资源文件的地址来实现。

本实例主要利用MediaPlayer实现从网络中播放音频,其具体实现步骤如下。

(1) 在Eclipse中创建一个Android应用项目,命名为MediaPlayer_3。

(2) 打开res\layout目录下的main.xml布局文件,在文件中声明一个TextView控件及

3 个 Button 控件。代码为：

```xml
<?xml version = "1.0" encoding = "utf-8"?>
<LinearLayout xmlns:android = "http://schemas.android.com/apk/res/android"
    android:layout_width = "fill_parent"
    android:layout_height = "fill_parent"
    android:orientation = "vertical"
    android:background = "#0e0">
    <TextView
        android:id = "@+id/mp3_name"
        android:layout_width = "wrap_content"
        android:layout_height = "wrap_content"
        android:text = "Large Text"
        android:textAppearance = "?android:attr/textAppearanceLarge" />
    <LinearLayout
        android:id = "@+id/linearLayout1"
        android:layout_width = "match_parent"
        android:layout_height = "wrap_content"
        android:orientation = "horizontal">
        <Button
            android:id = "@+id/button_start"
            android:layout_width = "wrap_content"
            android:layout_height = "wrap_content"
            android:text = "播放" />
        <Button
            android:id = "@+id/button_pause"
            android:layout_width = "wrap_content"
            android:layout_height = "wrap_content"
            android:text = "暂停" />
        <Button
            android:id = "@+id/button_stop"
            android:layout_width = "wrap_content"
            android:layout_height = "wrap_content"
            android:text = "停止" />
    </LinearLayout>
</LinearLayout>
```

(3) 打开 src\fs.mediaplayer_3 包下的 MainActivity.java 文件，在文件中实现从网络中播放音频。代码为：

```java
package fs.mediaplayer_3;
import java.io.File;
import android.app.Activity;
import android.media.MediaPlayer;
import android.os.Bundle;
import android.os.Environment;
import android.view.View;
import android.widget.Button;
import android.widget.TextView;
public class MainActivity extends Activity {
    //声明控件
    private MediaPlayer mediaPlayer = null;      //创建一个空 MediaPlayer 对象
    private Button startButton = null;           //播放 Button 组件对象
    private Button pauseButton = null;           //暂停 Button 组件对象
    private Button stopButton = null;            //停止 Button 组件对象
    private TextView nameTextView = null;        //文件名称 TextView 组件对象
```

```java
        private boolean isPause = false;                      //是否暂停
        /** 第一次调用 Activity 活动 */
        @Override
        public void onCreate(Bundle savedInstanceState) {
            super.onCreate(savedInstanceState);
            setContentView(R.layout.main);
            //实例化文件名称 TextView 组件对象
            nameTextView = (TextView) findViewById(R.id.mp3_name);
            nameTextView.setText("网络");                                            //设置文件名称
            startButton = (Button) findViewById(R.id.button_start);    //实例化播放 Button 组件对象
            //添加播放按钮单击事件监听
            startButton.setOnClickListener(new Button.OnClickListener() {
                @Override
                public void onClick(View arg0) {
                    start();                        //调用 MP3 播放方法
                }
            });
            pauseButton = (Button) findViewById(R.id.button_pause);    //实例化暂停 Button 组件对象
            //添加暂停按钮单击事件监听
            pauseButton.setOnClickListener(new Button.OnClickListener() {
                @Override
                public void onClick(View arg0) {
                    pause();                        //调用 MP3 暂停播放方法
                }
            });
            stopButton = (Button) findViewById(R.id.button_stop);{     //实例化停止 Button 组件对象
            //添加停止按钮单击事件监听
            stopButton.setOnClickListener(new Button.OnClickListener() {
                @Override
                public void onClick(View arg0) {
                    stop();                         //调用 MP3 停止播放方法
                }
            });
        }
        /**
         * MP3 开始播放方法
         */
        public void start() {
            try {
                if (mediaPlayer != null) {          //判断 MediaPlayer 对象不为空
            //判断 MediaPlayer 对象正在播放中,并不执行以下程序
                    if (mediaPlayer.isPlaying()) {
                        return;
                    }
                }
                stop();                             //调用停止播放方法
                if (isPause) {
            //判断 MediaPlayer 对象是否暂停,如果暂停就不重新播放
                    return;
                }
                /*
                 * 网络资源
                 */
                mediaPlayer = new MediaPlayer();
                String path = "http://zhangmenshiting2.baidu.com/data2/music/10547672/10547672.
```

```
mp3?xcode = 4013468857a89a277cf2f0741d8293f2&mid = 0.62331205608975";
            mediaPlayer.setDataSource(path);        //为MediaPlayer设置数据源
            mediaPlayer.prepare();                  //准备播放
            mediaPlayer.start();                    //开始播放
            //文件播放完毕监听事件
            mediaPlayer.setOnCompletionListener(new MediaPlayer.OnCompletionListener() {
                @Override
                public void onCompletion(MediaPlayer arg0) {
                    //覆盖文件播出完毕事件
                    //解除资源与MediaPlayer的赋值关系,让资源可以为其他程序利用
                            mediaPlayer.release();
                            startButton.setText("播放");
                            isPause = false;         //取消暂停状态
                            mediaPlayer = null;
                }
            });
            //文件播放错误监听
            mediaPlayer.setOnErrorListener(new MediaPlayer.OnErrorListener() {
                @Override
                public boolean onError(MediaPlayer arg0, int arg1, int arg2) {
                    //解除资源与MediaPlayer的赋值关系,让资源可以为其他程序利用
                            mediaPlayer.release();
                            return false;
                }
            });
            startButton.setText("正在播放");
            pauseButton.setText("暂停");
        } catch (Exception e) {
            e.printStackTrace();
        }
    }
    /**
     * MP3播放暂停方法
     */
    public void pause() {
        try {
            if (mediaPlayer != null) {               //判断MediaPlayer对象不为空
                if (mediaPlayer.isPlaying()) {       //判断MediaPlayer对象正在播放中
                    mediaPlayer.pause();             //暂停播放
                    pauseButton.setText("取消暂停");
                    isPause = true;                  //暂停状态
                } else {
                    mediaPlayer.start();             //开始播放
                    pauseButton.setText("暂停");
                    isPause = false;
                }
            }
        } catch (Exception e) {
            e.printStackTrace();
        }
    }
    /**
     * MP3停止播放方法
     */
    public void stop() {
```

```
        try {
            if (mediaPlayer != null) {                //判断 MediaPlayer 对象不为空
                mediaPlayer.stop();                   //停止播放
                startButton.setText("播放");
                pauseButton.setText("暂停");
                isPause = false;                      //取消暂停状态
                mediaPlayer.release();
                mediaPlayer = null;
            }
        } catch (Exception e) {
            e.printStackTrace();
        }
    }
}
```

图 7-3　从网络中播放音频

7.2　SoundPool 播放音频实例

前面几个实例介绍了利用 MediaPlayer 从不同地址中播放音频，但 MediaPlayer 占用资源较高，且不支持同时播放多个音频，所以 Android 还提供了另一个播放音频的 SoundPool。SoundPool 也是音频池，它可以同时播放多个短促的音频，而且占用的资源少。SoundPool 适合在应用程序中的播放按键音或消息提示音等，也适合在游戏中实现密集而短暂的声音，例如，多个飞机的爆炸声等。使用 SoundPool 播放音频，首先需要创建 SoundPool 对象，然后加载所要播放的音频，最后再调用 play() 方法播放音频。

本实例利用 SoundPool 实现同时播放多个音频，其具体实现步骤如下。

(1) 在 Eclipse 中创建一个 Android 应用项目，命名为 SoundPool_test。

(2) 打开 res\layout 目录下的 main.xml 布局文件，在文件中添加 4 个 Button 控件。代码为：

```
<?xml version = "1.0" encoding = "utf - 8"?>
<LinearLayout xmlns:android = "http://schemas.android.com/apk/res/android"
```

```xml
        android:layout_width = "match_parent"
        android:layout_height = "match_parent"
        android:background = " # ccddee"
        android:orientation = "vertical" >
<Button
        android:id = "@ + id/button1"
        android:layout_width = "wrap_content"
        android:layout_height = "wrap_content"
        android:text = "风铃声" />
<Button
        android:id = "@ + id/button2"
        android:layout_width = "wrap_content"
        android:layout_height = "wrap_content"
        android:layout_gravity = "clip_horizontal"
        android:text = "布谷鸟叫声" />
<Button
        android:id = "@ + id/button3"
        android:layout_width = "wrap_content"
        android:layout_height = "wrap_content"
        android:text = "门铃声" />
<Button
        android:id = "@ + id/button4"
        android:layout_width = "wrap_content"
        android:layout_height = "wrap_content"
        android:text = "电话声" />
</LinearLayout>
```

(3) 在 res 中创建一个 raw 文件夹，在文件夹中放置 4 种类型的音频文件。

(4) 打开 src\fs.soundpool_test 包下的 MainActivity.java 文件，在文件中利用 soundPool 类实现音频播放。代码为：

```java
package fs.soundpool_test;
import java.util.HashMap;
import android.app.Activity;
import android.media.AudioManager;
import android.media.SoundPool;
import android.os.Bundle;
import android.view.KeyEvent;
import android.view.View;
import android.view.View.OnClickListener;
import android.widget.Button;
public class MainActivity extends Activity {
    private SoundPool soundpool;                              //声明一个 SoundPool 对象
    //创建一个 HashMap 对象
    private HashMap < Integer, Integer > soundmap = new HashMap < Integer, Integer > ( );
    @Override
    public void onCreate(Bundle savedInstanceState) {
        super.onCreate(savedInstanceState);
        setContentView(R.layout.main);
        Button chimes = (Button) findViewById(R.id.button1);      //获取"风铃声"按钮
        Button enter = (Button) findViewById(R.id.button2);       //获取"布谷鸟叫声"按钮
        Button notify = (Button) findViewById(R.id.button3);      //获取"门铃声"按钮
        Button ringout = (Button) findViewById(R.id.button4);     //获取"电话声"按钮
        //创建一个 SoundPool 对象，该对象可以容纳 5 个音频流
        soundpool = new SoundPool(5,AudioManager.STREAM_SYSTEM, 0);
        //将要播放的音频流保存到 HashMap 对象中
        soundmap.put(1, soundpool.load(this, R.raw.chimes, 1));
```

```java
        soundmap.put(2, soundpool.load(this, R.raw.enter, 1));
        soundmap.put(3, soundpool.load(this, R.raw.notify, 1));
        soundmap.put(4, soundpool.load(this, R.raw.ringout, 1));
        soundmap.put(5, soundpool.load(this, R.raw.ding, 1));
        //为各按钮添加单击事件监听器
        chimes.setOnClickListener(new OnClickListener() {
            @Override
            public void onClick(View v) {
                soundpool.play(soundmap.get(1), 1, 1, 0, 0, 1);             //播放指定的音频
            }
        });
        enter.setOnClickListener(new OnClickListener() {
            @Override
            public void onClick(View v) {
                soundpool.play(soundmap.get(2), 1, 1, 0, 0, 1);             //播放指定的音频
            }
        });
        notify.setOnClickListener(new OnClickListener() {
            @Override
            public void onClick(View v) {
                soundpool.play(soundmap.get(3), 1, 1, 0, 0, 1);             //播放指定的音频
            }
        });
        ringout.setOnClickListener(new OnClickListener() {
            @Override
            public void onClick(View v) {
                soundpool.play(soundmap.get(4), 1, 1, 0, 0, 1);             //播放指定的音频
    soundpool.play(soundpool.load(MainActivity.this, R.raw.notify, 1), 1, 1, 0, 0, 1);
            }
        });
    }
    //重写键被按下的事件
    @Override
    public boolean onKeyDown(int keyCode, KeyEvent event) {
        soundpool.play(soundmap.get(5), 1, 1, 0, 0, 1);                     //播放按键音
        return true;
    }
}
```

运行程序,效果如图 7-4 所示。当单击界面中的按钮时,即播放相应的声音。

图 7-4 SoundPool 播放音频

7.3 MediaRecorder 录制音频实例

在前面两节中实现了在 Android 播放音频,已经了解了在 Android 系统中多媒体文件播放的实现。由于 Android 提供了对多媒体的播放,自然会提供对多媒体的采样录制功能。当然,这需要手机本身的硬件支持,Android 中的多媒体录制由 MediaRecorder 类提供了相关方法。

本实例利用 MediaRecorder 实现录制声音和播放声音。其具体实现步骤如下。

(1) 在 Eclipse 中创建一个 Android 应用项目,命名为 MediaRecorder_test。

(2) 打开 res\layout 目录下的 main.xml 布局文件,在文件中声明两个 Button 控件及一个 ListView 控件。代码为:

```xml
<RelativeLayout xmlns:android = "http://schemas.android.com/apk/res/android"
    xmlns:tools = "http://schemas.android.com/tools"
    android:layout_width = "match_parent"
    android:layout_height = "match_parent"
    android:paddingBottom = "@dimen/activity_vertical_margin"
    android:paddingLeft = "@dimen/activity_horizontal_margin"
    android:paddingRight = "@dimen/activity_horizontal_margin"
    android:paddingTop = "@dimen/activity_vertical_margin"
    tools:context = ".MainActivity"
    android:background = "#b00">
    <LinearLayout
        android:id = "@+id/li1"
        android:orientation = "horizontal"
        android:layout_width = "match_parent"
        android:layout_height = "wrap_content">
        <Button
            android:id = "@+id/start"
            android:layout_width = "wrap_content"
            android:layout_height = "wrap_content"
            android:layout_weight = "1"
            android:text = "录音"/>
        <Button
            android:id = "@+id/stop"
            android:layout_width = "wrap_content"
            android:layout_height = "wrap_content"
            android:layout_weight = "1"
            android:text = "停止/播放"/>
    </LinearLayout>
    <ListView
        android:id = "@+id/list"
        android:layout_below = "@id/li1"
        android:layout_width = "match_parent"
        android:layout_height = "wrap_content"></ListView>
</RelativeLayout>
```

(3) 在 res\layout 目录下新建一个 list_show.xml 布局文件,该文件用于显示 ListView 的内容。代码为:

```xml
<?xml version = "1.0" encoding = "utf-8"?>
```

```xml
<LinearLayout xmlns:android = "http://schemas.android.com/apk/res/android"
    android:layout_width = "match_parent"
    android:layout_height = "match_parent"
    android:orientation = "horizontal" >
    <TextView
        android:layout_width = "wrap_content"
        android:layout_height = "wrap_content"
        android:id = "@ + id/show_file_name" />
    <Button
        android:id = "@ + id/bt_list_play"
        android:layout_width = "wrap_content"
        android:layout_height = "wrap_content"
        android:text = "播放录音"/>
    <Button android:id = "@ + id/bt_list_stop"
        android:layout_width = "wrap_content"
        android:layout_height = "wrap_content"
        android:text = "停止播放"/>
</LinearLayout>
```

（4）打开 src\fs.mediarecorder 包下的 MainActivity.java 文件,在文件中实现声明的录制、停止和播放等功能。代码为：

```java
package fs.mediarecorder_test;
import java.io.File;
import java.io.IOException;
import java.text.SimpleDateFormat;
import android.app.Activity;
import android.app.AlertDialog;
import android.app.AlertDialog.Builder;
import android.app.Dialog;
import android.content.DialogInterface;
import android.media.MediaPlayer;
import android.media.MediaRecorder;
import android.os.Bundle;
import android.os.Environment;
import android.view.LayoutInflater;
import android.view.Menu;
import android.view.View;
import android.view.View.OnClickListener;
import android.view.ViewGroup;
import android.widget.BaseAdapter;
import android.widget.Button;
import android.widget.EditText;
import android.widget.ListView;
import android.widget.TextView;
public class MainActivity extends Activity implements OnClickListener {
    private Button start;
    private Button stop;
    private ListView listView;
    //录音文件播放
    private MediaPlayer myPlayer;
    //录音
    private MediaRecorder myRecorder;
    //音频文件保存地址
    private String path;
    private String paths = path;
```

```java
    private File saveFilePath;
    //所录音的文件
    String[] listFile = null;
    ShowRecorderAdpter showRecord;
    AlertDialog aler = null;
    @Override
    protected void onCreate(Bundle savedInstanceState) {
        super.onCreate(savedInstanceState);
        setContentView(R.layout.main);
        start = (Button) findViewById(R.id.start);
        stop = (Button) findViewById(R.id.stop);
        listView = (ListView) findViewById(R.id.list);
        myPlayer = new MediaPlayer();
        myRecorder = new MediaRecorder();
        //从麦克风源进行录音
        myRecorder.setAudioSource(MediaRecorder.AudioSource.DEFAULT);
        //设置输出格式
        myRecorder.setOutputFormat(MediaRecorder.OutputFormat.DEFAULT);
        //设置编码格式
        myRecorder.setAudioEncoder(MediaRecorder.AudioEncoder.DEFAULT);
        showRecord = new ShowRecorderAdpter();
        if (Environment.getExternalStorageState().equals(
                Environment.MEDIA_MOUNTED)) {
            try {
                path = Environment.getExternalStorageDirectory()
                        .getCanonicalPath().toString()
                        + "/XIONGRECORDERS";
                File files = new File(path);
                if (!files.exists()) {
                    files.mkdir();
                }
                listFile = files.list();
            } catch (IOException e) {
                e.printStackTrace();
            }
        }
        start.setOnClickListener(this);
        stop.setOnClickListener(this);
        if (listFile != null) {
            listView.setAdapter(showRecord);
        }
    }
    @Override
    public boolean onCreateOptionsMenu(Menu menu) {
        getMenuInflater().inflate(R.menu.main, menu);
        return true;
    }
    class ShowRecorderAdpter extends BaseAdapter {
        @Override
        public int getCount() {
            return listFile.length;
        }
        @Override
        public Object getItem(int arg0) {
            return arg0;
```

```java
        }
        @Override
        public long getItemId(int arg0) {
            return arg0;
        }
        @Override
        public View getView(final int postion, View arg1, ViewGroup arg2) {
            View views = LayoutInflater.from(MainActivity.this).inflate(
                    R.layout.list_show, null);
            TextView filename = (TextView) views
                    .findViewById(R.id.show_file_name);
            Button plays = (Button) views.findViewById(R.id.bt_list_play);
            Button stop = (Button) views.findViewById(R.id.bt_list_stop);
            filename.setText(listFile[postion]);
            //播放录音
            plays.setOnClickListener(new OnClickListener() {
                @Override
                public void onClick(View arg0) {
                    try {
                        myPlayer.reset();
                        myPlayer.setDataSource(path + "/" + listFile[postion]);
                        if (!myPlayer.isPlaying()) {
                            myPlayer.prepare();
                            myPlayer.start();
                        } else {
                            myPlayer.pause();
                        }
                    } catch (IOException e) {
                        e.printStackTrace();
                    }
                }
            });
            //停止播放
            stop.setOnClickListener(new OnClickListener() {
                @Override
                public void onClick(View arg0) {
                    if (myPlayer.isPlaying()) {
                        myPlayer.stop();
                    }
                }
            });
            return views;
        }
    }
    @Override
    public void onClick(View v) {
        switch (v.getId()) {
        case R.id.start:
            final EditText filename = new EditText(this);
            Builder alerBuidler = new Builder(this);
            alerBuidler
                    .setTitle("请输入要保存的文件名")
                    .setView(filename)
                    .setPositiveButton("确定",
                            new DialogInterface.OnClickListener() {
```

```java
                        @Override
                        public void onClick(DialogInterface dialog,
                                int which) {
                            String text = filename.getText().toString();
                            try {
                                paths = path
                                        + "/"
                                        + text
                                        + new SimpleDateFormat("yyyyMMddHHmmss").format(System.currentTimeMillis())
                                        + ".amr";
                                saveFilePath = new File(paths);
                                myRecorder.setOutputFile(saveFilePath
                                        .getAbsolutePath());
                                saveFilePath.createNewFile();
                                myRecorder.prepare();
                                //开始录音
                                myRecorder.start();
                                start.setText("正在录音中……");
                                start.setEnabled(false);
                                aler.dismiss();
                                //重新读取文件
                                File files = new File(path);
                                listFile = files.list();
                                //刷新 ListView
                                showRecord.notifyDataSetChanged();
                            } catch (Exception e) {
                                e.printStackTrace();
                            }
                        }
                    });
            aler = alerBuidler.create();
            aler.setCanceledOnTouchOutside(false);
            aler.show();
            break;
        case R.id.stop:
            if (saveFilePath.exists() && saveFilePath != null) {
                myRecorder.stop();
                myRecorder.release();
                //判断是否保存,如果不保存则删除
                new AlertDialog.Builder(this)
                        .setTitle("是否保存该录音")
                        .setPositiveButton("确定", null)
                        .setNegativeButton("取消",
                                new DialogInterface.OnClickListener() {
                                    @Override
                                    public void onClick(DialogInterface dialog,
                                            int which) {
                                        saveFilePath.delete();
                                        //重新读取文件
                                        File files = new File(path);
                                        listFile = files.list();
                                        //刷新 ListView
                                        showRecord.notifyDataSetChanged();
                                    }
                                }).show();
```

```
                    }
                    start.setText("录音");
                    start.setEnabled(true);
                default:
                    break;
            }
        }
        @Override
        protected void onDestroy() {
            //释放资源
            if (myPlayer.isPlaying()) {
                myPlayer.stop();
                myPlayer.release();
            }
            myPlayer.release();
            myRecorder.release();
            super.onDestroy();
        }
    }
```

（5）打开 AndroidManifest.xml 文件,为录制声音设置权限。代码为:

```
…
    </application>
<!-- 添加录制多媒体权限 -->
<uses-permission android:name = "android.permission.MOUNT_FORMAT_FILESYSTEMS"/>
<uses-permission android:name = "android.permission.WRITE_EXTERNAL_STORAGE"/>
<uses-permission android:name = "android.permission.RECORD_AUDIO"/>
</manifest>
```

运行程序,效果如图 7-5(a)所示,单击界面中的"录音"按钮,则弹出为录制的声音命名对话框,效果如图 7-5(b)所示。

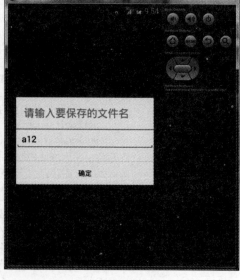

(a) 默认界面　　　　　　　　　(b) 保存录制对话框

图 7-5　录制多媒体

7.4 VideoView 播放视频实例

为了在 Android 中播放视频，Android 提供了 VideoView 组件。它的作用与 ImageView 类似，只是 ImageView 用于显示图片，而 VideoView 用于播放视频。

实际上与 VideoView 一起结合使用的还有一个 MediaController 类，它的作用是提供一个友好的图形控制界面，通过该控制界面来控制视频的播放。

本实例利用 VideoView 来播放视频。其具体实现步骤如下。

(1) 在 Eclipse 中创建一个 Android 应用项目，命名为 VideoView_test。

(2) 打开 res\layout 目录下的 main.xml 布局文件，在文件中定义一个 TextView 控件、一个 VideoView 控件及 3 个 Button 控件。代码为：

```xml
<?xml version = "1.0" encoding = "utf - 8"?>
<AbsoluteLayout xmlns:android = "http://schemas.android.com/apk/res/android"
    android:layout_width = "fill_parent"
    android:layout_height = "fill_parent"
    android:orientation = "vertical"
    android:background = "#ccddee" >
    <TextView
        android:layout_width = "fill_parent"
        android:layout_height = "wrap_content"
        android:text = "@string/hello_world" />
    <VideoView
        android:id = "@ + id/VideoView01"
        android:layout_width = "320px"
        android:layout_height = "240px" />
    <Button
        android:id = "@ + id/PlayButton"
        android:layout_width = "90dp"
        android:layout_height = "wrap_content"
        android:layout_x = "111dp"
        android:layout_y = "204dp"
        android:text = "播放" />
    <Button
        android:id = "@ + id/LoadButton"
        android:layout_width = "84dp"
        android:layout_height = "wrap_content"
        android:layout_x = "17dp"
        android:layout_y = "202dp"
        android:text = "装载" />
    <Button
        android:id = "@ + id/PauseButton"
        android:layout_width = "93dp"
        android:layout_height = "wrap_content"
        android:layout_x = "215dp"
        android:layout_y = "208dp"
        android:text = "暂停" />
</AbsoluteLayout>
```

(3) 打开 src\fs.videoview_test 包下的 MainActivity.java 文件，在文件中利用 VideoView 及 MediaController 控件实现播放视频。代码为：

```java
package fs.videoview_test;
import android.app.Activity;
import android.os.Bundle;
import android.view.View;
import android.view.View.OnClickListener;
import android.widget.Button;
import android.widget.MediaController;
import android.widget.VideoView;
public class MainActivity extends Activity
{
    /** 第一次调用 Activity 活动 */
    public void onCreate(Bundle savedInstanceState)
    {
        super.onCreate(savedInstanceState);
        setContentView(R.layout.main);
        final VideoView videoView = (VideoView) findViewById(R.id.VideoView01);
        Button PauseButton = (Button) this.findViewById(R.id.PauseButton);
        Button LoadButton = (Button) this.findViewById(R.id.LoadButton);
        Button PlayButton = (Button) this.findViewById(R.id.PlayButton);
        //载入
        LoadButton.setOnClickListener(new OnClickListener() {
            public void onClick(View arg0)
            {
                videoView.setVideoPath("/sdcard/bady.mp4");
                videoView.setMediaController(new MediaController(MainActivity.this));
                videoView.requestFocus();
            }
        });
        //播放
        PlayButton.setOnClickListener(new OnClickListener() {
            public void onClick(View arg0)
            {
                videoView.start();
            }
        });
        //暂停
        PauseButton.setOnClickListener(new OnClickListener() {
            public void onClick1(View arg0)
            {
                videoView.pause();
            }
            @Override
            public void onClick(View arg0) {
                //TODO 自动生成的方法存根
            }
        });
    }
}
```

运行程序,利用 VideoView 控件实现播放视频的效果如图 7-6 所示。

图 7-6 VideoView 播放视频

7.5 SurfaceView 播放视频实例

使用 VideoView 播放视频简单、方便,但有些早期的开发者还是更喜欢使用 MediaPlayer 来播放视频,但由于 MediaPlayer 主要用于播放音频,因此它没有提供图像输出界面,此时就要借助于 SurfaceView 来显示 MediaPlayer 播放的图像输出。

本实例利用 MediaPlayer 和 SurfaceView 播放视频。本实例的具体实现步骤如下。

(1) 在 Eclipse 中创建一个 Android 应用项目,命名为 Media_Surface_test。

(2) 打开 res\layout 目录下的 main.xml 布局文件,在文件中声明一个 SurfaceView 控件及 3 个 ImageButton 控件。代码为:

```xml
<?xml version = "1.0" encoding = "utf-8"?>
<LinearLayout
    xmlns:android = "http://schemas.android.com/apk/res/android"
    android:orientation = "vertical"
    android:layout_width = "fill_parent"
    android:layout_height = "fill_parent"
    android:background = "#0a0">
<SurfaceView
    android:id = "@ + id/surfaceView"
    android:layout_width = "fill_parent"
    android:layout_height = "360px"/>
<LinearLayout
    android:orientation = "horizontal"
    android:layout_width = "fill_parent"
    android:layout_height = "wrap_content"
    android:gravity = "center_horizontal">
<ImageButton
    android:id = "@ + id/play"
    android:layout_width = "wrap_content"
    android:layout_height = "wrap_content"
```

```xml
            android:src = "@drawable/b12"/>
    <ImageButton
            android:id = "@+id/pause"
            android:layout_width = "wrap_content"
            android:layout_height = "wrap_content"
            android:src = "@drawable/b13"/>
    <ImageButton
            android:id = "@+id/stop"
            android:layout_width = "wrap_content"
            android:layout_height = "wrap_content"
            android:src = "@drawable/b11"/>
</LinearLayout>
</LinearLayout>
```

（3）打开 src\fs.media_surface_test 包下的 MainActivity.java 文件，在文件中利用 MediaPalyer 及 SurfaceView 实现视频播放。代码为：

```java
package fs.media_surface_test;
import java.io.IOException;
import android.app.Activity;
import android.media.AudioManager;
import android.media.MediaPlayer;
import android.os.Bundle;
import android.view.SurfaceHolder;
import android.view.SurfaceView;
import android.view.View;
import android.view.View.OnClickListener;
import android.widget.ImageButton;
public class MainActivity extends Activity implements OnClickListener
{
    //定义控件
    SurfaceView surfaceView;
    ImageButton play, pause, stop;
    MediaPlayer mPlayer;
    //记录当前视频的播放位置
    int position;
    @Override
    public void onCreate(Bundle savedInstanceState)
    {
        super.onCreate(savedInstanceState);
        setContentView(R.layout.main);
        //获取界面中的3个按钮
        play = (ImageButton) findViewById(R.id.play);
        pause = (ImageButton) findViewById(R.id.pause);
        stop = (ImageButton) findViewById(R.id.stop);
        //为3个按钮的单击事件绑定事件监听器
        play.setOnClickListener(this);
        pause.setOnClickListener(this);
        stop.setOnClickListener(this);
        //创建 MediaPlayer
        mPlayer = new MediaPlayer();
        surfaceView = (SurfaceView) this.findViewById(R.id.surfaceView);
        //设置 SurfaceView 自己不管理的缓冲区
```

```java
            surfaceView.getHolder().setType(
                SurfaceHolder.SURFACE_TYPE_PUSH_BUFFERS);
        //设置播放时打开屏幕
        surfaceView.getHolder().setKeepScreenOn(true);
        surfaceView.getHolder().addCallback(new SurfaceListener());
    }
    @Override
    public void onClick(View source)
    {
        try
        {
            switch (source.getId())
            {
                //播放按钮被单击
                case R.id.play:
                    play();
                    break;
                //暂停按钮被单击
                case R.id.pause:
                    if (mPlayer.isPlaying())
                    {
                        mPlayer.pause();
                    }
                    else
                    {
                        mPlayer.start();
                    }
                    break;
                //停止按钮被单击
                case R.id.stop:
                    if (mPlayer.isPlaying())
                        mPlayer.stop();
                    break;
            }
        }
        catch (Exception e)
        {
            e.printStackTrace();
        }
    }
    private void play() throws IOException
    {
        mPlayer.reset();
        mPlayer.setAudioStreamType(AudioManager.STREAM_MUSIC);
        //设置需要播放的视频
        mPlayer.setDataSource("/sdcard/sibling.mp4");
        //把视频画面输出到 SurfaceView
        mPlayer.setDisplay(surfaceView.getHolder());
        mPlayer.prepare();
        mPlayer.start();
    }
    private class SurfaceListener implements SurfaceHolder.Callback
```

```java
    {
        @Override
        public void surfaceChanged(SurfaceHolder holder, int format, int width, int height)
        {
        }
        @Override
        public void surfaceCreated(SurfaceHolder holder)
        {
            if (position > 0)
            {
                try
                {
                    //开始播放
                    play();
                    //直接从指定位置开始播放
                    mPlayer.seekTo(position);
                    position = 0;
                }
                catch (Exception e)
                {
                    e.printStackTrace();
                }
            }
        }
        @Override
        public void surfaceDestroyed(SurfaceHolder holder)
        {
        }
    }
    //当其他Activity被打开,暂停播放
    @Override
    protected void onPause()
    {
        if (mPlayer.isPlaying())
        {
            //保存当前的播放位置
            position = mPlayer.getCurrentPosition();
            mPlayer.stop();
        }
        super.onPause();
    }
    @Override
    protected void onDestroy()
    {
        //停止播放
        if (mPlayer.isPlaying())
            mPlayer.stop();
        //释放资源
        mPlayer.release();
        super.onDestroy();
    }
}
```

运行程序,效果如图 7-7 所示。

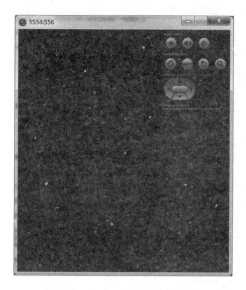

图 7-7　SurfaceView 播放视频

7.6　摄像头实例

拍照是现在智能手机普通存在的功能,现在手机上摄像头的像素越来越高,拍照的清晰度也越来越高。在 Android 中提供了专门用于处理相机相关事件的类,它就是 android.hardware 包中的 Camera 类,可以通过其提供的 open()方法打开相机。打开相机后,可以通过 Camera.Parameters 类处理相机的拍照参数,调用 startPreview()方法预览拍照画面,也可以调用 takePicture()方法进行拍照。而通过调用 stopPreview()方法结束预览,调用 Camera 类的 release()方法释放相机资源。

本实例利用 Camera 类实现拍照、预览和保存相片功能。其具体实现步骤如下。

(1) 在 Eclipse 中新建一个 Android 应用项目,命名为 Camera_test。

(2) 打开 res\layout 目录下的 main.xml 布局文件,在文件中声明一个 SurfaceView 控件及两个 Button 控件。代码为:

```xml
<?xml version = "1.0" encoding = "utf-8"?>
<LinearLayout xmlns:android = "http://schemas.android.com/apk/res/android"
    android:layout_width = "fill_parent"
    android:layout_height = "fill_parent"
    android:orientation = "horizontal"
    android:background = "#0a0" >
    <LinearLayout
        android:id = "@+id/linearLayout1"
        android:layout_width = "72dp"
        android:layout_height = "match_parent"
        android:orientation = "vertical" >
        <Button
            android:id = "@+id/preview"
            android:layout_width = "wrap_content"
```

```
                android:layout_height = "wrap_content"
                android:text = "预览" />
            <Button
                android:id = "@ + id/takephoto"
                android:layout_width = "wrap_content"
                android:layout_height = "wrap_content"
                android:text = "拍照" />
        </LinearLayout>
        <SurfaceView
            android:id = "@ + id/surfaceView1"
            android:layout_width = "match_parent"
            android:layout_height = "match_parent" />
</LinearLayout>
```

（3）在 res\layout 目录下新建一个 save.xml 布局文件，用于实现保存界面，在文件中声明一个 ImageView 控件、一个 TextView 控件及一个 EditText 控件。代码为：

```
<?xml version = "1.0" encoding = "utf - 8"?>
<LinearLayout xmlns:android = "http://schemas.android.com/apk/res/android"
    android:orientation = "vertical"
    android:layout_width = "fill_parent"
    android:layout_height = "fill_parent"
    android:background = "#00a">
<LinearLayout
    android:orientation = "horizontal"
    android:layout_width = "fill_parent"
    android:layout_height = "wrap_content">
<TextView
    android:layout_width = "wrap_content"
    android:layout_height = "wrap_content"
    android:layout_marginRight = "8dp"
    android:text = "相片名称：" />
<EditText
    android:id = "@ + id/phone_name"
    android:layout_width = "fill_parent"
    android:layout_height = "wrap_content"/>
</LinearLayout>
<ImageView
    android:id = "@ + id/show"
    android:contentDescription = "用于显示相片预览"
    android:layout_width = "320dp"
    android:layout_height = "240dp"
    android:scaleType = "fitCenter"
    android:layout_marginTop = "10dp"/>
</LinearLayout>
```

（4）打开 src\fs.carema_test 包下的 MainActivity.java 文件，在文件中实现拍照、预览及保存相片功能。代码为：

```
package fs.camera_test;
import java.io.File;
import java.io.FileOutputStream;
import java.io.IOException;
import android.app.Activity;
import android.app.AlertDialog;
```

```java
import android.content.DialogInterface;
import android.graphics.Bitmap;
import android.graphics.BitmapFactory;
import android.graphics.PixelFormat;
import android.hardware.Camera;
import android.hardware.Camera.PictureCallback;
import android.os.Bundle;
import android.view.SurfaceHolder;
import android.view.SurfaceView;
import android.view.View;
import android.view.Window;
import android.widget.Button;
import android.widget.EditText;
import android.widget.ImageView;
import android.widget.Toast;
public class MainActivity extends Activity {
    //定义控件
    private Camera camera;                          //相机对象
    private boolean isPreview = false;              //是否为预览模式
    //第一次调用 Activity 活动
    @Override
    public void onCreate(Bundle savedInstanceState) {
        super.onCreate(savedInstanceState);
        requestWindowFeature(Window.FEATURE_NO_TITLE);                    //设置全屏显示
        setContentView(R.layout.main);
        //判断是否安装 SD 卡
        if (!android.os.Environment.getExternalStorageState().equals(
                android.os.Environment.MEDIA_MOUNTED)) {
            //弹出消息提示框显示提示信息
            Toast.makeText(this, "请安装 SD 卡!", Toast.LENGTH_SHORT).show();
        }
        //获取 SurfaceView 组件,用于显示相机预览
        SurfaceView sv = (SurfaceView) findViewById(R.id.surfaceView1);
        final SurfaceHolder sh = sv.getHolder();
        //设置该 SurfaceHolder 自己不维护缓冲
        sh.setType(SurfaceHolder.SURFACE_TYPE_PUSH_BUFFERS);
        Button preview = (Button) findViewById(R.id.preview);             //获取"预览"按钮
        preview.setOnClickListener(new View.OnClickListener() {
            @Override
            public void onClick(View v) {
                //如果相机为非预览模式,则打开相机
                if (!isPreview) {
                    camera = Camera.open();       //打开相机
                }
                try {
                    camera.setPreviewDisplay(sh);    //设置用于显示预览的 SurfaceView
                    Camera.Parameters parameters = camera.getParameters();   //获取相机参数
                    parameters.setPictureSize(640, 480);              //设置预览画面的尺寸
                    parameters.setPictureFormat(PixelFormat.JPEG);    //指定图片为 JPEG 图片
                    parameters.set("jpeg-quality", 80);               //设置图片的质量
                    parameters.setPictureSize(640, 480);              //设置拍摄图片的尺寸
                    camera.setParameters(parameters);                 //重新设置相机参数
```

```java
                    camera.startPreview();              //开始预览
                    camera.autoFocus(null);             //设置自动对焦
                } catch (IOException e) {
                    e.printStackTrace();
                }
            }
        });
        Button takePhoto = (Button) findViewById(R.id.takephoto);      //获取"拍照"按钮
        takePhoto.setOnClickListener(new View.OnClickListener() {
            @Override
            public void onClick(View v) {
                if(camera!= null){
                    camera.takePicture(null, null, jpeg);               //进行拍照
                }
            }
        });
    }
    //实现拍照的回调接口
    final PictureCallback jpeg = new PictureCallback() {
        @Override
        public void onPictureTaken(byte[] data, Camera camera) {
            //根据拍照所得的数据创建位图
            final Bitmap bm = BitmapFactory.decodeByteArray(data, 0,
                    data.length);
            //加载 layout/save.xml 文件对应的布局资源
            View saveView = getLayoutInflater().inflate(R.layout.save, null);
            final EditText photoName = (EditText) saveView
                    .findViewById(R.id.phone_name);
            //获取对话框上的 ImageView 组件
            ImageView show = (ImageView) saveView.findViewById(R.id.show);
            show.setImageBitmap(bm);                    //显示刚刚拍好的照片
            camera.stopPreview();                       //停止预览
            isPreview = false;
            //使用对话框显示 saveDialog 组件
            new AlertDialog.Builder(MainActivity.this).setView(saveView)
                .setPositiveButton("保存", new DialogInterface.OnClickListener() {
                    @Override
                    public void onClick(DialogInterface dialog, int which) {
                        File file = new File("/sdcard/pictures/" + photoName
                                .getText().toString() + ".jpg");         //创建文件对象
                        try {
                            file.createNewFile();                        //创建一个新文件
                            //创建一个文件输出流对象
                            FileOutputStream fileOS = new FileOutputStream(file);
                            //将图片内容压缩为 JPEG 格式输出到输出流对象中
                            bm.compress(Bitmap.CompressFormat.JPEG, 100, fileOS);
                            //将缓冲区中的数据全部写出到输出流中
                            fileOS.flush();
                            fileOS.close();                              //关闭文件输出流对象
                            isPreview = true;
                            resetCamera();
                        } catch (IOException e) {
```

```java
                            e.printStackTrace();
                        }
                    }
                }).setNegativeButton("取消", new DialogInterface.OnClickListener() {
                    public void onClick(DialogInterface dialog, int which) {
                        isPreview = true;
                        resetCamera();              //重新预览
                    }
                }).show();
            }
        };
        //重新预览
        private void resetCamera(){
            if(isPreview){
                camera.startPreview();              //开启预览
            }
        }
        //停止预览并释放资源
        @Override
        protected void onPause() {
            if(camera!= null){
                camera.stopPreview();               //停止预览
                camera.release();                   //释放资源
            }
            super.onPause();
        }
}
```

(5) 打开 AndroidManifest.xml 布局文件,在文件中设置权限。代码为:

```xml
…
<uses-sdk
    android:minSdkVersion = "8"
    android:targetSdkVersion = "18" />
<!-- 授予程序可以向 SD 卡中保存文件的权限 -->
<uses-permission android:name = "android.permission.MOUNT_UNMOUNT_FILESYSTEMS"/>
<uses-permission android:name = "android.permission.WRITE_EXTERNAL_STORAGE"/>
<!-- 授予程序使用摄像头的权限 -->
<uses-permission android:name = "android.permission.CAMERA" />
<uses-feature android:name = "android.hardware.camera" />
<uses-feature android:name = "android.hardware.camera.autofocus" />
<application
    android:allowBackup = "true"
    android:icon = "@drawable/ic_launcher"
    android:label = "@string/app_name"
    android:theme = "@style/AppTheme" >
…
```

运行程序,默认界面如图 7-8(a)所示,当单击界面中的"拍照"按钮时,即可实现拍照图片并弹出图片保存对话框,当单击界面中的"预览"按钮时,可以实现图片的预览,效果如图 7-8(b)所示。

（a）默认界面　　　　　　　　　　（b）预览界面

图 7-8　摄像头实例

7.7　录制视频实例

在 Android 提供的 MediaRecorder 除了可用于录制音频外，还可用于录制视频。使用 MediaRecorder 录制视频与录制音频的步骤基本相同。只是录制视频时不仅需要采集声音，还需要采集图像。为了让 MediaRecorder 录制时采集图像，应该在调用 setAudioSource()方法时再调用 setVideoSource()方法来设置图像来源。除此之外，还需要调用 setOutputFormat()设置输出文件格式。

本实例用一个摄像按钮，实现单击该按钮进入视频录制的 Activity，在这个 Activity 右侧的画面随着手机而变换，这就是预览效果，在这个 Activity 的右侧有两个按钮，一个"录制"按钮，一个"停止"按钮，单击"录制"按钮，程序开始录制视频，单击"停止"按钮，程序结束录制，并且弹出提示框，提示视频已经录制完成，是否保存，单击"确定"即可将录制的视频保存到 SD Card 中。

本实例的具体实现步骤如下：

（1）在 Eclipse 中创建一个 Android 应用项目，命名为 Recorder_Video。

（2）打开 res\layout 目录下的 main.xml 主布局文件，在文件中声明一个 TextView 控件及一个 Button 控件。代码为：

```
<?xml version = "1.0" encoding = "utf - 8"?>
<LinearLayout xmlns:android = "http://schemas.android.com/apk/res/android"
    android:layout_width = "fill_parent"
    android:layout_height = "fill_parent"
    android:orientation = "vertical"
    android:background = "#0a0">
    <TextView
```

```xml
        android:layout_width = "fill_parent"
        android:layout_height = "wrap_content"
        android:text = "@string/hello_world" />
    <Button
        android:id = "@+id/camera_button"
        android:layout_width = "wrap_content"
        android:layout_height = "wrap_content"
        android:layout_gravity = "center_horizontal"
        android:text = "摄像" />
</LinearLayout>
```

（3）在 res\layout 目录下新建一个 video.xml 录制布局文件，在文件中声明一个 SurfaceView 控件及两个 Button 控件。代码为：

```xml
<LinearLayout xmlns:android = "http://schemas.android.com/apk/res/android"
        android:id = "@+id/linearLayout1"
        android:layout_width = "fill_parent"
        android:layout_height = "fill_parent"
        android:background = "#b00">
    <SurfaceView
        android:id = "@+id/surface_view"
        android:layout_width = "wrap_content"
        android:layout_height = "fill_parent"
        android:layout_weight = "0.65" />
    <LinearLayout
        android:id = "@+id/linearLayout2"
        android:layout_width = "wrap_content"
        android:layout_height = "match_parent"
        android:orientation = "vertical" >
        <Button
            android:id = "@+id/start"
            android:layout_width = "wrap_content"
            android:layout_height = "wrap_content"
            android:text = "录制" />
        <Button
            android:id = "@+id/stop"
            android:layout_width = "wrap_content"
            android:layout_height = "wrap_content"
            android:text = "停止" />
    </LinearLayout>
</LinearLayout>
```

（4）打开 src\fs.recorder_video 包下的 MainActivity.java 文件，在文件中实现启动 Activity 类、实例化 Button 控件及跳转到主界面的功能。代码为：

```java
package fs.recorder_video;
import android.app.Activity;
import android.content.Intent;
import android.os.Bundle;
import android.view.View;
import android.widget.Button;
public class MainActivity extends Activity {
    /** 第一次调用 Activity 活动 */
    @Override
    public void onCreate(Bundle savedInstanceState) {
        super.onCreate(savedInstanceState);
```

```java
        setContentView(R.layout.main);
        Button button = (Button) findViewById(R.id.camera_button);    //实例化 Button 组件对象
        button.setOnClickListener(new Button.OnClickListener() {
            //为 Button 添加单击监听
            @Override
            public void onClick(View arg0) {
                Intent intent = new Intent();                          //初始化 Intent
                intent.setClass(MainActivity.this, ActivityRecording.class);  //指定 intent 对象启动的类
                startActivity(intent);                                 //启动新的 Activity
            }
        });
    }
}
```

（5）在 src\fs.recorder_video 包下新建一个 ActivityRecording.java 文件，用于实现视频的录制、保存录制的视频及停止视频录制。代码为：

```java
package fs.recorder_video;
import java.io.File;
import java.io.IOException;
import android.app.Activity;
import android.app.AlertDialog;
import android.content.DialogInterface;
import android.content.Intent;
import android.content.pm.ActivityInfo;
import android.graphics.PixelFormat;
import android.hardware.Camera;
import android.media.MediaRecorder;
import android.os.Bundle;
import android.os.Environment;
import android.view.KeyEvent;
import android.view.SurfaceHolder;
import android.view.SurfaceView;
import android.view.View;
import android.view.View.OnClickListener;
import android.view.Window;
import android.view.WindowManager;
import android.widget.Button;
public class ActivityRecording extends Activity implements SurfaceHolder.Callback {
    //声明控件
    private SurfaceView surfaceView = null;              //创建一个空 SurfaceView 对象
    private SurfaceHolder surfaceHolder = null;          //创建一个空 SurfaceHolder 对象
    private Button startButton = null;                   //创建开始录制按钮的 Button 组件对象
    private Button stopButton = null;                    //创建停止录制按钮的 Button 组件对象
    private MediaRecorder mediaRecorder = null;          //创建一个空 MediaRecorder 对象
    private Camera camera = null;                        //创建一个空 Camera 对象
    private boolean previewRunning = false;              //预览状态
    private File videoFile = null;                       //录制视频文件的 File 对象
    /** 第一次调用 Activity 活动 */
    @Override
    public void onCreate(Bundle savedInstanceState) {
        super.onCreate(savedInstanceState);
        getWindow().setFormat(PixelFormat.TRANSLUCENT);  //窗口设置为半透明
```

```java
        requestWindowFeature(Window.FEATURE_NO_TITLE);                    //窗口去掉标题
getWindow().setFlags(WindowManager.LayoutParams.FLAG_FULLSCREEN,WindowManager.LayoutParams.
FLAG_FULLSCREEN);                                  //窗口设置为全屏
        //调用 setRequestedOrientation 来翻转 PreviewsetRequestedOrientation(ActivityInfo.
        //SCREEN_ORIENTATION_LANDSCAPE);
        setContentView(R.layout.video);
surfaceView = (SurfaceView) findViewById(R.id.surface_view);              //实例化 SurfaceView
        surfaceHolder = surfaceView.getHolder();    //获取 SurfaceHolder
        surfaceHolder.addCallback(this);              //注册实现好的 Callback
        //设置缓存类型
        surfaceHolder.setType(SurfaceHolder.SURFACE_TYPE_PUSH_BUFFERS);
        //实例化开始录制按钮的 Button 组件对象
        startButton = (Button) findViewById(R.id.start);
        //实例化停止录制按钮的 Button 组件对象
        stopButton = (Button) findViewById(R.id.stop);
        startButton.setEnabled(true);               //摄像按钮生效
        stopButton.setEnabled(false);               //停止按钮失效
        //添加摄像按钮单击事件监听
        startButton.setOnClickListener(new OnClickListener() {
            @Override
            public void onClick(View v) {
                startRecording();              //调用开始摄像方法
            }
        });
        //添加停止按钮单击事件监听
        stopButton.setOnClickListener(new OnClickListener() {
            @Override
            public void onClick(View v) {
                stopRecording();              //调用停止摄像方法
            }
        });
    }
    //开始摄像方法
    public void startRecording() {
        try {
            stopCamera();                      //调用停止 Camera 方法
            if (!getStorageState()) {          //判断是否有存储卡,如果没有就关闭页面
                ActivityRecording.this.finish();  //结束应用程序
            }
            //获取存储卡(SD Card)的根目录
            String sdCard = Environment.getExternalStorageDirectory().getPath();
            //获取相片存放位置的目录
            String dirFilePath = sdCard + File.separator + "AVideo";
            File dirFile = new File(dirFilePath);//获取录制文件夹的路径的 File 对象
            if (!dirFile.exists()) {           //判断文件夹是否存在
                dirFile.mkdir();               //创建文件夹
            }
 videoFile = File.createTempFile("video", ".3gp", dirFile);     //创建录制视频临时文件
            mediaRecorder = new MediaRecorder();              //初始化 MediaRecorder 对象
            mediaRecorder.setPreviewDisplay(surfaceHolder.getSurface());   //预览
    mediaRecorder.setVideoSource(MediaRecorder.VideoSource.DEFAULT);       //视频源
mediaRecorder.setAudioSource(MediaRecorder.AudioSource.MIC);               //录音源为麦克风
```

```java
            //输出格式为3gp
            mediaRecorder.setOutputFormat(MediaRecorder.OutputFormat.THREE_GPP);
            mediaRecorder.setVideoSize(480, 320);  //视频尺寸
            mediaRecorder.setVideoFrameRate(15);   //视频帧频率
mediaRecorder.setVideoEncoder(MediaRecorder.VideoEncoder.H263);//视频编码
mediaRecorder.setAudioEncoder(MediaRecorder.AudioEncoder.AMR_NB);        //音频编码
            mediaRecorder.setMaxDuration(10000);   //最大期限
mediaRecorder.setOutputFile(videoFile.getAbsolutePath());                //保存路径
            mediaRecorder.prepare();               //准备录制
            mediaRecorder.start();                 //开始录制
            //文件录制错误监听
            mediaRecorder
                    .setOnErrorListener(new MediaRecorder.OnErrorListener() {
                        @Override
                        public void onError(MediaRecorder arg0, int arg1,
                                int arg2) {
                            stopRecording();       //调用停止摄像方法
                        }
                    });
            startButton.setText("录制中");
            startButton.setEnabled(false);         //摄像按钮失效
            stopButton.setEnabled(true);           //停止按钮生效
        } catch (IOException e) {
            e.printStackTrace();
        }
    }
    //停止摄像方法
    public void stopRecording() {
        if (mediaRecorder != null) {               //判断 MediaRecorder 对象是否为空
            mediaRecorder.stop();                  //停止摄像
            mediaRecorder.release();               //释放资源
            mediaRecorder = null;                  //置空 MediaRecorder 对象
            startButton.setEnabled(true);          //摄像按钮生效
            stopButton.setEnabled(false);          //停止按钮失效
            startButton.setText("录制");
            isSave();                              //调用是否保存方法保存
        }
        stopCamera();                              //调用停止 Camera 方法
        prepareCamera();                           //调用初始化 Camera 方法
        startCamera();                             //调用开始 Camera 方法
    }
    //初始化摄像
    public void prepareCamera() {
        camera = Camera.open();                    //初始化 Camera
        try {
            camera.setPreviewDisplay(surfaceHolder);                       //设置预览
        } catch (IOException e) {
            camera.release();                      //释放相机资源
            camera = null;                         //置空 Camera 对象
        }
    }
    //开始摄像
```

```java
        public void startCamera() {
            if (previewRunning) {                          //判断预览开启
                camera.stopPreview();                      //停止预览
            }
            try {
                //设置用 SurfaceView 作为承载镜头取景画面的显示
                camera.setPreviewDisplay(surfaceHolder);
                camera.startPreview();                     //开始预览
                previewRunning = true;                     //设置预览状态为 true
            } catch (IOException e) {
                e.printStackTrace();
            }
        }
        //停止摄像
        public void stopCamera() {
            if (camera != null) {                          //判断 Camera 对象不为空
                camera.stopPreview();                      //停止预览
                camera.release();                          //释放摄像头资源
                camera = null;                             //置空 Camera 对象
                previewRunning = false;                    //设置预览状态为 false
            }
        }
        //手机按键监听事件
        @Override
        public boolean onKeyDown(int keyCode, KeyEvent event) {
            //判断手机键盘按下的是否是返回键
            if (keyCode == KeyEvent.KEYCODE_BACK) {
                stopRecording();                           //调用停止摄像方法
                Intent intent = new Intent();              //初始化 Intent
    intent.setClass(ActivityRecording.this, MainActivity.class);    //指定 intent 对象启动的类
                //清除该进程空间的所有 Activity
                intent.setFlags(Intent.FLAG_ACTIVITY_CLEAR_TOP);
            startActivity(intent);                         //启动新的 Activity
                ActivityRecording.this.finish();           //销毁这个 Activity
            }
            return super.onKeyDown(keyCode, event);
        }
        //是否保存录制的视频文件
        public void isSave() {
AlertDialog alertDialog = new AlertDialog.Builder(this).create();       //创建 AlertDialog 对象
            alertDialog.setTitle("提示信息");              //设置信息标题
            //设置信息内容
            alertDialog.setMessage("是否保存 " + videoFile.getName() + " 视频文件?");
            //设置确定按钮,并添加按钮监听事件
            alertDialog.setButton("确定",
                    new android.content.DialogInterface.OnClickListener() {
                        @Override
                        public void onClick(DialogInterface arg0, int arg1) {
                        }
                    });
            //设置取消按钮,并添加按钮监听事件
            alertDialog.setButton2("取消",
```

```java
                        new android.content.DialogInterface.OnClickListener() {
                            @Override
                            public void onClick(DialogInterface arg0, int arg1) {
                                if (videoFile.exists()) {       //判断文件是否存在
                                    videoFile.delete();         //删除该文件
                                }
                            }
                        });
                alertDialog.show();                              //设置弹出提示框
    }
    /**
     * 获取手机 SD Card 的存储状态
     * @return 手机 SD Card 的存储状态(true/false)
     */
    public boolean getStorageState() {
        if (Environment.getExternalStorageState().equals(
                Environment.MEDIA_MOUNTED)) {        //判断手机 SD Card 的存储状态
            return true;
        } else {
            AlertDialog alertDialog = new AlertDialog.Builder(this).create();   //创建 AlertDialog 对象
            alertDialog.setTitle("提示信息");                    //设置信息标题
            alertDialog.setMessage("未安装 SD 卡,请检查你的设备");            //设置信息内容
            //设置确定按钮,并添加按钮监听事件
            alertDialog.setButton("确定",
                    new android.content.DialogInterface.OnClickListener() {
                        @Override
                        public void onClick(DialogInterface arg0, int arg1) {
                            ActivityRecording.this.finish();             //结束应用程序
                        }
                    });
            alertDialog.show();                          //设置弹出提示框
            return false;
        }
    }
    //当预览界面的格式和大小发生改变时,该方法被调用
    @Override
    public void surfaceChanged(SurfaceHolder arg0, int arg1, int arg2, int arg3) {
        startCamera();                                   //调用开始 Camera 方法
    }
    //初次实例化,预览界面被创建时,该方法被调用
    @Override
    public void surfaceCreated(SurfaceHolder arg0) {
        prepareCamera();                                 //调用初始化 Camera 方法
    }
    //当预览界面被关闭时,该方法被调用
    @Override
    public void surfaceDestroyed(SurfaceHolder arg0) {
        stopCamera();                                    //调用停止 Camera 方法
    }
}
```

运行程序,默认界面如图 7-9(a)所示,单击界面中的"摄像"按钮,跳转到"录制"预览界面,

效果如图 7-90(b)所示。

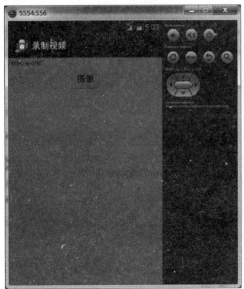

(a) 默认界面　　　　　　　　　　　　(b) 录制界面

图 7-9　录制视频

7.8　SensorManager 开发传感器实例

　　Android 系统提供了对传感器的支持,如果手机设备的硬件提供了这些传感器,Android 应用可以通过传感器来获取设备的外界条件,包括手机设备的运行状态、当前摆放方向、外界的磁场、温度和压力等。Android 系统提供了驱动程序去管理这些传感器硬件,当传感器硬件感知到外部环境发生改变时,Android 系统负责管理这些传感器数据。

　　在 Android 平台上开发传感器应用十分简单,下面通过一个简单的加速传感器来介绍传感器应用的开发。本实例的具体实现步骤如下。

　　(1) 在 Eclipse 中创建一个 Android 应用项目,命名为 Develop_Sendsor。

　　(2) 打开 res\layout 目录下的 main.xml 布局文件,在文件中声明一个 TextView 控件及一个 EditText 控件。代码为:

```
<?xml version = "1.0" encoding = "utf - 8"?>
<LinearLayout
    xmlns:android = "http://schemas.android.com/apk/res/android"
    android:orientation = "vertical"
    android:layout_width = "fill_parent"
    android:layout_height = "fill_parent"
    android:background = "#a00">
<TextView
    android:layout_width = "fill_parent"
    android:layout_height = "wrap_content"
    android:text = "返回的值" />
<EditText
    android:id = "@ + id/txt1"
```

```xml
        android:layout_width = "fill_parent"
        android:layout_height = "wrap_content"
        android:editable = "false"
        android:cursorVisible = "false" />
</LinearLayout>
```

(3) 打开 src\fs.develop_sendsor 包下的 MainActivity.java 文件，在文件中开发一个加速度传感器，在界面中提供一个文本框显示加速度值。代码为：

```java
package fs.develop_sendsor;
import android.app.Activity;
import android.content.Context;
import android.hardware.Sensor;
import android.hardware.SensorEvent;
import android.hardware.SensorEventListener;
import android.hardware.SensorManager;
import android.os.Bundle;
import android.widget.EditText;
public class MainActivity extends Activity implements SensorEventListener
{
    //定义系统的Sensor管理器
    SensorManager sensorManager;
    EditText etTxt1;
    @Override
    public void onCreate(Bundle savedInstanceState)
    {
        super.onCreate(savedInstanceState);
        setContentView(R.layout.main);
        //获取程序界面上的文本框组件
        etTxt1 = (EditText) findViewById(R.id.txt1);
        //获取系统的传感器管理服务
        sensorManager = (SensorManager) getSystemService(
            Context.SENSOR_SERVICE);
    }
    @Override
    protected void onResume()
    {
        super.onResume();
        //为系统的加速度传感器注册监听器
        sensorManager.registerListener(this,
            sensorManager.getDefaultSensor(Sensor.TYPE_ACCELEROMETER),
            SensorManager.SENSOR_DELAY_GAME);
    }
    @Override
    protected void onStop()
    {
        //取消注册
        sensorManager.unregisterListener(this);
        super.onStop();
    }
    //以下是实现SensorEventListener接口必须实现的方法
    //当传感器的值发生改变时回调该方法
    @Override
    public void onSensorChanged(SensorEvent event)
    {
        float[] values = event.values;
```

```
            StringBuilder sb = new StringBuilder();
            sb.append("X方向上的加速度: ");
            sb.append(values[0]);
            sb.append("\nY方向上的加速度: ");
            sb.append(values[1]);
            sb.append("\nZ方向上的加速度: ");
            sb.append(values[2]);
            etTxt1.setText(sb.toString());
        }
        //当传感器精度改变时回调该方法.
        @Override
        public void onAccuracyChanged(Sensor sensor, int accuracy)
        {
        }
}
```

运行程序,效果如图 7-10 所示,当拿着手机移动时,即加速度返回的值会出现不同(真机模拟器)。

图 7-10 加速度传感器

7.9 Android 定位实例

Android 支持 GPS 和网络地图,通常将各种不同的定位技术称为 LBS。LBS 是基于位置的服务(Location Based Service)的简称,它是通过电信移动运营商的无线电通信网络(如 GSM 网和 CDMA 网)或外部定位方式(如 GPS)获取移动终端用户的位置信息(地理坐标或大地坐标),在地理信息系统(Geographic Information System,GIS)平台的支持下,为用户提供相应服务的一种增值业务。

在 Android 中提供了相关函数用于获取 GPS 位置信息,包括精度、方位、经纬度、海拔、速度、高度和运营商收费等信息。

本实例通过手机实时获取定位信息,包括用户所在的经度、纬度、高度、方向和移动速度等。其具体实现步骤如下。

（1）在 Eclipse 中创建一个 Android 应用项目，命名为 GPS_Data。

（2）打开 res\layout 目录下的 main.xml 文件，在文件中定义一个 EditText 控件。代码为：

```xml
<?xml version = "1.0" encoding = "utf - 8"?>
<LinearLayout xmlns:android = "http://schemas.android.com/apk/res/android"
    android:orientation = "vertical"
    android:layout_width = "fill_parent"
    android:layout_height = "fill_parent"
    android:background = "#aabbcc">
<EditText
    android:id = "@ + id/show"
    android:layout_width = "fill_parent"
    android:layout_height = "wrap_content"
    android:editable = "false"
    android:cursorVisible = "false"/>
</LinearLayout>
```

（3）打开 src\fs.gps_data 包下的 MainActivity.java 文件，在文件中实现手机实时获取定位信息。代码为：

```java
package fs.gps_data;
import android.app.Activity;
import android.content.Context;
import android.location.Location;
import android.location.LocationListener;
import android.location.LocationManager;
import android.os.Bundle;
import android.widget.EditText;
public class MainActivity extends Activity
{
    //定义 LocationManager 对象
    LocationManager locManager;
    //定义程序界面中的 EditText 组件
    EditText show;
    @Override
    public void onCreate(Bundle savedInstanceState)
    {
        super.onCreate(savedInstanceState);
        setContentView(R.layout.main);
        //获取程序界面上的 EditText 组件
        show = (EditText) findViewById(R.id.show);
        //创建 LocationManager 对象
        locManager = (LocationManager) getSystemService(Context.LOCATION_SERVICE);
        //从 GPS 获取最近的定位信息
        Location location = locManager.getLastKnownLocation(
            LocationManager.GPS_PROVIDER);
        //使用 location 对象更新 EditText 中的内容
        updateView(location);
        //设置每 3 秒获取一次 GPS 的定位信息
        locManager.requestLocationUpdates(LocationManager.GPS_PROVIDER
            , 3000, 8, new LocationListener()
        {
```

```java
            @Override
            public void onLocationChanged(Location location)
            {
                //当 GPS 定位信息发生改变时,更新位置
                updateView(location);
            }
            @Override
            public void onProviderDisabled(String provider)
            {
                updateView(null);
            }
            @Override
            public void onProviderEnabled(String provider)
            {
                //当 GPS LocationProvider 可用时,更新位置
                updateView(locManager
                    .getLastKnownLocation(provider));
            }
            @Override
            public void onStatusChanged(String provider, int status,
                Bundle extras)
            {
            }
        });
    }
    //更新 EditText 中显示的内容
    public void updateView(Location newLocation)
    {
        if (newLocation != null)
        {
            StringBuilder sb = new StringBuilder();
            sb.append("实时的位置信息: \n");
            sb.append("经度: ");
            sb.append(newLocation.getLongitude());
            sb.append("\n 纬度: ");
            sb.append(newLocation.getLatitude());
            sb.append("\n 高度: ");
            sb.append(newLocation.getAltitude());
            sb.append("\n 速度: ");
            sb.append(newLocation.getSpeed());
            sb.append("\n 方向: ");
            sb.append(newLocation.getBearing());
            show.setText(sb.toString());
        }
        else
        {
            //如果传入的 Location 对象为空则清空 EditText
            show.setText("");
        }
    }
}
```

(4) 为程序添加 GPS 信号的访问权限,打开 AndroidManifest.xml 文件,添加权限代码为:

```
            </application>
            <!-- 授权获取定位信息 -->
            <uses-permission android:name="android.permission.ACCESS_FINE_LOCATION" />
</manifest>
```

运行程序，打开 DDMS 的 Emulator Control 面板，如图 7-11 所示，填写相关数据，并单击界面中的 Send 按钮，即 Android 中接收到定位信息，效果如图 7-12 所示。

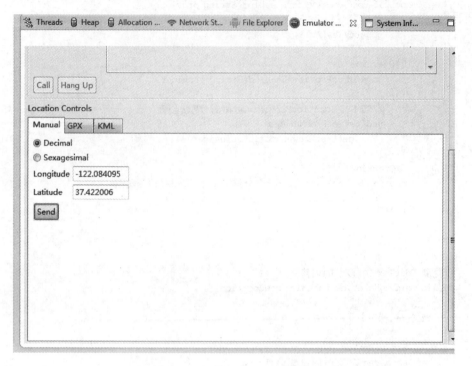

图 7-11　Emulator Control 面板

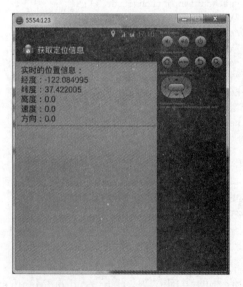

图 7-12　实时的定位信息

7.10 城市定位实例

城市的定位相信大家都比较熟悉,对于城市定位主要实现查询城市的位置和天气情况等内容。对于天气预报,主要实现对当前天气的详细描述和未来几天天气情况的描述以及更加直观的天气变化趋势图和需要查询天气的地址。

本节将实现一个城市天气预报的查询管理,实现一个对城市的当前天气的详细描述以及未来温度变化趋势图的直观呈现,并实现对城市管理进行查询功能。

本实例的具体实现步骤如下。

(1) 在 Eclipse 中创建一个 Android 应用项目,命名为 City_Manager。

(2) 打开 res\layout 目录下的 main.xml 主布局文件,在文件中主要布局城市天气预报及查询的主页面。代码为:

```xml
<?xml version = "1.0" encoding = "UTF-8"?>
<RelativeLayout xmlns:android = "http://schemas.android.com/apk/res/android"
    android:layout_width = "fill_parent"
    android:layout_height = "fill_parent"
    android:background = "#a00">
    <LinearLayout
        android:id = "@+id/weather_condition"
        android:layout_width = "fill_parent"
        android:layout_height = "wrap_content"
        android:layout_alignParentTop = "true"
        android:layout_marginLeft = "10.0dip"
        android:layout_marginRight = "10.0dip"
        android:layout_marginTop = "10.0dip"
        android:orientation = "horizontal" >
        <LinearLayout
            android:layout_width = "fill_parent"
            android:layout_height = "wrap_content"
            android:layout_weight = "1.0"
            android:gravity = "left"
            android:orientation = "vertical" >
            <LinearLayout
                android:layout_width = "wrap_content"
                android:layout_height = "wrap_content"
                android:gravity = "center_vertical"
                android:orientation = "horizontal" >
                <fs.city_manager.SelfPView
                    android:id = "@+id/progress"
                    android:layout_width = "wrap_content"
                    android:layout_height = "wrap_content" />
                <TextView
                    android:id = "@+id/city_name"
                    android:layout_width = "wrap_content"
                    android:layout_height = "wrap_content"
                    android:layout_marginLeft = "5.0dip"
                    android:ellipsize = "marquee"
                    android:focusable = "true"
                    android:marqueeRepeatLimit = "marquee_forever"
                    android:singleLine = "true"
```

```
                    android:text = "城市"
                    android:textSize = "30.0sp" />
            </LinearLayout>
            ...
        </FrameLayout>
</RelativeLayout>
```

(3) 在 res\layout 目录下新建一个 city.xml 布局文件,主要用于实现城市定位及确定查询城市页面。代码为:

```
<?xml version = "1.0" encoding = "UTF - 8"?>
<RelativeLayout xmlns:android = "http://schemas.android.com/apk/res/android"
    android:layout_width = "fill_parent"
    android:layout_height = "fill_parent"
    android:background = "#0b0" >
    <AutoCompleteTextView
        android:id = "@ + id/ed_com"
        android:layout_width = "match_parent"
        android:layout_height = "wrap_content"
        android:layout_alignParentLeft = "true"
        android:layout_below = "@ + id/imageView1"
        android:ems = "10"
        android:hint = "搜索城市名"/>
    <ImageView
        android:id = "@ + id/imageView1"
        android:layout_width = "wrap_content"
        android:layout_height = "wrap_content"
        android:layout_alignParentLeft = "true"
        android:layout_alignParentTop = "true"
        android:background = "@drawable/city" />
    <TextView
        android:id = "@ + id/textView1"
        android:layout_width = "wrap_content"
        android:layout_height = "wrap_content"
        android:layout_alignParentTop = "true"
        android:layout_alignRight = "@ + id/button1"
        android:text = "当前城市是: " />
    <Button
        android:id = "@ + id/button2"
        android:layout_width = "wrap_content"
        android:layout_height = "wrap_content"
        android:layout_alignBaseline = "@ + id/button1"
        android:layout_alignBottom = "@ + id/button1"
        android:layout_alignParentRight = "true"
        android:layout_marginRight = "29dp"
        android:text = "确定城市" />
    <Button
        android:id = "@ + id/button1"
        android:layout_width = "wrap_content"
        android:layout_height = "wrap_content"
        android:layout_below = "@ + id/ed_com"
        android:layout_marginTop = "38dp"
        android:layout_toRightOf = "@ + id/imageView1"
        android:text = "自动定位" />
</RelativeLayout>
```

(4) 在 res\layout 目录下新建一个 trend.xml 布局文件，主要用于实现温度变化趋势图界面。代码为：

```xml
<?xml version = "1.0" encoding = "UTF-8"?>
<LinearLayout xmlns:android = "http://schemas.android.com/apk/res/android"
    android:id = "@ + id/trendxml"
    android:layout_width = "fill_parent"
    android:layout_height = "fill_parent"
    android:background = "#d01"
    android:orientation = "vertical" >
    <RelativeLayout
        android:layout_width = "fill_parent"
        android:layout_height = "45.0dip"
        android:background = "@drawable/title_bg" >
        <TextView
            android:id = "@ + id/cityname"
            android:layout_width = "wrap_content"
            android:layout_height = "fill_parent"
            android:layout_centerHorizontal = "true"
            android:layout_centerVertical = "true"
            android:background = "@drawable/button_press"
            android:gravity = "center"
            android:text = "温度趋势图"
            android:textColor = "@color/white"
            android:textSize = "20.0sp"/>
    </RelativeLayout>
    <LinearLayout
        android:id = "@ + id/trend_top"
        android:layout_width = "fill_parent"
        android:layout_height = "wrap_content"
        android:layout_weight = "-10.0"
        android:orientation = "vertical">
        <LinearLayout
            android:layout_width = "fill_parent"
            android:layout_height = "wrap_content"
            android:layout_marginTop = "5.0dip"
            android:paddingLeft = "8.0dip"
            android:paddingRight = "8.0dip" >
            <LinearLayout
                android:layout_width = "fill_parent"
                android:layout_height = "fill_parent"
                android:layout_weight = "1.0"
                android:orientation = "horizontal"
                android:gravity = "center">
                ……
            </LinearLayout>
        </LinearLayout>
    </LinearLayout>
</LinearLayout>
```

(5) 在 res\values 目录下新建一个 color.xml 文件，用于实现颜色资源设置。代码为：

```xml
<?xml version = "1.0" encoding = "UTF-8"?>
<resources>
    <color name = "gray">#ffdddddd</color>
    <color name = "light_gray">#ff555555</color>
```

```xml
<color name = "white">#ffffffff</color>
<color name = "black">#ff000000</color>
<color name = "green">#ff669966</color>
<color name = "red">#ffcc6666</color>
<color name = "blue">#ff336699</color>
<color name = "brown">#ffcc9966</color>
<color name = "purple">#ff996699</color>
<color name = "orange">#ffff6633</color>
<color name = "none">#00000000</color>
<color name = "alpha_white">#88ffffff</color>
<color name = "alpha_gray">#55555555</color>
<color name = "alpha_black">#33000000</color>
<color name = "alpha_blue">#990066cc</color>
<color name = "alpha_gold">#88ffcc33</color>
<color name = "android_black">#55000000</color>
<color name = "android_white">#ffffffff</color>
<color name = "transparent">#00000000</color>
<color name = "common_buttonCharColor">#ff000000</color>
<color name = "common_charColor">#ffffffff</color>
</resources>
```

（6）打开 src/fs.city_manager 包下的 MainActivity.java 文件，该文件用于实现显示界面的内容，并获取查询城市、获取天气信息并对数据文件进行初始化。代码为：

```java
public class MainActivity extends Activity {
    /** 第一次调用 Activity 活动 */
    private CityWeatherInfo cityWeatherInfo = null;
    GetWeatherInfo weatherInfo = null;
    private Button btn_trend;
    private TextView tv_cityname, tv_synchtime, tv_date, tv_week, tv_condition,
            tv_nowtemp, tv_temp, tv_windconditionTextView;
    private TextView tv_next_1_day_date, tv_next_1_day_temperature,
            next_1_weather;
    …
    Handler updateHandler = new Handler() {
        public void handleMessage(Message paramMessage) {
            super.handleMessage(paramMessage);
            if (paramMessage.what == 1) {
                cityWeatherInfo = paramMessage.getData().getParcelable(
                        "weather");
                if (cityWeatherInfo != null) {
                    tv_cityname.setText(cityWeatherInfo.getCity());
                    tv_synchtime.setText(cityWeatherInfo.getnowtime() + "发布");
                    tv_date.setText(cityWeatherInfo.getnowdate());
                    tv_week.setText(cityWeatherInfo.getCurrentDayOfWeek());
                    tv_condition.setText(cityWeatherInfo.getCurrentCondition());
                    tv_nowtemp.setText(String.valueOf(cityWeatherInfo
                            .getnowTemp()) + "°C");
                    tv_temp.setText(cityWeatherInfo.getCurrentLowTemp() + "°C"
                            + "~" + cityWeatherInfo.getCurrentHighTemp() + "°C");
                    tv_windconditionTextView.setText(cityWeatherInfo
                            .getCurrentWindCondition());
                    imgv_now.setImageDrawable(getResources().getDrawable(
                            R.drawable.w1 + cityWeatherInfo.getnowIcon()));
                    tv_next_1_day_date.setText(cityWeatherInfo.getNext1date());
                    tv_next_1_day_temperature.setText(cityWeatherInfo
```

```java
                            .getNext1LowTemp()
                            + "°C"
                            + "~"
                            + cityWeatherInfo.getNext1HighTemp() + "°C");
                    next_1_weather.setText(cityWeatherInfo.getNext1Condition());
                    imgv_next1.setImageDrawable(getResources().getDrawable(
                            R.drawable.w1 + cityWeatherInfo.getNext1Icon()));
                    ...
                    imgv_next4.setImageDrawable(getResources().getDrawable(
                            R.drawable.w1 + cityWeatherInfo.getNext4Icon()));
                }
            }
        }
    };
    @Override
    protected void onActivityResult(int requestCode, int resultCode, Intent data) {
        //TODO 自动存根法
        super.onActivityResult(requestCode, resultCode, data);
        if (requestCode == 2) {
            if (resultCode == RESULT_CANCELED) {
                Bundle bundle = data.getExtras();
                String cityString = bundle.getString("CITY");
                if (cityString.equals("")) {
                    return;
                }
                tv_cityname.setText(cityString);
                if (weatherInfo == null) {
                    weatherInfo = new GetWeatherInfo(this);
                }
                weatherInfo.getInfo(cityString, updateHandler);
            }
        }
    }
    @Override
    public void onCreate(Bundle savedInstanceState) {
        super.onCreate(savedInstanceState);
        setContentView(R.layout.main);
        init_view();
        init_citydb();
        weatherInfo = new GetWeatherInfo(this);
        weatherInfo.getInfo("上海", updateHandler);
        btn_trend.setOnClickListener(new OnClickListener() {
            public void onClick(View v) {
                //TODO 自动存根法
                Intent intent = new Intent(MainActivity.this, TrendActivity.class);
                Bundle mBundle = new Bundle();
                mBundle.putParcelable("myweather", cityWeatherInfo);
                intent.putExtras(mBundle);
                startActivity(intent);
            }
        });
        tv_cityname.setOnClickListener(new OnClickListener() {
            public void onClick(View v) {
                //TODO 自动存根法
                Intent intent2 = new Intent(MainActivity.this, S_city.class);
```

```java
                startActivityForResult(intent2, 2);
            }
        });
    }
    private void init_view() {
        btn_trend = (Button) findViewById(R.id.btn_temperature);
        tv_cityname = (TextView) findViewById(R.id.city_name);
        tv_synchtime = (TextView) findViewById(R.id.synch_info);
        tv_date = (TextView) findViewById(R.id.forcast_date);
        tv_week = (TextView) findViewById(R.id.current_day_of_week);
        tv_condition = (TextView) findViewById(R.id.current_condition);
        tv_nowtemp = (TextView) findViewById(R.id.temperature);
        tv_temp = (TextView) findViewById(R.id.temperature_range);
        tv_windconditionTextView = (TextView) findViewById(R.id.wind_condition);
        imgv_now = (ImageView) findViewById(R.id.na);
        tv_next_1_day_date = (TextView) findViewById(R.id.next_1_day_date);
        tv_next_1_day_temperature = (TextView) findViewById(R.id.next_1_day_temperature);
        next_1_weather = (TextView) findViewById(R.id.next_1_weather_condition);
        imgv_next1 = (ImageView) findViewById(R.id.next_1_day_img);
        …

    }
    @SuppressWarnings("finally")
    private boolean init_citydb() {
        boolean is_suc = false;
        String dirPath = getApplicationContext().getFilesDir().getParentFile()
                .getAbsolutePath()
                + "/databases";
        File dir = new File(dirPath);
        if (!dir.exists()) {
            dir.mkdirs();
        }
        //数据库文件
        File dbfile = new File(dir, "chinacity.db");
        try {
            if (!dbfile.exists()) {
                dbfile.createNewFile();
                //加载欲导入的数据库
                InputStream is = this.getApplicationContext().getResources()
                        .openRawResource(R.raw.chinacity);
                FileOutputStream fos = new FileOutputStream(dbfile);
                byte[] buffere = new byte[is.available()];
                is.read(buffere);
                fos.write(buffere);
                is.close();
                fos.close();
            }
            is_suc = true;
        } catch (Exception e) {
            //TODO：异常处理
            is_suc = false;
        } finally {
            return is_suc;
        }
    }
}
```

(7) 在 src/fs.city_manager 包下新建一个 CityWeatherInfo.java 文件,用于定义天气信息类。主要包括城市名、城市 ID 以及当前天气的详细信息,如当前日期、发布天气预报时间、当前温度、当天最高最低温度、风速以及天气描述;还有未来五天的日期、星期、最高最低温度、天气描述以及天气图片。代码为:

```java
public class CityWeatherInfo implements Parcelable
{
    public static final Parcelable.Creator<CityWeatherInfo> CREATOR = new Parcelable.Creator()
    {
        public CityWeatherInfo createFromParcel(Parcel paramParcel)
        {
            return new CityWeatherInfo(paramParcel);
        }
        public CityWeatherInfo[] newArray(int paramInt)
        {
            return new CityWeatherInfo[paramInt];
        }
    };
    //定义变量
    private String city;
    private String cityId;
    private String nowdate;
    private String nowtime;
    …
    private int next5LowTemp;
    public CityWeatherInfo()
    {
        //为变量赋值
        this.city = "";
        this.cityId = "";
        …
        this.next5Icon = -1;
    }
    public CityWeatherInfo(Parcel paramParcel)           //从序列化数据读取到类中
    {
        this.city = paramParcel.readString();
        this.cityId = paramParcel.readString();
        …
        this.next5Condition = paramParcel.readString();
        this.next5Icon = paramParcel.readInt();
    }
    public int describeContents()
    {
        return 0;
    }
    public String getCity()                              //获得城市方法
    {
        return this.city;
    }
    public String getCityId()
    {
        return this.cityId;
    }
    …
```

```java
public void setNext5DayOfWeek(String paramString)
{
    this.next5DayOfWeek = paramString;
}
public void setNext5HighTemp(int paramInt)
{
    this.next5HighTemp = paramInt;
}
public void setNext5Icon(int paramString)
{
    this.next5Icon = paramString;
}
public void setNext5LowTemp(int paramInt)
{
    this.next5LowTemp = paramInt;
}
public void writeToParcel(Parcel paramParcel, int paramInt)      //序列化到数据中
{
    paramParcel.writeString(this.city);
    paramParcel.writeString(this.cityId);
    ...
    paramParcel.writeString(this.next5Condition);
    paramParcel.writeInt(this.next5Icon);
}
}
```

(8) 在 src/fs.city_manager 包下新建一个 GetWeatherInfo.java 文件,实现天气的查询、更新显示以及获取未来天气信息的操作。代码为:

```java
public class GetWeatherInfo {
    final String WEATHER_URL = "http://m.weather.com.cn/data/";
    final String WEATHER_NOW_URL = "http://www.weather.com.cn/data/sk/";
    Context context;
    public GetWeatherInfo(Context context) {
        //TODO 自动存根法
        this.context = context;
    }
    public void getInfo(final String loccity, final Handler handler) {
        new Thread(new Runnable() {
            public void run() {
                //TODO 自动存根法
                CityWeatherInfo locCityWeatherInfo = new CityWeatherInfo();
                Weather_n weather_num = new Weather_n(context);
                String weather_numString = weather_num.get_weatherNum(loccity);
                if (weather_numString == "" || weather_numString == null) {
                    locCityWeatherInfo = null;
                } else {
                    locCityWeatherInfo = getUrlInfo(weather_numString);
                }
                Message msg = new Message();
                msg.what = 1;
                Bundle mBundle = new Bundle();
                mBundle.putParcelable("weather", locCityWeatherInfo);
                msg.setData(mBundle);
                handler.sendMessage(msg);
            }
```

```java
            }).start();
    }
    private CityWeatherInfo getUrlInfo(String weather_num) {
        CityWeatherInfo locCityWeatherInfo = new CityWeatherInfo();
        boolean is_suc = false;
        try {
            DefaultHttpClient httpClient = new DefaultHttpClient();
            HttpGet httpGet = new HttpGet(WEATHER_NOW_URL + weather_num + ".html");
            int res = 0;
            res = httpClient.execute(httpGet).getStatusLine().getStatusCode();
            if (res == 200) {
                //当返回码为 200 时,做处理,得到服务器端返回 json 数据,并做处理
                HttpResponse httpResponse = httpClient.execute(httpGet);
                StringBuilder builder = new StringBuilder();
                BufferedReader bufferedReader2 = new BufferedReader(
                        new InputStreamReader(httpResponse.getEntity().getContent()));
                String str2 = "";
                for (String s = bufferedReader2.readLine(); s != null; s = bufferedReader2.readLine()) {
                    builder.append(s);
                }
                System.out.println(">>>>>>" + builder.toString());
                //解析 json
                JSONObject nowjsonObject = new JSONObject(builder.toString()).getJSONObject("weatherinfo");
                locCityWeatherInfo.setCity(nowjsonObject.getString("city"));
                locCityWeatherInfo.setCityId(nowjsonObject.getString("cityid"));
                locCityWeatherInfo.setnowTemp(Integer.valueOf(nowjsonObject
                        .getString("temp")));
                locCityWeatherInfo.setCurrentWindCondition(nowjsonObject
                        .getString("WD") + ":" + nowjsonObject.getString("WS"));
                locCityWeatherInfo.setnowtime(nowjsonObject.getString("time"));
            }
            httpGet = new HttpGet(WEATHER_URL + weather_num + ".html");
            res = 0;
            res = httpClient.execute(httpGet).getStatusLine().getStatusCode();
            if (res == 200) {
                //当返回码为 200 时,做处理,得到服务器端返回 json 数据,并做处理
                HttpResponse httpResponse = httpClient.execute(httpGet);
                StringBuilder builder = new StringBuilder();
                BufferedReader bufferedReader2 = new BufferedReader(
                        new InputStreamReader(httpResponse.getEntity().getContent()));
                String str2 = "";
                for (String s = bufferedReader2.readLine(); s != null; s = bufferedReader2.readLine()) {
                    builder.append(s);
                }
                System.out.println(">>>>>>" + builder.toString());
                //解析 json
                JSONObject jsonObject = new JSONObject(builder.toString()).getJSONObject("weatherinfo");
                locCityWeatherInfo.setnowdate(jsonObject.getString("date_y"));
                locCityWeatherInfo.setCurrentDayOfWeek(jsonObject.getString("week"));
                //温度
                locCityWeatherInfo.setCurrentHighTemp(getHighOrLowTemp(
                        jsonObject.getString("temp1"), "high"));
```

```java
            //描述
            String tmpString = "";
            tmpString = jsonObject.getString("weather1");
            locCityWeatherInfo.setCurrentCondition(tmpString);
            locCityWeatherInfo.setnowIcon(seticon(tmpString));
            ...
            setdate(locCityWeatherInfo);
        }
    } catch (Exception e) {
        //TODO 异常处理
        System.out.println(e.toString());
    }
    return locCityWeatherInfo;
}
private static int getHighOrLowTemp(String paramString1, String paramString2) {
    String str1 = paramString1.split("~")[0];
    String str2 = paramString1.split("~")[1];
    int i = Integer.parseInt(str1.substring(0, str1.length() - 1));
    int j = Integer.parseInt(str2.substring(0, str2.length() - 1));
    int k;
    if (paramString2.equals("high")) {
        if (i > j)
            k = i;
        else {
            k = j;
        }
    } else {
        if (i > j)
            k = j;
        else {
            k = i;
        }
    }
    return k;
}
//设置天气
private int seticon(String locCondition) {
    int icon_num = 0;
    if (locCondition.indexOf("云") != -1) {
        icon_num = 6;
    } else if (locCondition.indexOf("雨") != -1) {
        icon_num = 4;
    } else if (locCondition.indexOf("雪") != -1) {
        icon_num = 8;
    } else if (locCondition.indexOf("雷") != -1) {
        icon_num = 5;
    } else if (locCondition.indexOf("晴") != -1) {
        icon_num = 0;
    } else {
        icon_num = 6;
    }
    return icon_num;
}
//设置日期
private void setdate(CityWeatherInfo locCityWeatherInfo) {
```

```java
String dateString = locCityWeatherInfo.getnowdate();
int y = dateString.indexOf("年");
int m = dateString.indexOf("月");
int d = dateString.indexOf("日");
int mouth = Integer.valueOf(dateString.substring(y + 1, m));
int date = Integer.valueOf(dateString.substring(m + 1, d));
int[][] datearray = new int[6][2];
datearray[0][0] = date;
datearray[0][1] = mouth;
for (int i = 1; i < 6; i++) {
    datearray[i][0] = datearray[i - 1][0] + 1;
    datearray[i][1] = datearray[i - 1][1];
    switch (datearray[i][1]) {
    case 1:
    case 3:
    case 5:
    case 7:
    case 8:
    case 10:
    case 12:
        if (datearray[i][0] == 32) {
            datearray[i][0] = 1;
            datearray[i][1] = datearray[i][1] + 1;
        }
        break;
    case 4:
    case 6:
    case 9:
    case 11:
        if (datearray[i][0] == 31) {
            datearray[i][0] = 1;
            datearray[i][1] = datearray[i][1] + 1;
        }
        break;
    case 2:
        if (datearray[i][0] == 29) {
            datearray[i][0] = 1;
            datearray[i][1] = datearray[i][1] + 1;
        }
        break;
    default:
        break;
    }
}
locCityWeatherInfo.setnowdate(datearray[0][1] + "/" + datearray[0][0]);
locCityWeatherInfo.setNext1date(datearray[1][1] + "/" + datearray[1][0]);
locCityWeatherInfo.setNext2date(datearray[2][1] + "/" + datearray[2][0]);
locCityWeatherInfo.setNext3date(datearray[3][1] + "/" + datearray[3][0]);
locCityWeatherInfo.setNext4date(datearray[4][1] + "/" + datearray[4][0]);
locCityWeatherInfo.setNext5date(datearray[5][1] + "/" + datearray[5][0]);
}
}
```

(9) 在 src/fs.city_manager 包下新建一个 Weather_n.java 文件。将保存在数据库中的城市名与城市编码的对应值通过查询本地数据库文件来获得。代码为：

```java
public class Weather_n {
    Context context;
    public Weather_n(Context context) {
        this.context = context;
    }
    public String get_weatherNum(String city) {
        String Weather_num = "";
        String Path = context.getApplicationContext().getFilesDir()
                .getParentFile().getAbsolutePath()
                + "/databases/chinacity.db";
        File dir = new File(Path);
        if (!dir.exists()) {
            return Weather_num;
        }
        SQLiteDatabase mdb = SQLiteDatabase.openOrCreateDatabase(dir, null);
        Cursor cursor = null;
        try {
            cursor = mdb.query("city_table", null, "CITY = '" + city + "'",
                    null, null, null, null);
            if (cursor != null) {
                cursor.moveToFirst();
                Weather_num = cursor.getString(cursor
                        .getColumnIndex("WEATHER_ID"));
                cursor.close();
            }
        } catch (Exception e) {
            //TODO 异常处理
            System.out.println(e.toString());
        } finally {
            mdb.close();
        }
        return Weather_num;
    }
}
```

(10) 在 src/fs.city_manager 包下新建一个 WeatherTInfo.java 文件，因为在温度趋势表中，需要呈现的数据为星期、日期、气候描述以及最高、最低温度，为了方便信息的获取，需要定义温度趋势类。代码为：

```java
public class WeatherTInfo
{
    //定义并赋值变量
    public String mDate;
    public int mHighTemperature;
    public int mHightWeatherID = 44;
    public int mId;
    public boolean mIsEmpty = true;
    public String mLowTempDes;
    public int mLowTemperature;
    public int mLowWeatherID = 44;
    public String mWeek;
    public void clean()
```

```java
{
    this.mId = 0;
    this.mDate = "";
    this.mWeek = "";
    this.mHightWeatherID = -1;
    this.mLowWeatherID = -1;
    this.mLowTempDes = "";
    this.mIsEmpty = true;
}
}
```

(11) 在 src/fs.city_manager 包下新建一个 TrendActivity.java 文件,用于实现温度描述相关显示、绘制趋势图、绘制温度之间的变化折线以及温度的数字标记。代码为:

```java
public class TrendActivity extends Activity
{
    //定义并赋值变量
    private static final int ALLOW_SLIDE_DISTANCE = 50;
    private static final int ANI_DURATION = 500;
    …
    private TextView[] mWeekDays = new TextView[6];
    private List<WeatherTInfo> listTrendInfos;
    private void createTrendView() {                    //温度趋势视图
        resetTrendBuffer();                             //重置数据
        this.mContent.removeAllViews();                 //移除视图
        this.mTrendView = null;
        this.mTrendView = new TrendView(this);          //实例化 TrendView
        this.mContent.addView(this.mTrendView);         //添加视图
        init();                                         //初始化
    }
    private Bitmap getNewBuffer() {                     //缓存清空
        Bitmap localBitmap = Bitmap.createBitmap(this.mContent.getWidth(),
                this.mContent.getHeight(), Bitmap.Config.ARGB_8888);
        this.mCanvas = new Canvas(localBitmap);
        return localBitmap;
    }
    private TextView getTextViewByID(String paramString) {
        return (TextView) findViewById(getResources().getIdentifier(
                paramString, "id", getPackageName()));
    }
    private void init() {                               //数据的初始化
        for (int i = 0; i < 6; i++) {                   //获取星期、日期和气候描述
            int j = i;
            this.mWeekDays[i] = getTextViewByID("weekday" + j);
            this.mDates[i] = getTextViewByID("date" + j);
            this.mWeatherNights[i] = getTextViewByID("weathernight" + j);
        }
        this.mCityName = ((TextView) findViewById(R.id.cityname));
        this.mCityInfo = (CityWeatherInfo)getIntent().getParcelableExtra("myweather");
        setweek(mCityInfo);                             //转化为星期
        setDate(mCityInfo);                             //转化为日期
        this.listTrendInfos = new ArrayList<WeatherTInfo>();
        {
            WeatherTInfo wTrendInfo = new WeatherTInfo();
            ……
            listTrendInfos.add(wTrendInfo);             //添加到温度趋势类中
```

```java
            ...
        }
        setDescriptionTemp();                          //设置界面
    }
    private void resetTrendBuffer() {
        if (this.mTrendDrawBuffer != null) {
            this.mTrendDrawBuffer.recycle();
            this.mTrendDrawBuffer = null;
        }
    }
    private void setDescriptionTemp() {                //设置界面
        CityWeatherInfo localCityWeatherInfo = this.mCityInfo;
        //设置显示
        this.mCityName.setText(localCityWeatherInfo.getCity() + "温度趋势图");
        if (this.listTrendInfos.size() != 0) {
            for (int i = 0; i < 6; i++) {
                WeatherTInfo localWeatherTrendInfo = (WeatherTInfo) listTrendInfos.get(i);
                //设置星期显示
                this.mWeekDays[i].setText(localWeatherTrendInfo.mWeek);
                //设置日期显示
                this.mDates[i].setText(localWeatherTrendInfo.mDate);
                if (localWeatherTrendInfo.mLowTempDes.length() >= 4)
                    this.mWeatherNights[i].setTextSize(13.0F);
                this.mWeatherNights[i]
                    .setText(localWeatherTrendInfo.mLowTempDes);
            }
        }
    }
    //设置星期
    private void setweek(CityWeatherInfo locCityWeatherInfo){
        String locweekString = locCityWeatherInfo.getCurrentDayOfWeek();
        int weekid = 0;
        if (locweekString.equals("星期一")) {
            weekid = 1;
        }else if (locweekString.equals("星期二")) {
            weekid = 2;
        }else if (locweekString.equals("星期三")) {
            weekid = 3;
        }else if(locweekString.equals("星期四")){
            weekid = 4;
        }else if (locweekString.equals("星期五")) {
            weekid = 5;
        }else if (locweekString.equals("星期六")) {
            weekid = 6;
        }else if (locweekString.equals("星期天") || locweekString.equals("星期日")) {
            weekid = 7;
        }
        switch (weekid ) {
        case 1:
            locCityWeatherInfo.setNext1DayOfWeek("周二");
            locCityWeatherInfo.setNext2DayOfWeek("周三");
            locCityWeatherInfo.setNext3DayOfWeek("周四");
            locCityWeatherInfo.setNext4DayOfWeek("周五");
            locCityWeatherInfo.setNext5DayOfWeek("周六");
            locCityWeatherInfo.setCurrentDayOfWeek("今天");
```

```java
                break;
            ...
        }
    }
    //设置日期
    private void setDate(CityWeatherInfo locCityWeatherInfo){
    }
protected void onActivityResult(int paramInt1, int paramInt2,
        Intent paramIntent) {
    super.onActivityResult(paramInt1, paramInt2, paramIntent);
}
public void onAnimationEnd(Animation paramAnimation) {
    this.mIsDoingFadeAnimation = false;
}
public void onAnimationRepeat(Animation paramAnimation) {
}
public void onAnimationStart(Animation paramAnimation) {
    this.mIsDoingFadeAnimation = true;
}
protected void onCreate(Bundle paramBundle) {
    super.onCreate(paramBundle);
    instance = this;
    setContentView(R.layout.trend);
    this.mContent = ((LinearLayout) findViewById(R.id.trend_content));
    createTrendView();
    Bitmap localBitmap = BitmapFactory.decodeResource(getResources(),
            R.drawable.trend_line_content);
    this.mLineBmp = new NinePatch(localBitmap,
            localBitmap.getNinePatchChunk(), null);
    this.mScale = getResources().getDisplayMetrics().density;
    this.mLineTopMagin = (0.5F + 10.0F * this.mScale);
}
...
private class TrendView extends View {
    private static final int LINE_DISTANCE = 25;
    float moveX = 0.0F;
    public TrendView(Context arg2) {
        super(arg2);
    }
    @SuppressLint("WrongCall") private void prepareDrawBuffer() {
        System.out.println("prepareDrawBuffer");
        if (TrendActivity.this.mTrendDrawBuffer == null) {
            TrendActivity.this.mTrendDrawBuffer = TrendActivity.this.getNewBuffer();
        }
        if (TrendActivity.this.mIsTempType) {
            DrawTrendTView.onDraw(TrendActivity.this,
                    TrendActivity.this.mContent.getWidth(),
                    ...
        }
    }
    protected void onDraw(Canvas paramCanvas) {
        System.out.println("TrendView onDraw");
        RectF localRectF = new RectF(0.0F,
                TrendActivity.this.mLineTopMagin,
                TrendActivity.this.mContent.getWidth(),
```

```
                    1.0F + TrendActivity.this.mLineTopMagin);
            int i = TrendActivity.this.mContent.getHeight() / 25;
            for (int j = 0; j < i; j++) {
                TrendActivity.this.mLineBmp.draw(paramCanvas, localRectF);
                localRectF.offset(0.0F, 25.0F);
            }
            prepareDrawBuffer();
            paramCanvas.drawBitmap(TrendActivity.this.mTrendDrawBuffer, 0.0F,
                    0.0F, TrendActivity.this.mPaintBuffer);
        }
    }
}
```

（12）在 src/fs.city_manager 包下新建一个 S_city.java 文件,本文件用于实现查询城市的管理。在城市选择界面中,需要实现城市的输入以及自动定位到当前城市的功能。代码为:

```
public class S_city extends Activity {
    TextView tv_city;
    AutoCompleteTextView et_com;
    Button btn_ok, btn_loc;
    Context context;
    Get_l_city get_loc_city = null;
    @Override
    public void onCreate(Bundle savedInstanceState) {
        super.onCreate(savedInstanceState);
        setContentView(R.layout.city);
        init_view();                                    //初始化界面
        btn_loc.setOnClickListener(new OnClickListener() {   //自动定位城市
            public void onClick(View v) {
                //TODO 自动存根法
                get_loc_city = new Get_l_city(context);
                String cityString = get_loc_city.getcity();
                et_com.setText(cityString);
            }
        });
        btn_ok.setOnClickListener(new OnClickListener() {
            //输入城市,返回主界面
            public void onClick(View v) {
                String cityString = et_com.getText().toString();
                if (cityString.equals("")) {
                    Toast.makeText(S_city.this, "请输入查询的城市",
                            Toast.LENGTH_LONG).show();
                    return;
                }
                if (get_loc_city != null) {
                    //取消定位
                    get_loc_city.unre();
                }
                on_Previous();                          //回退
            }
        });
    }
    private void init_view() {
        //界面控件初始化
```

```java
        context = this;
        et_com = (AutoCompleteTextView) findViewById(R.id.ed_com);
        tv_city = (TextView) findViewById(R.id.textView1);
        btn_loc = (Button) findViewById(R.id.button1);
        btn_ok = (Button) findViewById(R.id.button2);
    }
    private void on_Previous() {                                //定义返回携带数据方法
        Bundle bundle = new Bundle();
        String cityString = et_com.getText().toString();        //获取输入的数据
        bundle.putString("CITY", cityString);                   //保存数据在 bundle 中
        S_city.this.setResult(RESULT_CANCELED, S_city.this
                .getIntent().putExtras(bundle));                //设置返回结果
        S_city.this.finish();                                   //结束当前 Activity B
    }
    @Override
    public boolean onKeyDown(int keyCode, KeyEvent event) {     //重写按钮单击监听
        if (keyCode == KeyEvent.KEYCODE_BACK) {                 //判断是否单击返回键
            on_Previous();                                      //调用返回数据方法
            return true;
        } else {
            return super.onKeyDown(keyCode, event);             //其他键时,不另处理
        }
    }
}
```

（13）在 src/fs.city_manager 包下新建一个 Get_l_city.java 文件，在文件中实现自动定位到所在城市的功能并从网络返回的数据中解析得到城市名。代码为：

```java
public class Get_l_city {
    //借助 Google MAP 通过用户当前经纬度获得用户当前城市
    static final String GOOGLE_MAPS_API_KEY = "abcdefg";
    private LocationManager locationManager;
    private Location currentLocation;
    private String city = "全国";
    LocationListener ll;
    public Get_l_city(Context context) {
        this.locationManager = (LocationManager) context.getSystemService(Context.LOCATION_SERVICE);
        ll = new LocationListener() {
            public void onLocationChanged(Location loc) {
    //当坐标改变时触发此函数,如果 Provider 传进相同的坐标,它就不会被触发
                currentLocation = loc;                          //保存最新的位置
                //更新经纬度并把值放置到 TextViews
                System.out.println("getCity()"
                        + (loc.getLatitude() + " " + loc.getLongitude()));
            }
            public void onProviderDisabled(String arg0) {
                System.out.println(".onProviderDisabled(关闭)" + arg0);
            }
            public void onProviderEnabled(String arg0) {
                System.out.println(".onProviderEnabled(开启)" + arg0);
            }
            public void onStatusChanged(String arg0, int arg1, Bundle arg2) {
                System.out.println(".onStatusChanged(Provider 的转态在可用、" + "暂时不可用
                    和无服务 3 个状态直接切换时触发此函数)" + arg0 + " " + arg1 + " " + arg2);
            }
```

```java
        };
        //只是简单地获取城市,不需要实时更新
        this.locationManager.requestLocationUpdates(
                LocationManager.GPS_PROVIDER, 1000, 0, ll);
        currentLocation = locationManager
                .getLastKnownLocation(LocationManager.GPS_PROVIDER);
        if (currentLocation == null)
            currentLocation = locationManager.getLastKnownLocation(LocationManager.NETWORK_
            PROVIDER);
}
//开始解析
public String getcity() {
    if (currentLocation != null) {
        String temp = reverseGeocode(currentLocation);
        if (temp != null && temp.length() >= 2)
            city = temp;
    } else {
        System.out.println("GetCity.start()未获得location");
    }
    return city;
}
public void unre(){
    this.locationManager.removeUpdates(ll);
}
//通过 Google map api 解析出城市
private String reverseGeocode(Location loc) {
    String localityName = "";
    HttpURLConnection connection = null;
    URL serverAddress = null;
    try {
        serverAddress = new URL("http://maps.google.com/maps/geo?q = "
                + Double.toString(loc.getLatitude()) + ","
                + Double.toString(loc.getLongitude())
                + "&output = xml&language = zh-CN&sensor = true"
                +
                "&key = "
                + GOOGLE_MAPS_API_KEY);
        connection = null;
        //设置的初始连接
        connection = (HttpURLConnection) serverAddress.openConnection();
        connection.setRequestMethod("GET");
        connection.setDoOutput(true);
        connection.setReadTimeout(10000);
        connection.connect();
        try {
            InputStreamReader isr = new InputStreamReader(
                    connection.getInputStream());
            System.out.println(isr.toString());
            InputSource source = new InputSource(isr);
            SAXParserFactory factory = SAXParserFactory.newInstance();
            SAXParser parser = factory.newSAXParser();
            XMLReader xr = parser.getXMLReader();
            GoogleReverseGeocodeXmlHandler handler = new GoogleReverseGeocodeXmlHandler();
            xr.setContentHandler(handler);
            xr.parse(source);
```

```java
                localityName = handler.getLocalityName();
                System.out.println("GetCity.reverseGeocode()" + localityName);
            } catch (Exception ex) {
                ex.printStackTrace();
            }
        } catch (Exception ex) {
            ex.printStackTrace();
            System.out.println("GetCity.reverseGeocode()" + ex);
        }
        return localityName;
    }
    private class GoogleReverseGeocodeXmlHandler extends DefaultHandler {
        private boolean inLocalityName = false;
        private boolean finished = false;
        private StringBuilder builder;
        private String localityName;
        public String getLocalityName() {
            return this.localityName;
        }
        @Override
        public void characters(char[] ch, int start, int length)
                throws SAXException {
            super.characters(ch, start, length);
            if (this.inLocalityName && !this.finished) {
                if ((ch[start] != '\n') && (ch[start] != ' ')) {
                    builder.append(ch, start, length);
                }
            }
        }
        @Override
        public void endElement(String uri, String localName, String name)
                throws SAXException {
            super.endElement(uri, localName, name);
            if (!this.finished) {
                if (localName.equalsIgnoreCase("LocalityName")) {
                    this.localityName = builder.toString();
                    this.finished = true;
                }
                if (builder != null) {
                    builder.setLength(0);
                }
            }
        }
        @Override
        public void startDocument() throws SAXException {
            super.startDocument();
            builder = new StringBuilder();
        }
        @Override
        public void startElement(String uri, String localName, String name,
                Attributes attributes) throws SAXException {
            super.startElement(uri, localName, name, attributes);
            if (localName.equalsIgnoreCase("LocalityName")) {
                this.inLocalityName = true;
            }
```

    }
  }
}

(14) 在 src/fs.city_manager 包下新建一个 SelfPView.java 文件,用于实现自动更新城市天气情况。代码为:

```java
public class SelfPView extends View {
    //定义并赋值变量
    private int height = 0;
    …
    public SelfPView(Context paramContext, AttributeSet paramAttributeSet)
    {
        super(paramContext, paramAttributeSet);
        this.progressBitmap = BitmapFactory.decodeResource(paramContext.getResources(), R.drawable.update );
        this.height = this.progressBitmap.getHeight();
        this.width = this.progressBitmap.getWidth();
    }
    @Override
    protected void onDraw(Canvas paramCanvas)
    {
        super.onDraw(paramCanvas);
        if (this.progressBitmap != null)
        {
            Paint localPaint = new Paint();
            localPaint.setAntiAlias(true);
            localPaint.setDither(true);
            localPaint.setFilterBitmap(true);
            paramCanvas.save();
            paramCanvas.rotate(this.rotateDegree, this.progressBitmap.getWidth() / 2 + this.offsetValue / 2, this.progressBitmap.getHeight() / 2 + this.offsetValue / 2);
            paramCanvas.drawBitmap(this.progressBitmap, this.offsetValue / 2, this.offsetValue / 2, localPaint);
            this.rotateDegree -= 10;
            if (this.rotateDegree >= 360)
                this.rotateDegree %= 360;
            paramCanvas.restore();
        }
    }
    @Override
    protected void onMeasure(int paramInt1, int paramInt2)
    {
        super.onMeasure(paramInt1, paramInt2);
        setMeasuredDimension(this.width + this.offsetValue, this.height + this.offsetValue);
    }
    public void startProgress()
    {
        this.isProgress = true;
        new Thread()
        {
```

```java
            @Override
            public void run()
            {
                while (true)
                {
                    if (!SelfPView.this.isProgress)
                        return;
                    SelfPView.this.postInvalidate();
                    try
                    {
                        sleep(50L);
                    }
                    catch (InterruptedException localInterruptedException)
                    {
                    }
                }
            }
        }
        .start();
    }
    public void stopProgress()
    {
        this.isProgress = false;
    }
}
```

(15) 打开AndroidManifest.xml布局文件,在文件中声明相关权限以及界面。代码为:

```xml
…
    <uses-permission android:name="android.permission.INTERNET"/>
    <uses-permission android:name="android.permission.ACCESS_FINE_LOCATION"/>
    <uses-permission android:name="android.permission.ACCESS_COARSE_LOCATION"/>"
    <application
        android:allowBackup="true"
        android:icon="@drawable/ic_launcher"
        android:label="@string/app_name"
        android:theme="@style/AppTheme" >
        <activity
            android:name="fs.city_manager.MainActivity"
            android:label="@string/app_name" >
            <intent-filter>
                <action android:name="android.intent.action.MAIN" />
                <category android:name="android.intent.category.LAUNCHER" />
            </intent-filter>
        </activity>
        <activity android:name="fs.city_manager.TrendActivity" />
        <activity android:name="fs.city_manager.S_city" />
    </application>
</manifest>
```

运行程序,得到城市的天气图,效果如图7-13(a)所示,当单击界面中的"气温趋势"按钮,即弹出城市的温度趋势图,如图7-13(b)所示,返回上一步即可实现城市的定位,如图7-13(c)所示。

(a）城市天气图　　　　（b）城市温度趋势图　　　　（c）城市定位

图 7-13　城市定位

附录A 网上参考资源

http://blog.csdn.net/x605940745/article/details/18405951.

http://blog.csdn.net/x605940745.

http://blog.csdn.net/cxf7394373/article/details/8313980.

http://www.2cto.com/kf/201410/343274.html.

http://weizhulin.blog.51cto.com/1556324/311694.

http://www.open-open.com/lib/view/open1333418857983.html.

http://www.oschina.net/code/snippet_107931_17160.

http://translate.google.cn/#en/zh-CN/Flags.

http://www.cnblogs.com/linjiqin/archive/2011/02/25/1965319.html.

http://blog.csdn.net/lee576/article/details/7994910.

http://www.l99.com/EditText_view.action?textId=541245.

http://blog.chinaunix.net/attachment/201106/12/25422700_1307864514joDf.jpg.

http://blog.chinaunix.net/uid-25422700-id-368672.html.

http://www.cnblogs.com/menlsh/archive/2012/12/09/2810372.html.

http://byandby.iteye.com/blog/831011.

http://blog.csdn.net/lee576/article/details/7900228.

http://ipjmc.iteye.com/blog/1290170.

http://blog.chinaunix.net/uid-24129645-id-3609120.html.

http://www.2cto.com/kf/201204/128926.html.

http://blog.csdn.net/cjjky/article/details/6881582.

http://www.cnblogs.com/linjiqin/archive/2011/03/08/1977579.html.

http://blog.csdn.net/cjjky/article/details/7065356.

http://blog.csdn.net/bailu66/article/details/7054232.

http://www.cnblogs.com/mengdd/archive/2013/05/08/3065156.html.

http://blog.csdn.net/x605940745/article/details/11981049.

http://www.cnblogs.com/linjiqin/archive/2011/02/23/1962535.html.

http://blog.csdn.net/loongggdroid/article/details/7581236.

http://blog.csdn.net/sjf0115/article/details/7254409.

http://www.iteye.com/topic/540423.
http://blog.csdn.net/lanjianhun/article/details/8198108.
http://blog.csdn.net/pathuang68/article/details/6561380.
http://www.jb51.net/article/52529.htm.
http://www.apkbus.com/android-131096-1-1.html.
http://blog.csdn.net/tianjf0514/article/details/7526421.
http://www.cnblogs.com/zhuawang/p/3675381.html.
http://blog.csdn.net/boyupeng/article/details/6213466.
http://www.jb51.net/article/33109.htm.
http://www.cnblogs.com/over140/archive/2010/11/04/1869316.html.
http://www.oschina.net/question/54100_34553.
http://www.cnblogs.com/wt616/archive/2011/06/20/2085531.html.
http://www.2cto.com/kf/201110/109222.html.
http://www.cnblogs.com/tornadomeet/archive/2012/07/29/2614251.html.
http://www.2cto.com/kf/201201/117915.html.
http://www.cnblogs.com/hnrainll/archive/2012/03/28/2420901.html.
http://wenku.baidu.com/link?url=bmcrTXg1d-0AwOdHX_fIqGyRh5o6efr2d1jYRVR0cYzgElLFQGnq7e0b2dpmG59Wq3t-CZkc3T_NNHJsrYjfjyBqvnHDFdbBqSTsaTQKFa7.
http://www.2cto.com/kf/201304/204571.html.
http://www.jb51.net/article/42349.htm.
http://byandby.iteye.com/blog/815212.
http://blog.csdn.net/jaycee110905/article/details/8964090.

参 考 文 献

[1] 李刚. 疯狂 Android 讲义[M]. 北京：电子工业出版社,2011.
[2] 张余. Android 网络开发从入门到精通[M]. 北京：清华大学出版社,2014.
[3] 欧阳零. Android 编程兵书[M]. 北京：电子工业出版社,2014.
[4] 李佐彬,等. Android 开发入门与实战体验[M]. 北京：机械工业出版社,2011.
[5] 软件开发技术联盟. Android 开发实战[M]. 北京：清华大学出版社,2013.
[6] 楚无咎. Android 编程经典 200 例[M]. 北京：电子工业出版社,2013.
[7] 高洪岩. Android 学习精要[M]. 北京：清华大学出版社,2012.
[8] 明日科技. Android 从入门到精通[M]. 北京：清华大学出版社,2012.
[9] 李波,史江萍,王祥凤. Android 4.X 从入门到精通[M]. 北京：清华大学出版社,2012.